Lecture Notes of the Institute for Computer Sciences, Social Informatics and Telecommunications Engineering 638

Editorial Board Members

Ozgur Akan, *Middle East Technical University, Ankara, Türkiye*
Paolo Bellavista, *University of Bologna, Bologna, Italy*
Jiannong Cao, *Hong Kong Polytechnic University, Hong Kong, Hong Kong*
Geoffrey Coulson, *Lancaster University, Lancaster, UK*
Falko Dressler, *University of Erlangen, Erlangen, Germany*
Domenico Ferrari, *Università Cattolica Piacenza, Piacenza, Italy*
Mario Gerla, *UCLA, Los Angeles, USA*
Hisashi Kobayashi, *Princeton University, Princeton, USA*
Sergio Palazzo, *University of Catania, Catania, Italy*
Sartaj Sahni, *University of Florida, Gainesville, USA*
Xuemin Shen, *University of Waterloo, Waterloo, Canada*
Mircea Stan, *University of Virginia, Charlottesville, USA*
Xiaohua Jia, *City University of Hong Kong, Kowloon, Hong Kong*
Albert Y. Zomaya, *University of Sydney, Sydney, Australia*

The LNICST series publishes ICST's conferences, symposia and workshops.
LNICST reports state-of-the-art results in areas related to the scope of the Institute.
The type of material published includes

- Proceedings (published in time for the respective event)
- Other edited monographs (such as project reports or invited volumes)

LNICST topics span the following areas:

- General Computer Science
- E-Economy
- E-Medicine
- Knowledge Management
- Multimedia
- Operations, Management and Policy
- Social Informatics
- Systems

Xianchao Zhang · Joey Tianyi Zhou · Hong Sun
Editors

Advanced Hybrid Information Processing

8th International Conference, ADHIP 2024
Jiaxing, China, September 20–22, 2024
Proceedings, Part IV

Springer

Editors
Xianchao Zhang
Provincial Key Laboratory of Multimodal
Perceiving and Intelligent Systems
Jiaxing, China

Joey Tianyi Zhou
Technology and Research (A*STAR)
Fusionopolis Way, Singapore

Hong Sun
Provincial Key Laboratory of Multimodal
Perceiving and Intelligent Systems
Jiaxing, China

ISSN 1867-8211　　　　　　ISSN 1867-822X (electronic)
Lecture Notes of the Institute for Computer Sciences, Social Informatics
and Telecommunications Engineering
ISBN 978-3-032-00305-8　　　ISBN 978-3-032-00306-5 (eBook)
https://doi.org/10.1007/978-3-032-00306-5

© ICST Institute for Computer Sciences, Social Informatics and Telecommunications Engineering 2026

This work is subject to copyright. All rights are solely and exclusively licensed by the Publisher, whether the whole or part of the material is concerned, specifically the rights of translation, reprinting, reuse of illustrations, recitation, broadcasting, reproduction on microfilms or in any other physical way, and transmission or information storage and retrieval, electronic adaptation, computer software, or by similar or dissimilar methodology now known or hereafter developed.
The use of general descriptive names, registered names, trademarks, service marks, etc. in this publication does not imply, even in the absence of a specific statement, that such names are exempt from the relevant protective laws and regulations and therefore free for general use.
The publisher, the authors and the editors are safe to assume that the advice and information in this book are believed to be true and accurate at the date of publication. Neither the publisher nor the authors or the editors give a warranty, expressed or implied, with respect to the material contained herein or for any errors or omissions that may have been made. The publisher remains neutral with regard to jurisdictional claims in published maps and institutional affiliations.

This Springer imprint is published by the registered company Springer Nature Switzerland AG
The registered company address is: Gewerbestrasse 11, 6330 Cham, Switzerland

If disposing of this product, please recycle the paper.

Preface

We are delighted to introduce the proceedings of the Eighth European Alliance for Innovation (EAI) International Conference on Advanced Hybrid Information Processing (ADHIP 2024). This conference brought together researchers, developers and practitioners around the world who are leveraging and developing deeper and wider use of hybrid information processing, especially weak signal processing. The theme of ADHIP 2024 was "weak signal detection and estimation, multi-source information fusion and hybrid intelligent computing".

The technical program of ADHIP 2024 consisted of 115 full papers. The conference tracks were: Track 1- Signal Processing and Enhancement; Track 2- Information Fusion and Integration; Track 3- Intelligent Computing and Machine Learning; and Track 4- Applications and Intelligent Systems.

Aside from the high-quality technical paper presentations, the technical program also featured four keynote speeches.

The four keynote speakers were Xindong Wu from Hefei University of Technology, China, Chenggang Yan from Hangzhou Dianzi University, China, Joey Tianyi Zhou from Agency for Science, Technology and Research (A*STAR), Singapore and Guangjie Han from Hehai University, China.

Coordination with the general and program chairs, Xianchao Zhang, Yaodong Zhu, Wei Wang and Junjie Hu was essential for the success of the conference. We sincerely appreciate their constant support and guidance.

It was also a great pleasure to work with such an excellent organizing committee team for their hard work in organizing and supporting the conference. In particular, the Technical Program Committee, led by our TPC Co-Chairs, Lei Chen, Hong Sun, Fengli Huang, Yu Wang, Ming Yan and Zheng Ma completed the peer-review process of technical papers and made a high-quality technical program. We are also grateful to the Conference Managers, Ivana Bujdakova, Ly Dao and Lin Zhong, for their support and to all the authors who submitted their papers to the ADHIP 2024 conference.

We strongly believe that ADHIP 2024 provided a good forum for all researchers, developers and practitioners to discuss all science and technology aspects that are relevant to advanced hybrid information processing. We also expect that future ADHIP conferences will be as successful and stimulating, as indicated by the contributions presented in this volume.

Xianchao Zhang
Joey Tianyi Zhou
Hong Sun

Organization

Organizing Committee

General Chair

Xianchao Zhang — Jiaxing University, China

General Co-chairs

Yaodong Zhu — Jiaxing University, China
Wei Wang — National Key Laboratory of Electromagnetic Space Security, China

Program Chairs

Xianchao Zhang — Jiaxing University, China
Joey Tianyi Zhou — Agency for Science, Technology and Research (A*STAR), Singapore

TPC Chair and Co-chairs

Xianchao Zhang — Jiaxing University, China
Junjie Hu — National Key Laboratory of Electromagnetic Space Security, China
Lei Chen — Georgia Southern University, USA
Hong Sun — Jiaxing University, China
Fengli Huang — Jiaxing University, China
Yu Wang — Nanjing University of Posts and Telecommunications, China
Ming Yan — Agency for Science, Technology and Research (A*STAR), Singapore
Zheng Ma — University of Southern Denmark, Denmark

Sponsorship and Exhibit Chair

Lin Zhong — Jiaxing University, China

Local Chairs

Zhiguo Jiang Jiaxing University, China
Huan Li National Key Laboratory of Electromagnetic Space Security, China

Workshops Chairs

Hong Sun Jiaxing University, China
Ali Kashif Manchester Metropolitan University, UK

Publicity and Social Media Chairs

Chaochao Wang Jiaxing University, China
Ruolin Zhou University of Massachusetts Dartmouth, USA

Publications Chair

Tong Wu Jiaxing University, China

Web Chairs

Tao Li Jiaxing University, China
Zheng Ma University of Southern Denmark, Denmark

Panels Chair

Chen Cen Institute for Infocomm Research, Singapore

Demos Chair

Wei Xue Harbin Engineering University, China

Technical Program Committee

Ali Kashif Bashir Manchester Metropolitan University, UK
Chao Li RIKEN-AIP, Japan
Dörthe Arndt Technische Universitat Dresden, Germany
Feng Xu Fudan University, China
Gang Xu Southeast University, Bangladesh

Gokce Banu Laleci Erturkmen	SRDC, Turkey
Jian Wang	Fudan University, China
Jianfeng Li	Nanjing University of Aeronautics and Astronautics, China
Jianhua Tang	Nanyang Technological University, China
Jinming Wen	McGill University, Canada
Ke Guan	Beijing Jiaotong University, China
Lan Lan	Xidian University, China
Lei Chen	Georgia Southern University, USA
Michael Lones	Heriot-Watt University, UK
Mingqian Liu	Xidian University, China
Mohamed Rezk	Dedalus Healthcare, Belgium
Peihan Qi	Xidian University, China
Qingzhong Liu	Sam Houston State University, USA
Qisong Wu	Southeast University, China
Ruolin Zhou	University of Massachusetts Dartmouth, USA
Sai Huang	Beijing University of Posts and Telecommunication, China
Tian Jin	National University of Defense Technology, China
Vincenzo De Florio	Vrije Universiteit Brussel, Belgium
Yan Ming	A*STAR, Singapore
Zhenyu Na	Dalian Maritime University, China
Zhi Sun	Tsinghua University, China

Contents – Part IV

Applications and Intelligent Systems

A Method for Monitoring the Operational Status of Distribution Network Equipment by Integrating Digital Twin Technology and Association Rules 3
   Xintao Li, Degao Li, Tao Wang, Jiangtao Guo, Junqiang Jia, and Lulu Liu

Real-time Analysis of an Embedded CNC System Based on Information Fusion in the Internet of Things 18
   Niyan Wu

A Multi-dimensional Early Warning Method for Digital E-commerce Supply Chain Risk Based on IoT Information Fusion 32
   Liwen Zuo

Cloud Monitoring Technology of Fish Pond Water Quality Based on Internet of Things Multi-source Data Collection 51
   Shaoyong Cao, Huanqing Han, and Cao Song

A Transient Response Control Method for Hybrid Wind and Solar Power Microgrids Based on Composite Information Fusion 65
   Jingtao Zhao, Chao Song, and Shaoyong Song

Data Intrusion Detection Method for Embedded Components of Power Terminal Based on Machine Learning 81
   Hongwei Wang and Xuedong Dai

The Intelligent Monitoring System of University Personnel File Falsification Data Based on Wireless Network 95
   Hui Li and Weiwei Zhang

A Study of Resource Sharing Methods for Teaching English Reading Based on Blockchain Technology 112
   Man Zhan

A Hybrid Teaching Resource Sharing Method for Higher Vocational English Based on Cloud Storage 127
   Man Zhan

A Personalized and Accurate Push Method for Online Teaching Resources Based on Social Media Information Integration 143
   Jianhua Jiang and Ziyu Ai

Evaluating the Effectiveness of the Application of an Interactive Teaching Model for English Courses in Undergraduate Colleges and Universities Based on Graph Neural Networks 157
   Yalin Sun and Ping Huang

Research on Software Test Data Optimization Using Adaptive Differential Evolution Algorithm 174
   Zheheng Liang, Wuqiang Shen, and Chaosheng Yao

A Hybrid Swarm Intelligence Algorithm Based Approach for Information Integration of English Language Database Calls 186
   Xue Che and Yuefei Wang

A Study on Fuzzy Comprehensive Evaluation of Blended Teaching Quality Based on Multi-source Information Fusion 201
   Lei Ma, Hongxue Yang, Jingyu Li, and Jiatong Wei

A Study on the Job Information Recommendation Method Based on Social Network Information Fusion 214
   Shan Gao and Xiangjun Shi

A Study of Algorithms for Deep Integration of Information on Teaching Resources of Aerobics Course in the Context of Curriculum Thinking and Politics 229
   Xiuyan Hong and Yuanyuan Zhang

Research on the Optimal Route Recommendation Method for Regional Tourism Based on Ant Colony Algorithm in Digital Economy Era 243
   Ce Liu and Fang Wang

An Intelligent Recommendation Method for Cross-Border E-commerce with Multiple Collaborative Information Based on Deep Learning and Graph Neural Network 258
   Liwen Zuo

Research on English Teaching Data Location Based on Multi-agent Hierarchical Reinforcement Learning 272
   Tingwen Wang

Optimal Allocation Method of College Students' Ideological and Political
Education Resources Based on PSO Algorithm 285
   He Kong, Zhengfang Lu, and Xiaodi Li

A Study of Collaborative Filtering-Based Recommendation Algorithms
for University Aesthetic Education Teaching Resources 300
   Xiaodi Li and He Kong

Data Topic Mining Method of Online English Teaching in Higher
Vocational Colleges Based on LDA Model 317
   Yuanyuan Zhang

A Study on the Clustering Method of Digital English Teaching Resources
Based on Deep Learning 330
   Hui Xu and Xiaorong Zhu

An Approach to Integrating Micro-video English Teaching Resources
Based on Improved Deep Learning 344
   Ping Huang and Yalin Sun

Intelligent Optimization Method for Charging Power of Electric Vehicle
Charging Station Based on VSM 358
   Jiatong Wei, Chunhua Kong, Shujiang Song, and Lei Ma

Author Index 373

Applications and Intelligent Systems

A Method for Monitoring the Operational Status of Distribution Network Equipment by Integrating Digital Twin Technology and Association Rules

Xintao Li[✉], Degao Li, Tao Wang, Jiangtao Guo, Junqiang Jia, and Lulu Liu

State Grid Xinjiang Electric Power Co., Ltd. Information and Communication Company, Urumqi 830001, China
15022891329@163.com

Abstract. In order to reduce the average correct deviation of the online condition monitoring method of distribution network equipment and improve the monitoring recall rate, the operation condition monitoring method of distribution network equipment integrating digital twin technology and association rules is studied. Innovatively utilizing digital twin technology to construct a data interaction model between physical entities and virtual models of distribution network equipment; Based on association rules, extract key state parameters of device operation, and further explore the correlation between device items and attributes in the digital twin model. The K-means algorithm is used to discretize the amplitude of the running data, and the initial threshold is obtained through residual calculation model. By comparing the relationship between the prediction error of offline models and the preset initial threshold, state monitoring is completed. The experimental results show that the proposed method has an average correct monitoring deviation of 1–1.2 and a monitoring recall rate of over 96%, which improves the accuracy of online monitoring of distribution network equipment and helps maintain the continuous, effective, safe and stable operation of the power system.

Keywords: Digital twin technology · Association rules · Distribution network · Condition monitoring

1 Introduction

The continuous growth of China's economy for many years has led to a rapid growth in both electricity demand and consumer groups. In view of this, power grid enterprises are continuously increasing their investment and support in ultra-high voltage projects, AC/DC hybrid power supply, and new energy technologies, making the power grid structure increasingly diversified and innovative, and the complexity is gradually increasing. During peak and low periods of electricity supply, the load value is gradually increasing. To ensure the continuous, effective, efficient, energy-saving, and safe and stable operation of the power system, and to bring benefits to the country and society, power grid enterprises are making unremitting efforts. The power system strongly depends on the reliable power supply capacity. Continuous promotion of people's quality of life and living standards for efficient power demand increases year by year, the national power grid company in this context must continue to rapid development and rapid expansion, so that such as the west to the east, north-south connectivity, the power network of the big pattern has also been formed. The product of the power network is the power system of all kinds of components of the type and number of exponential increase and expansion, more complex transmission and transformation components make the grid operating environment presents complexity, diversification of trends, so to ensure that the power system components of the effective and reliable operation of the whole network to achieve the stable operation of the cornerstone, but also is the current hot research topics [1]. Therefore, if the power system equipment can be found in a timely manner the slightest changes in the components, scientific and reasonable condition monitoring and overhaul, and efficiently solve the corresponding faults, so that the life of the power system equipment has been qualitatively improved, and ultimately to improve the detection mechanism and the management level of the purpose of the power system to ensure the continued and effective safe and stable operation.

Power system equipment often exhibits no discernible signs of impending failure initially, yet over time, the cumulative effects of aging amplify, expediting the emergence of failure probabilities, which in turn escalates equipment malfunction rates and undermines system reliability. Researchers advocate for an optimal maintenance approach: monitoring equipment operations preemptively, before the threshold of failure is reached. Maintenance in this way can not only greatly extend the operation life and reliability of the equipment, but also effectively reduce the operation risk of the system. However, it is accompanied by the high standard requirements of information electronic technology brought by real-time condition based maintenance. There is still a long way to go before it can reach the application state of the whole network, which is also a hot topic of current research. Study effective monitoring methods for key components, and give early warning when the operating conditions and status of equipment change beyond the threshold. Then, according to the equipment condition evaluation report, we will focus on equipment repair and maintenance and condition monitoring, carry out special inspection for equipment health and quality improvement, strengthen the monitoring of vulnerable points of equipment components again at the end of the period, carry out structural maintenance and troubleshooting of corresponding parts, and determine the replacement strategy for overhaul or minor repair at any time in the future. Various facility components in the power grid can clearly reflect the current loss and operating status through real-time monitoring data. Combined with the health assessment index, we can maintain and monitor unstable equipment to ensure safe and efficient work, and effectively prevent the occurrence of major accidents. Therefore, the researcher Zhao YL et al. [2] proposed a real-time monitoring method based on the least mean square for the operation of distribution network terminal equipment. The LMS algorithm is used to iterate the channel model parameters to obtain the amplitude attenuation characteristics and phase shift; The LMS algorithm adopted reflects the real-time monitoring information of distribution network equipment. Wang SW et al. [3] designed an automatic monitoring method for the operation status of intelligent distribution networks. Collect distribution network operation data and set multidimensional monitoring nodes; Build an adaptive big data power grid operation status monitoring model, and complete the operation monitoring of power grid equipment by using the modified evaluation method. Li FJ et al. [4] proposed a fault monitoring model for distribution network equipment operation based on fuzzy frequent term mining algorithm. Firstly, the fault feature data is extracted from the multivariate information database, and the redundant features are eliminated by Relief-F algorithm; Build equipment fault correlation feature library through data preprocessing and data integration; Based on the fuzzy set theory, a fault monitoring model of distribution network equipment operation is constructed to complete the equipment operation monitoring. He T et al. [5] proposed a leakage current monitoring method for low-voltage distribution networks based on extended Kalman filtering. Using wavelet transform method to extract low-frequency information of leakage current, using first-order Taylor expansion extended Kalman filtering technology to achieve monitoring of leakage current in low-voltage distribution network. Song Jialei et al. [6] proposed an online monitoring method for series compensation capacitor faults in 10 kV distribution networks based on the FP Growth algorithm. Identify the series branch status of the 10 kV distribution network; Using the FP Growth algorithm to fit the fault amplitude envelope curve, testing the correlation between the received signal and the fault location characteristic parameters, and achieving monitoring of the hazard level of the capacitor state.

Based on the above research results, this paper proposes a method for monitoring the operation status of distribution network equipment by integrating digital twin technology and association rules. Innovatively constructing a distribution network equipment model based on digital twin technology; Extract key state parameters of equipment operation based on association rules, use K-means algorithm to discretize the amplitude of key state parameter data, and obtain the initial threshold through residual calculation model. By comparing the prediction error of offline models with the preset initial threshold, equipment status monitoring can be achieved. This paper enhances the application effect of digital twin technology and association rule theory in fault monitoring of distribution network equipment operation.

2 Condition Risk Analysis of Distribution Network Equipment

2.1 Constructing Distribution Network Equipment Models Based on Digital Twins

The focus of digital twin technology has gradually been on the interaction of data between physical entities and virtual models, while two dimensions have been added to the digital twin three-dimensional model: twin data and services. A "virtual entity" is a virtual mapping of a physical entity, describing the physical entity in four dimensions: geometric, physical, behavioral, and rules. The geometric model summarizes the shape, size, location and other parameters of the physical entity; the physical model contains the physical properties of the physical object such as stress-strain, fatigue deformation, etc., which is the basis for processing and analyzing the data of the physical object. Behavioral model and rule model describe the real-time response and dynamic behavior of the model according to the experience and norms that exist in the objective world.

The "twin data" is the driving force of the entire digital twin five-dimensional model, integrating and fusing the data related to the physical model and dynamic information data, supporting the operation of the entire system, and through real-time iterative updates to fully and accurately respond to the model's state changes. Service is a collection of different functions provided by the digital twin system for users. Various types of data, models, simulation results generated by the system are displayed to the user in the form of platforms or software to meet the different needs of users. "Connection" is a physical entity, virtual entity, services, the three communication bridge, so that the three in the system operation to maintain consistent synchronization, and to ensure that the twin data and the three interact to drive the operation. This is illustrated in the following Fig. 1.

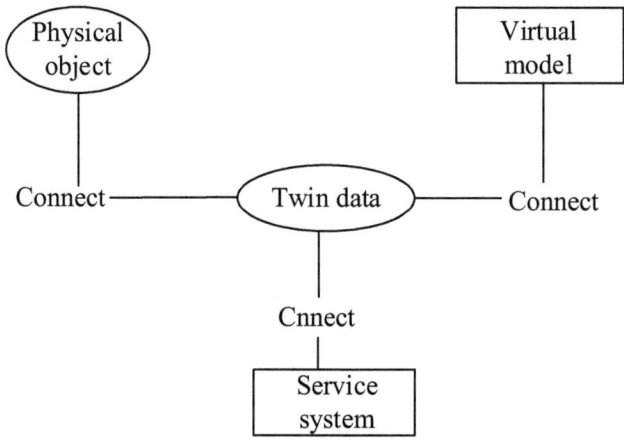

Fig. 1. Five-dimensional structure of the digital twin

To streamline the production process while enhancing modeling efficiency and authenticity, the key approach involves leveraging Catia software to meticulously design the component model of the physical device, then import it into 3ds Max software, assign material maps to it and lightweight the number of triangular patches of the model, and finally integrate and import it into Unity3D software to complete the rendering and loading of the device virtual model. Actual measurement of physical size parameters of the equipment, followed by modeling of various components of the equipment in the Catia software environment. After completing the drawing of the part model, integrate and assemble it based on its hierarchical architecture and parent-child relationship, convert the equipment into the corresponding file format, and export it. At this point, a preliminary 3D model of the device has been obtained. Import the made device model into 3ds Max and assign corresponding maps and materials, which can improve the accuracy and authenticity of modeling. At the same time, lightweight the overall model in this process, reducing the burden of subsequent rendering on the host image processing function. Import the model into Unity3D for device rendering, and improve the rendering effect by adding appropriate light and shadow components. The corresponding components are added to the virtual device, and the model-based dynamic behavior research enables it to achieve synchronous operation with the physical device.

2.2 Collecting Data on the Operating Status of the Equipment

In the distribution network, use situational awareness to focus on relevant elements in a special and complex environment in a specific time. Through the real-time acquisition, understanding, display and analysis of the key security technology element data that can quickly cause major changes in the future network situation, and the prediction of its recent development and trend, the relevant decision analysis and emergency action can be timely carried out. During data analysis, an extensive compilation of operational data from pertinent network equipment is imperative. Post-processing ensures the harmonization of diverse data types into a consistent, unified format. The converted data unit contains a vast amount of information, but distinguishing between real data and fake data is extremely challenging. Therefore, The preprocessing of interference data is paramount for attaining precise insights into equipment operational status. Confronted with vast data volumes, leveraging big data mining technology for in-depth processing serves as the initial step, mining valuable data hidden within it, and exploring its intrinsic value in depth. Subsequently, through further transformation, these data are transformed into a more easily interpretable and analytical form. This processing process includes two core steps: predictive mining, which mainly relies on the analysis and inference of past data; Describing mining focuses on describing the fundamental characteristics of existing data in a database. The mode driving mode used by DMT technology is mainly based on the relevant security system dynamics. The driver of the security system dynamics is regarded as a more integrated application mode. Through this mode, the real-time network state environment can be considered more fully, and some network data information collected can be processed and analyzed, In this process, identify some threats and risks existing in the transmission data, as well as related threat paths, and locate and record the threats after determining them. In this process, it is also able to accurately clear the threat instructions of a certain network security data information.

Since the processing of real-time data has high requirements, and real-time data has the characteristics of massive, time sequence, strong correlation and high coupling, there are often data redundancy and uneven problems, which cause a lot of difficulties to the data acquisition. Therefore, the storage mechanism and continuous query access technology of real-time data are studied. Finally, a real-time data processing method is proposed for real-time data characteristics to provide high-quality data support for the subsequent virtual-real data fusion. In view of the continuity of real-time data, the operation state data flow is usually defined as a collection of infinite sequences arranged in the order of time increment, that is,

$$S = \{ S1, S2, ..., St\} \quad (1)$$

In the formula, S is the sequence appearing at time t; A sequence, in turn, is defined as a tuple of consecutive and ordered points, with the expectation of data timeliness $E(I)$ introduced in order to reflect the real-time nature of the data, to represent the life cycle of real-time data. In the following, the data timeliness expectation will be derived according to the relevant formulas and processes: Define the data collection that $S = \{s1, s2, ..., sn\}$, defining the probability density distribution function $P = \{p1, p2, ..., pn\}$, $F(t)$ indicates the probability that the data Si is still valid in the time frame of [0, t]. First, get the value of (Ii) for each data item in the data set Si, (Ii) can be expressed as follows:

$$EIi = \int_0^{+\infty} tFi(t)dt \qquad (2)$$

In the formula: $Fi(t)$ is to use the time-dependent probability function; real-time data in the process of generating, the time-dependent probability function should satisfy the normalization, and obtain the aging expectation for the data set S. In order to achieve real-time data collection, it is necessary to preprocess and redundantly process the real-time data through the data processing layer. And visualize the data through the data application layer, using a time-based dynamic window model to meet the continuous query requirements in streaming data, with all query results located within that window. As new data arrives, the window shifts, continually yielding pertinent results until query completion Assuming that the size of the window is M, which requires a continuous query timeframe between $[0, T]$, only the most recent M elements will be considered in the query process. Then the range of real-time data obtained from successive queries will be in the range of $[\max(0, T - M + 1), T]$. That is, the data in that range are located inside the sliding window, indicating the result of a continuous query of the real-time data in the $[0, T]$ time. Stream data, generated in real-time, enters the data stream processor, which encompasses a window controller, query executor, and query results. The window controller governs window formation, while the query executor regulates the sliding granularity of the window. The continuous query sliding window time interval confines the results to within the window's execution, and ultimately, the window outputs the data sequence's results. Streaming data are continuously queried in a window model, when a continuous query is performed at $T1$ time, the defined real-time data $T1$ will be obtained from the continuous real-time data within the time period is displayed in the window. Since it is a continuous query, when it comes to $T2$ time, the sliding window will slide forward to get another new continuous data in the window and update the original data in the window. According to the above process, take t as a sliding interval, periodically update the sliding distance of the sliding window, replace the original data with new window data, get the data of continuous query while sliding, and output the results in the form of set sequence.

2.3 Association Rule State Volume Parameter Extraction

Collect the operation status data, preprocess the data, extract the effective operation status data, analyze the status data by using reasonable methods, derive the operation status of the equipment, understand the possible failure of the equipment in advance, calculate the failure rate of the equipment, and finally conduct risk assessment of the equipment, comprehensive economic efficiency, optimize the maintenance strategy model analysis, and formulate a reasonable and effective maintenance plan [7]. Distribution network equipment in actual operation, distribution points are numerous and extensive, equipment decentralization, therefore, the change of its operating state is easily affected by internal performance, external environmental conditions and other factors. Therefore, the index factors of its health state become very complicated, and the number of parameters and the construction of the system also become very complicated. If all the basic parameters are extracted, it will not only reduce the monitoring efficiency, but also lead to the variability of the monitoring results. Moreover, with the intelligent development of the power grid, the state data of the equipment operation will grow continuously. It is necessary to extract the state quantities that can accurately reflect the real operating state of the equipment, and obtain the scientific key parameters of the state. The establishment of the parameter system can not only reduce the complexity of condition monitoring, but also ensure the accuracy of monitoring, and eliminate the parameters that are not related to the faults and retain the original state quantities. Therefore, this chapter proposes a method of extracting key state parameters of equipment operation based on correlation rules. The association rules can reflect the correlation between things, through the association rules algorithm can be further mined out the correlation between transactions, select the frequent items and attributes of the event of the entire subset, and find out the correlation between them, and even can be one of the things to predict the occurrence of other things. In the association rule algorithm, the database of things is written as I, the subset of things in I is i, then $I = \{i1, i2, .., in\}$, N is the number of all subset transactions in the database, recorded as $|I|$. If a subset of things is $ik = \{\sigma 1, \sigma 2, ..., \sigma k\}$, k is the number of all terms in i, set Y as the collection of all items contained in I, any subset X in Y is called the term set, if $|X| = k$, then the set X is called $k-$ item set. In the transaction database, the number of transactions with a particular item set X is called the frequency or support count of the item X, denoted as $f(X)$. Then the proportion of the total number of transactions containing the item set X in the transaction database is called the support degree of the item set X, denoted as sup(X), can obtain the probability $p(X)$ of the set of terms X is:

$$\sup(X) = p(X) = \frac{f(X)}{|I|} \qquad (3)$$

In the formula: sup(X) is the threshold for minimum support; when, the degree of support for X is greater than the minimum value of that of sup(X), it becomes the frequent term sets of X. If the set of terms, $X \subset I$, $Y \subset I$ and $X \cap Y \neq \phi$, then $X \rightarrow Y$ may form an association rule, and X is the antecedent, denoting the premise. Y is the posterior term, indicating the conclusion. The association rule embodies the likelihood of an association between the item set X and Y corresponding to the two sets of variables. Association rule support: itemset $(X \cup Y)$ indicates a database I contains the number of transactions at the same time in X and Y, the support degree of $(X \cup Y)$ is the support for the $X \rightarrow Y$, is the number of transactions in the database $(X \cup Y)$ [8]. Association rule confidence: Where the confidence of $X \rightarrow Y$ is the ratio of X, Y in I, is the conditional probability $p(X|Y)$, denoted as $con(X \rightarrow Y)$, then its expression is:

$$con(X \rightarrow Y) = \frac{p(X \cup Y)}{p(X)} \qquad (4)$$

In the formula: $X \rightarrow Y$ is the confidence level. To build a rule base, usually set the minimum value of support, the $\sup(X \rightarrow Y)_{\min}$ and placement confidence minimum $con(X \rightarrow Y)_{\min}$. The purpose of the association rule is to mine from a specified database greater than all items of $con(X \rightarrow Y)_{min}$ and $\sup(X \rightarrow Y)_{min}$, to establish the associated rule base. Support and confidence metrics quantify the validity and certainty of association rules within a database set. Support gauges the rule's prevalence and significance in the transaction database, with higher values signifying stronger correlations between X and Y. Meanwhile, confidence assesses the reliability of the correlation rule, with increased confidence reflecting a firmer relationship between Y and X within the database. According to the association rule algorithm, the key influence parameters can be selected from the many state quantities of the equipment, and using the concepts of support degree and confidence degree of the association rule, the intrinsic connection between the equipment and the basic parameter can be calculated, and the state model between the equipment and the basic parameter can be established. According to the principle of judging frequent itemsets of association rules, which is to adjust the threshold based on the quality and quantity of generated rules, a balance point can be found, which can generate meaningful rules without generating too many useless rules. Finally, the hyperparameters of association rules are selected as follows: frequent itemsets with association rule values greater than 0.7 and confidence levels greater than 0.5 are selected as key hyperparameters. In order to facilitate the calculation, the representation of the state quantities is therefore simplified by using, for example Y_i to represent the line components. X_i denotes the set of state quantity terms, where i represents the part number. j is the key state parameters of the components are analyzed and calculated. Obtain the key state parameters of the distribution network equipment, and process them to establish the key state hierarchical model of the distribution network equipment.

2.4 Modeling the State-Volume Hierarchy

This section mainly elaborates on the theory of grid risk, firstly introduces the basic concept of risk, secondly clarifies the grid risk assessment process according to the concept of risk in the field of electric power, explains in detail the details of the process, and finally introduces the grid risk index system according to the requirements of the risk calculation, which lays the foundation of the risk calculation of the power system later on. Hierarchical analysis is applied and combined with the results of expert consultation to determine the weighting coefficients of the state quantities and to score the health degree of the equipment.

Taking the distribution network transformer as an example, a hierarchical model of transformer state quantities is established, and its eigenvectors are obtained to determine the weight coefficients of each state quantity. The basic idea of hierarchical analysis is firstly to hierarchize the complex problem, according to the nature of the problem and the overall goal of the problem will be decomposed into the indicator layer, the criterion layer and the target layer, for the same level of the amount of two comparisons, the use of relative scales to quantify the importance of its occupancy in the level [9]. Analytic Hierarchy Process (AHP) is a practical method used to transform qualitative quantities into quantitative analysis, while also being able to objectively describe and subjectively make judgments. Based on the constructed hierarchical indicator system, experts will compare the importance of each indicator at the same level and construct a judgment matrix for each indicator at each level. The eigenvectors and eigenvalues of each judgment matrix are solved. According to the structure and characteristics of the equipment to be evaluated, it is decomposed into three levels. The target level is the health degree H of the equipment, according to the physical structure and function of the equipment, which describes the health status P of each component of the equipment; The state layer is the state parameter C that describes the health of the different components. The breakdown of the structure is shown in Fig. 2.

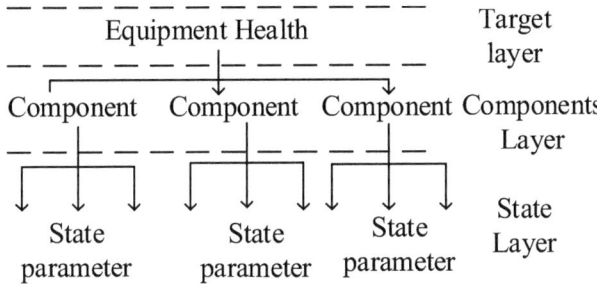

Fig. 2. Hierarchical analysis structure of power equipment state quantities

From the Fig. 2, it can be seen that the health of the equipment, the H can be done through a series of state parameters. Using x to express, the expression is defined as follows:

$$H = f(x1, x2, ..., xn) = \sum_{i=1}^{m} \sum_{j=1}^{n} Sij \times Pij \quad (5)$$

In the formula: Sij is the deduction value for each state parameter; the Pij is the weight value of each state parameter relative to the overall goal. Subsequent AHP calculations are carried out around the hierarchical structure diagram. According to the state parameters listed by the power grid company, the hierarchical structure model is formed by sorting and classifying the components. In the model, the transformer state parameters are shown in the Table 1:

The sum-product method is used to calculate the approximate eigenvectors. Firstly, the columns of the matrix are normalized, and the expression of normalization is as follows:

$$aij = \frac{aij}{\sum_{1}^{n} aij} \quad (6)$$

In the formula: aij are the eigenvectors. After normalization, each column of the matrix is summed by rows and the resulting vectors are normalized again to compute the resulting vectors, the $(w1, w2, ..., wn)^T$, which is the eigenvector approximation solution of the judgment matrix. The expression for the computation of its maximum eigenvalue is given by:

$$\lambda \max = \sum_{i=1}^{n} \frac{(AW)i}{n\omega i} \quad (7)$$

Table 1. State parameters of the transformer

Serial Number	State parameter	Serial Number	State parameter
1	Oil level in the main oil storage tank	13	Overexcitation
2	Oil leakage situation	14	External insulation condition
3	Noise and vibration situation	15	Oil level indicator
4	Operating oil temperature	16	Motor operation status
5	Winding DC resistance	17	Cooling device heat dissipation effect
6	Winding deformation situation	18	Oil level indicator
7	Winding insulation resistance and absorption ratio	19	Tapping position
8	Winding dielectric loss factor	20	Oil leakage situation
9	Number of short circuits	21	Online oil filtration device
10	Overload condition	22	Oil pressure resistance
11	Insulation of iron core and clamps	23	Pressure relief valve
12	Iron core deformation	24	Gas relay

In the formula: $n\omega i$ is the weight value of the feature matrix; the A is the judgment matrix, the formula is:

$$aij = \frac{aik}{ajk} \quad (8)$$

In the formula: aik is a positive matrix; the ajk is a negative matrix. According to matrix theory, the judgment matrix has a unique characteristic root when the consistency condition is satisfied, that is to say, except that the $\lambda \max = n$. By analogy, ranking is carried out from bottom to top according to the hierarchical analysis structure. The calculation formula is:

$$w_i^{k+1} = \sum_{j=1}^{n} w_{ij}^k w_j^k \quad (9)$$

In the formula: k is the number of states; W w_{ij}^k is the relative weight of the state i i on a property A A in layer k $k+1$. n is the number of sub-attributes A. Through the analysis of distribution network electrical equipment status and maintenance strategy, an optimization model for equipment status monitoring is formed based on power system equipment status monitoring and evaluation, based on transmission and substation equipment status analysis and taking into account the risk analysis of power equipment.

Condition monitoring based on digital twin technology, not only to establish the actual organization based on the twin model, but also need to carry out institutional data processing. As one of the three core elements of digital twin technology, how to collect the twin data from the real mechanism that can characterize the complete operating state of the equipment and drive the twin model to realize the condition monitoring of the equipment is a problem that needs to be solved.

3 Distribution Network Equipment Operation Condition Monitoring

Import the equipment model into 3ds MAX, and directly eliminate redundant points and lines in the geometric model. Simplify faces and points through Polygon Cruncher tool or simplify complex models through shell extraction. Assemble the lightweight model in SolidWorks. According to the analysis of the established state hierarchical model, the influencing factors are analyzed from multiple aspects, thus forming a multidimensional analysis method. Each analysis method includes data analysis and model analysis. Finally, realize multi-dimensional and multi-mode stereo analysis of equipment. The formula for constructing the analysis model is:

$$E = (m_a, m_m) \tag{10}$$

In the formula: m_a denotes the analytical model, the m_m represents a data-driven fault analysis model. The collected operation data of distribution network equipment is used as input to carry out intelligent learning through the model. This enables different types of fault data to form corresponding rules for operation status diagnosis and analysis. The model driven fault analysis mainly relies on CAE software. On the basis of establishing environmental constraints, the real-time movement mechanism simulation and vibration characteristics analysis of the equipment can be carried out, and the fault can be finally determined through the analysis results. As the running time increases and the surrounding environment changes, the virtual model of the equipment entity at different times in the iteration process [10]. Long time operation will lead to some failures of equipment. Through real-time failure analysis and prediction, it can be judged whether there will be failures at a certain time. Once the equipment fails at a certain time, first analyze the fault type, fault location, fault model and other information, then conduct fault alarm on the interactive interface, and update the model in the information space through the analysis of the fault information, so as to achieve the synchronization of the running state of the equipment in the physical space and the virtual model in the information space. K-means algorithm is introduced to discretize the amplitude of operation data. Secondly, the initial threshold is obtained through the residual calculation model $R1$. When modeled offline, the E its prediction error is greater than a preset initial threshold $R1$, when the discretized time series operation data is used as input, the association rule pattern is used as a criterion of whether to update the model online or not. If there is a matching association rule pattern, it means that the model failure is caused by the change of the rule pattern, then the weights of the fully connected layer in the model are updated online, and then the prediction is performed to complete the state monitoring. If there is no matching association rule pattern, the model mismatch is due to anomalies and no online update is performed.

4 Example Analysis

4.1 Experimental Preparation

In the experiment, the ultrasonic online monitoring of transformer partial discharge faults fixed the sensor on the outer wall of the transformer oil tank ultrasonic sensor, received and converted its internal partial discharge electrical signal, and used it as the partial discharge monitoring sample for this experiment. In order to control potential deviations, the measurement value of the electrical signal is converted to a set warning value for the number of partial discharges in the transformer. If the warning value is less than the set warning value, the measurement of the partial discharge signal is not far from the transmission and can only be carried out in the case of partial discharge. Only when the partial discharge is severe enough, the on-site measurement signal is transmitted to the monitoring room of the distribution substation to achieve fault alarm and control the internal partial discharge monitoring sample error in a relatively simple way. The wireless temperature measurement system for transformers uses traditional thermocouples as temperature sensing elements, supplemented by microelectronic processors, to achieve the collection, storage, and remote transmission of on-site measurement signals, which are used as temperature monitoring samples for transformers in this experiment. The ambient temperature at the transformer work site has a significant impact on the actual measured working temperature. In order to control the potential relative error of working temperature measurement under this influence, the following operations are carried out: first, thermocouples are installed in multiple positions of the radiator and outlet temperature sensors inside the transformer box, and the measurement error of relative temperature difference is controlled through multi-point monitoring. Secondly, with the help of wireless technology, online monitoring technology for transformer temperature in local area networks has been completed, which is well adapted to the integration of new distribution equipment status monitoring capacity. The new passive wireless device SAW chip is fixed on the switch contacts inside the switchgear busbar through circular metal fasteners, and the connection points are coated with thermal conductive adhesive to improve the heat transfer monitoring effect. Due to the wireless transmission of sensor signals, there is no need to consider the insulation issue of the equipment. The sensor antenna is installed in the instrument room of the box, without the need for any modifications to the switchgear. The signal line is led out through the side hole of the box, achieving on-site display of temperature monitoring and remote transmission of data.

In the experimental section of this article, better monitoring results were achieved through comparison with traditional monitoring methods.

In the experimental part of this paper, by comparing with the traditional monitoring methods, more adequate monitoring effect is obtained. The experimental sampling time of the monitoring data samples was 250s. The hardware and software environments for the experiments in this paper are summarized in the Table 2.

In the test, distribution network equipment units are selected for research. The unit is mainly composed of motor, transformer and other parts. Based on Unity3D platform, the digital Li Sheng model is built. In this test platform construction. C# script based on Unity3D and SQL Server database realize the back-end development of the platform. Combining the advantages of ECharts for data visualization, ECharts is integrated with

Table 2. Experimental equipment

Project	Configuration	Parameter
1	Operating system	NVIDIA Titan Xp
2	Development language	Python2.6
3	Third Development Language	Anaconda 4.3.1
4	Graphics card	Inter 1080 i7
5	Memory	3.40G
6	CPU	48G
8	Programming tools	Pycharm
9	GPU	NVIDIA

Unity3D platform to complete the front-end development of the platform. In addition, the function is developed in Matlab software. The construction measurement is also more accurate.

4.2 Results and Conclusions

After the preparation of the above experiments, different equipment online condition monitoring methods can simultaneously monitor the online condition of multiple equipment., ensuring the robustness and applicability of the experimental results through 3000 experimental samples. Calculate the average correct error value of the monitoring, so as to determine the optimal monitoring method. The results are shown in the Fig. 3.

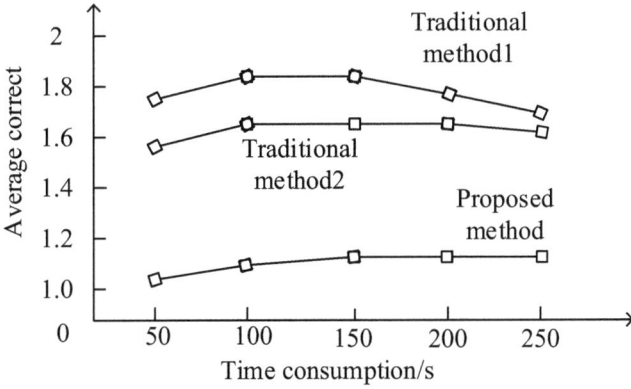

Fig. 3. Comparison of the results of the deviation

As can be seen from the experimental results, the average correct deviation of the proposed method is between 1 to 1.2, which is a large difference between the two, which is due to the fact that the size and position of the buffer zone and the size and position of the sliding window may be different in the processing of data at the same position of the data segment under different methods. The average correct deviation of monitoring for traditional method 1 and traditional method 2 is higher than 1.4. The comparative data proves that the monitoring deviation of our method is the smallest, which verifies the effectiveness of our monitoring method. At the same time, the monitoring recall rate was analyzed and the specific results were obtained in the Fig. 4.

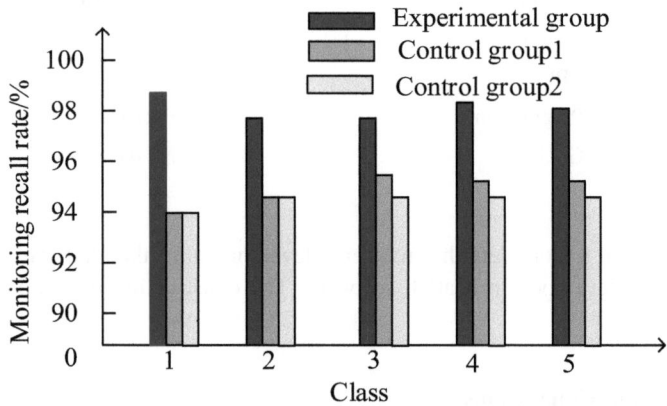

Fig. 4. Comparison of results of monitoring recall rates

In the experiment, the recall rate of the experimental group is high, reaching more than 96%. The monitoring recall rate of traditional method 1 and traditional method 2 is less than 95%. The above comparative data proves that this method has obvious advantages, especially in terms of recall rate.

In summary, the monitoring method proposed in this paper performs well in terms of recall and can effectively improve the monitoring accuracy of the target object. This conclusion provides a valuable reference for research and practice in related fields. This is because compared with other methods, this method adopts a monitoring method that combines digital twin technology and association rules, which can comprehensively and multi angle monitor the operation status of equipment. The digital twin model can reflect the real-time operation status of equipment, providing more comprehensive and accurate information support for operation and maintenance personnel. At the same time, the association rules are used to analyze the monitoring data, and the correlation relationship between the equipment operation status and various factors is found. These correlation rules not only help to understand the operation law of the equipment more deeply, but also can provide strong support for predictive maintenance of the equipment. In the future, the method will be further optimized to improve the monitoring accuracy and efficiency, so as to provide more comprehensive and efficient technical support for the operation and maintenance management of distribution network equipment.

5   Conclusion

To mitigate deviations in online monitoring of distribution network equipment status, a hybrid method integrating digital twin technology and association rules was explored. This approach constructs a virtual model to mimic equipment operation, leveraging association rules to uncover intricate device interdependencies, enhancing data analysis efficiency and precision. Experimental findings confirm reduced monitoring bias and improved recall. To amplify the study's impact in monitoring distribution network equipment status, optimizing Analytic Hierarchy Process model parameters and boosting monitoring efficiency are crucial avenues for future advancements.

References

1. Rajasekaran, A., Kalyanchakravarthi, P., Subudhi, P.S.: Anomaly detection of smart grid equipment using machine learning applications. Distrib. Gener. Altern. Energy J. **1**(7), 37518–37523 (2022)
2. Zhao, Y.L., Wang, Y.F., Ji, J., et al.: Real time acquisition of terminal equipment operation data of distribution network based on LMS algorithm. Electron. Des. Eng. **30**(18), 189–193 (2022)
3. Wang, S.W., Fu, B.: Design of an automatic monitoring system for the operation status of intelligent distribution networks. Electron. Technol. **52**(4), 313–315 (2023)
4. Li, F.J., Wang, Z.X., Sun, Q., et al.: A model for mining association of equipment faults in distribution network considering meteorological factors. Modern Power **40**(4), 605–613 (2023)
5. He, T., Zhai, W.L., Feng, Y.H.: Leakage current monitoring of low voltage distribution network based on extended Kalman filter. Inf. Technol. **46**(1), 148–152 (2022)
6. Jialei, S., Lijun, J.: Online monitoring method for series compensation capacitor fault in 10 kV distribution network based on FP-growth algorithm. Electr. Autom. **44**(1), 41–43 (2022)
7. Kousar, S., Zafar, N.A., Ali, T., et al.: Formal modeling of IoT-based distribution management system for smart grids. Sustainability **14**(8), 1–25 (2022)
8. Liu, S., Wang, S., Liu, X., et al.: Human memory update strategy: a multi-layer template update mechanism for remote visual monitoring. IEEE Trans. Multimedia **23**(3), 2188–2198 (2021)
9. Zjavka, L.: Power quality statistical predictions based on differential, deep and probabilistic learning using off-grid and meteo data in 24-hour horizon. Int. J. Energy Res. **46**(8), 10182–10196 (2022)
10. Jawthari, M., Stoffa, V.: Relation between student engagement and demographic characteristics in distance learning using association rules. Electronics **11**(5), 724–729 (2022)

Real-time Analysis of an Embedded CNC System Based on Information Fusion in the Internet of Things

Niyan Wu(✉)

Intelligent Manufacturing College, Yibin Vocational and Technical College, Yibin 644000, China
15892521108@163.com

Abstract. Given the challenge that the advancement of numerical control systems remains confined to the traditional PC+control card architecture, which is no longer adequate for contemporary demands, we incorporate the concept of the Internet of Things for information fusion to conduct real-time assessments and explorations within embedded numerical control systems. Through the Internet of Things technology, realize the fusion of multi-source and heterogeneous information of embedded CNC system operation. Process scheduling for embedded CNC system. Design CFS scheduler to realize real-time process scheduling. The real-time task scheduling delay test and context switching test prove that the optimized embedded CNC system has high real-time performance.

Keywords: Internet of Things · Embedded · Real-time · CNC system · Information fusion

1 Introduction

As one of the national economic lifeline industries, manufacturing plays an important role in measuring the national economic influence. In the process of optimizing and upgrading traditional NC machining equipment, embedded system, a new computer technology, has gradually become the focus of the current manufacturing transformation research because of its characteristics of "customization for specific applications".

As the core of modern intelligent equipment, embedded system plays a great role in various industries and shows great potential because of its rich development resources, strong professionalism, small volume and power consumption, and simplified system. It is a key step to improve the embedded system and create a platform that meets the requirements of NC machining to replace the traditional PC platform. In the field of embedded development, Linuz operating system is open source and free [1], which is used for the maintenance and development of open source community. At the same time, developers can modify the kernel configuration and compile it according to their own wishes. Although Linuz system has obvious advantages and great development potential, it is not a real-time system. This shortcoming makes it unable to be well applied to key fields that require high real-time performance, such as numerical control processing. The numerical control system is highly sensitive to time, and once its real-time performance is not met, it will cause irreparable consequences. Therefore, by improving the real-time performance of Linuz, an embedded system that can meet the real-time performance of CNC equipment in extreme conditions is constructed [2]. The combination of embedded platform and WEDM CNC system plays an important role in promoting the intelligent upgrade of WEDM machine tools, which can open up a broader market for CNC system and create higher commercial value for society and enterprises.

2 Multi-source Heterogeneous Information Fusion for the Operation of Embedded Numerical Control Systems Based on the Internet of Things

In an embedded numerical control system, the Internet of Things technology can facilitate device interconnection and communication, thereby enhancing the system's intelligence and automation. Additionally, information fusion techniques can be applied to integrate multi-source heterogeneous data, further boosting the system's sensing capabilities and decision-making abilities. Consequently, the integration of multi-source heterogeneous information within an IoT-based embedded CNC system serves as the foundation for a real-time embedded CNC system [3]. By achieving this integration, can enhance the system's sensing and decision-making capabilities, providing a more robust data foundation for real-time system analysis. At the same time, real-time analysis can also evaluate and optimize the effect of information fusion to further improve the performance and stability of the system. Combined with the Internet of Things technology, the embedded CNC system is established to run a multi-source heterogeneous information fusion model, as shown in Fig. 1.

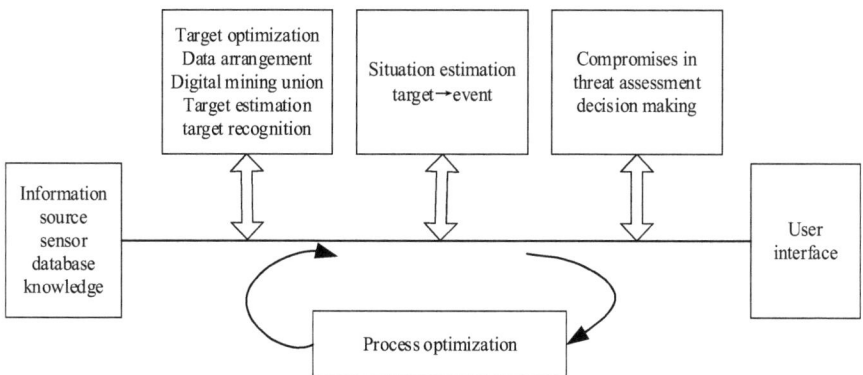

Fig. 1. Multi-source heterogeneous information fusion model for embedded CNC operation

Multi-source heterogeneous information fusion in the context of the Internet of Things (IoT) builds upon the multi-sensor information fusion approach, tailored to the unique characteristics of the IoT [4]. This fusion process not only retains the attributes of multi-sensor network data fusion but also primarily aims to address the challenges posed by multi-source heterogeneity. Due to the complex operating environment of the system and the different sources of relevant parameters, the data obtained belong to high attribute dimension data. The specific method of feature fusion for this kind of data is as follows: an efficient algorithm is given. This algorithm is based on the idea of division, the high attribute dimension data into multiple relatively low attribute dimension data, first of all, these relatively low attribute dimension data processing, and then use the results of these processes to calculate the necessary characteristics of the original high attribute dimension data, to ensure that the results obtained are the same as the results obtained by directly calculating the high attribute dimension data. The algorithm is divided into four steps: data preprocessing and modeling, high-dimensional data division, calculation of the set of kernel attributes at all levels, and feature selection calculation, as shown in Fig. 2.

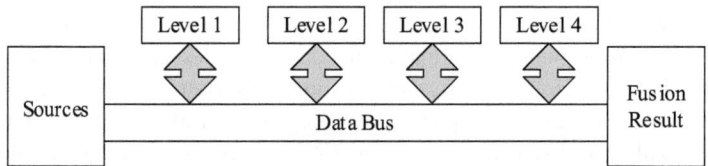

Fig. 2. Flowchart of the feature fusion algorithm for high attribute dimensions

Step 1: Data Preprocessing and Modeling

Data preprocessing primarily involves techniques such as quantization mapping and discretization of continuous data to ensure that sensor-collected data within the same attribute type are comparable, enabling the data to exhibit corresponding discriminative abilities.

Step 2: Division of High-Dimensional Attributes

To enhance the efficiency of feature fusion calculations for high-dimensional IoT data, this algorithm adopts a partitioning approach [5, 6]. It segments the data into multiple subsets with relatively lower dimensional attributes based on the division of high-dimensional attributes. These subsets are processed individually, and their results are utilized to compute the essential features of the original high-dimensional data, ensuring that the outcomes align with those obtained from direct calculations.

Data can be partitioned based on attributes following certain principles, such as grouping attributes by sensor type, where data from the same sensor type is assigned to the same subset, creating relatively low-dimensional information systems within each partitioned subset [7].

Step 3: Calculation of Core Attribute Set for the Full Attribute Set

First, the discernibility matrix for each partitioned subset is computed, and the core attribute sets for each subset are identified through the matrices. Then, theorems are applied to determine whether the core attributes of each partitioned subset are also core attributes of the full attribute set, thereby identifying the core attribute set for the entire attribute set.

Step 4: Attribute Reduction for the Full Attribute Set

The objective of attribute reduction is to eliminate redundant data while maintaining a certain level of data quality and the underlying knowledge, simplifying subsequent data processing computations and enhancing processing efficiency. The heuristic search method applied in this paper uses attribute significance as the heuristic function, starting from the core attribute set and ending when the partition of the universe of discourse based on equivalence relations is achieved. The algorithm checks if the current attribute set is a reduction. If yes, it terminates; if not, the attribute with the highest significance among the remaining attributes is added, and the check is repeated until the termination condition is met, ultimately yielding a reduction of the full attribute set. Typically, this reduction represents the simplest form possible.

3 Process Scheduling for Embedded CNC Systems

A process is an execution process of a program on the system. In an embedded numerical control system, a program uses system resources through a process, including memory space required for program execution, peripheral input, other programs or files and other data, such as opening a video editor or a text editor when using a computer. Since the CPU can only run one process at the same time, when multiple processes are opened at the same time, some processes need to occupy the CPU for a long time, while some processes need to wait for the user's operation before responding immediately to occupy the CPU, so the problem to be solved by process scheduling is to decide when to run which process and how long to run the process. Process scheduling is an important guarantee for the normal operation of the system, and its advantages and disadvantages are also an important indicator to measure the excellence of an operating system. The system allocates the time that processes occupy system resources reasonably through the scheduler, and provides maximum fairness to each process in the system according to the required computing power. At the same time, it fulfills the demands for prompt responsiveness, high processing capacity, and minimal power usage, among others. or from another perspective, it tries to ensure that no process is being treated unfairly. According to the different needs of the process, the process is divided into real-time process, interactive process and batch process, and the latter two are ordinary processes. For real-time processes with strong scheduling requirements, the response time should be as short as possible to complete the task within the specified time. Interactive processes often need to communicate with users. After receiving the data sent by the user, the process must respond quickly, or it will create a bad experience for the user. Batch processes are generally executed later, and do not need to respond quickly. For example, compilers of program languages, database search engines, and scientific computing pay more attention to throughput. The core of embedded CNC system uses the core scheduler to schedule the process, that is, the main scheduler and the periodic scheduler. The main scheduler is used to switch the front and rear processes on the CPU; The periodic scheduler periodically updates scheduling related information according to the clock interrupt sent by the timer, and is not responsible for process scheduling. The scheduler manages processes through operation scheduling classes, and the kernel associates them to different structure instances (members are pointers to functions) according to different types of processing process functions, which are called scheduling classes. The scheduling class is used to implement different scheduling algorithms. One scheduling class can implement multiple algorithms. Its core scheduling architecture is shown in Fig. 3.

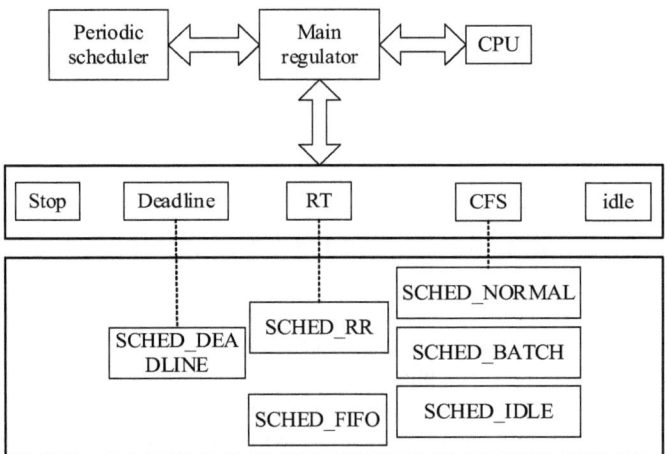

Fig. 3. Kernel scheduling architecture for embedded CNC

The system needs to process a lot of processes, which may be in runnable and sleeping states. Process scheduling first needs to add runnable processes to the CPU's running queue. The process currently executed by the CPU is known as the active process and is not present in the execution queue. Process scheduling primarily involves alternating the active process. Switching the active process involves two key procedures:

Step 1: Set the rescheduling flag;

Step 2: Check whether the current has set the rescheduling flag when the system call returns or the interrupt returns. If yes, call schedule to switch processes.

For the first step, the hardware timer will generate periodic clock interrupts as the beat of system operation. When timer interrupts are generated, the CPU will execute the interrupt handler and judge whether the current process needs to be preempted. This process requires a specific scheduling algorithm. The embedded CNC system will implement the scheduling algorithm through the specific function of the scheduling class. The corresponding scheduling class will calculate in the timer interrupt processing to execute the above judgment. If preemption is required, set the rescheduling flag for the current process. In the second step, the function in the main scheduler is used to switch processes. When the system call returns or the interrupt returns, check the current rescheduling flag. If set, the calling function selects the process to run from the running queue through the scheduling class.

In the specific scheduling strategy, the meaning of each structure member is shown in Table 1. Because different processes require different scheduling strategies, the kernel implements a variety of scheduling classes. However, not every scheduling strategy corresponds to a unique scheduling class. These scheduling classes, from top to bottom, are prioritized from high to low.

Table 1. Record of the meaning of each structure member in the scheduling strategy

Number	Function Name	Function functionality
1	Next	Connect to the next priority scheduling class
2	Enqueue_task	Add runnable processes to the run queue
3	Dequeue_to_task	Remove a process from the running queue
4	Yield_to_task	Make the process actively abandon the CPU
5	Check_preempt_curr	Check if the process entering the queue is preempting the current process and selecting the next suitable process from the running queue
6	Pick_next_task	Put the running process back into the queue
7	Put_prev_task	Call this function when the process scheduling strategy changes
8	Set_curr_task	Statistical information of periodic update processes
9	Update_curr	Statistics for updating the queue

The kernel uses the task_struct structure to describe a process, making it easier to manage processes. This structure contains the detailed information required to describe the process and the resources required for the program to run, such as the status of the process, the files used, the suspended signals, and the relevant address information. Table 2 lists the fields related to scheduling.

Table 2. Process description constructs

Number	Variable or structure	Explain
1	Volatile long	Process status, indicating whether it is running
2	Const struct sched_class*sched_class	Scheduling class pointer, pointing to the scheduling class to which the process belongs
3	Struct sched_entity se	Process scheduling entity structure members
4	Struct task_group*sched_task_group	Group scheduling related
5	Pid_t pid	Process Distinguished Variables

Real-time systems need to meet as many task timeframe requirements as possible when scheduling the system, and then meet the minimum response time under the requirements of the task timeframe. The timeframe of the task can be delineated by its cycle, initiation time, expiry date, and duration of implementation.

Cycle time indicates how often a task needs to be executed, and its value is the difference between the arrival times of two neighboring real-time tasks, the T_i denotes a periodic task, the τ_i is denoted as its cycle, and the arrival time is the first day of the cycle of the mission i the moment when a job arrives in the system, denoted as ϕ_i. Deadlines are divided into relative and absolute deadlines. A relative deadline is the length of the time period between the moment of arrival of the task and the absolute deadline. The absolute deadline is the last moment when the task must be completed. Relative deadline is a period of time, absolute deadline is a point in time. The relative deadline and the absolute deadline are denoted, respectively as D_i and d_i, generally default $D_i = \tau_i$, d_i can be derived from Eq. (1).

$$d_i = \phi_i + D_i \tag{1}$$

Execution time is the length of time it takes for the processor to complete the process requirements, and is usually considered to be the execution time of the most unfavorable state of the task, i.e., the maximum time that the processor needs to execute in the worst case scenario, denoted as C_i. In scheduling design analysis, the system status is a factor that must be considered. It can be measured by CPU utilization, which is recorded as U, periodic tasks represent CPU utilization through task sets, as shown in Eq. (2).

$$U = \sum_{i=1}^{m} \frac{C_i}{T_i} \tag{2}$$

According to the CPU utilization of the current task set, it can be divided into light load status and overload status [8]: when the CPU utilization is less than 1, the system is in light load status, that is, the system can meet the resource requirements of all tasks; When the CPU utilization rate is greater than 1, the task demand exceeds the upper limit of the system's service, and the system is overloaded.

According to whether the priority of the scheduling task is fixed, scheduling algorithms can be divided into static priority scheduling algorithm (RM monotone rate algorithm) and dynamic priority scheduling algorithm (EDF earliest deadline priority scheduling algorithm).

(1) RM regards the frequency of task arrival as the most important scheduling factor. The higher the frequency, the more important it is. It should be scheduled first. The priority of all tasks is determined by their frequency, that is, the priority is inversely proportional to the cycle. For a given task set $S(n)$, if the CPU utilization of the centralized task meets the following conditions, the task set is considered to be scheduled using RM scheduling algorithm, which is derived from Eq. (3).

$$U < n(2^{\frac{1}{n}} - 1) = L(n) \tag{3}$$

In the equation, $L(n)$ represents the upper limit where the task set $S(n)$ can be scheduled by the RM algorithm. If the number of tasks in the task set gradually increases to infinity, the limit value of $L(n)$ is about 0.69, that is, the CPU utilization limit value under RM algorithm scheduling is 69%. RM algorithm is widely used in real-time systems because of its simple structure and low scheduling overhead. However, due to the constraints of its schedulability conditions, the CPU utilization is not high, and it is difficult to give full play to the maximum performance of the CPU, which leads to its inability to be well applied to CNC system tasks. At the same time, the algorithm stipulates that the longer the cycle, the lower the priority. If multiple short cycle tasks are scheduled, these tasks may occupy the CPU of low priority processes for many times, resulting in frequent context switching, which brings more running overhead and consumes a lot of time, resulting in low priority processes missing their time limit requirements, which has a bad impact on the real-time performance of the system [9]. On the other hand, the real-time periodic tasks in the EDM numerical control system run alternately. For example, the gap voltage detection task cooperates with the gap adjustment task and the motion control task. If only the static priority cycle is short when switching tasks, the alternating error will occur. At the same time, some important real-time sudden tasks may exceed the task time limit because they cannot be scheduled, It brings irreparable losses to production.

(2) The EDF algorithm needs to dynamically adjust the priority of tasks according to the deadline, so the implementation of the algorithm requires a large system overhead [10]. At the same time, when the system is overloaded, the scheduling ability of the algorithm will drop sharply, which will cause chaos of a large number of periodic tasks. All the above conditions restrict the performance of the algorithm. The following will illustrate the scheduling process of the algorithm and the possible states when the system is overloaded to analyze and discuss the EDF algorithm.

To illustrate the working principle of EDF scheduling algorithm, it is assumed that there are three $t_i(C_i, \tau_i)$ the set of tasks that comprise the S_1 for EDF algorithm scheduling: $t_1(1, 3)$, $t_2(2, 4)$, $t_3(1, 6)$, CPU utilization rate is $1/3 + 2/4 + 1/6 = 1$, three tasks arrive at the same time at 0, according to the EDF algorithm, t_1 has the smallest absolute deadline, so it is scheduled first, and scheduling t_2 begins at moment 1 when the task is completed, t_2 run to moment 3 t_3 is dispatched, at moment 3 t_1 is ready but its absolute deadline is less than that of the first job until moment 4 t_1 is able to dispatch. Again, at moment 4 t_2 the second one is ready to work but its absolute deadline is greater than t_1. Therefore, its scheduling is delayed. This scheduling situation continues until moment 12, which is the least common multiple of the three task cycles, and then the previous scheduling situation is repeated.

In order to facilitate the analysis of the regular changes of EDF algorithm when scheduling task sets, the theory of super period is cited to illustrate that: for n set of tasks, with the hypercycle being the least common multiple of the cycles of all the tasks in the set. If the tasks can arrive simultaneously at the beginning of each cycle, then at the end of the supercycle, each task in the task set can end simultaneously. The task scheduling situation in the next hypercycle is the same as the previous one, and the arrival time and execution order of tasks in each hypercycle are duplicated in other hypercycles, so the scheduling situation can be analyzed in one hypercycle.

4 Scheduler Design and Real-Time Process Scheduling

In the same priority of the process is scheduled by the order of its order in the chain table, its structure is shown in Fig. 4.

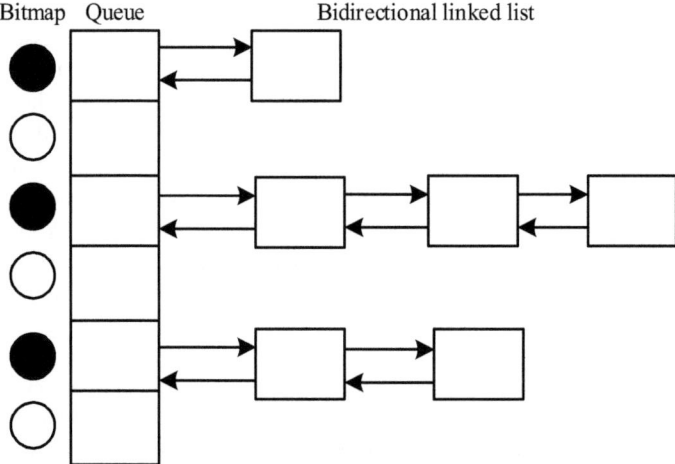

Fig. 4. Diagram of the structure of the run queue

The system overhead includes the scheduling overhead and running overhead of task scheduling and execution. Generally, the scheduling overhead of static scheduling algorithm is better than that of dynamic scheduling algorithm. For example, the RM algorithm has determined the priority of tasks and arranged their ready queues before scheduling, so that the calculation and operation time for queue maintenance are fixed, and it will not occupy too much resources, and the scheduling cost is small. For the EDF algorithm, because the priority of a task is dynamically determined by its arrival time and relative deadline, the system needs a lot of calculation and storage results to adjust the queue when managing the dynamic queue. When the number of tasks approaches a large number, the time spent on queue management increases, and the scheduling cost increases.

The running cost mainly refers to the system cost caused by task switching, which is directly related to task preemption. The running cost of dynamic scheduling algorithm is less than that of static scheduling algorithm. As previously analyzed, the readiness of high priority tasks in the RM algorithm will cause frequent task switching, while the later tasks in the EDE generally have lower priority and will not cause task switching, which is determined by its scheduling rules, so it is better than static scheduling in terms of running costs.

When the system is overloaded, in order to ensure the real limit requirements of real-time tasks and reduce a series of scheduling disorders caused by overload, the impact of the "domino effect" on all tasks can be limited to some unimportant tasks, that is, select the more important tasks within the CPU processing capacity to form a task set for priority scheduling, and organize the blocked and timeout tasks separately, When the CPU is idle, it can be scheduled to ensure the timeliness of hard real-time tasks in the NC machining system. This processing method can not only take advantage of the high CPU utilization of the EDE algorithm, but also meet the scheduling requirements of important tasks when the system is overloaded. It is suitable for industrial control work environments with a large number of hard real-time tasks.

EDF algorithm judges the schedulability of each priority task. In case of overload, a work request of an important real-time task is not responded in time due to overload. As its deadline approaches, the urgency of the process's request for execution increases. The system increases its load. When the task deadline is equal to the cycle, At this time, the time limit requirement is equal to the periodic task whose cycle is equal to its remaining time, which also reflects the need for more CPU resources to process the task. Therefore, the remaining time should be used to replace the period in the schedulability judgment.

The specific process of the improved EDF scheduling algorithm is described as follows:

Step 1: After the real-time task arrives at the system, judge the system state, if the system load U is less than 1, skip to step 2, and if the system load is more than 1, skip to step 3.
Step 2: Compare the absolute deadline of each priority chain header task in the priority queue, select the task with the smallest absolute deadline to schedule and hand it to the CPU for running.
Step 3: The system is in overload, according to the priority order and scheduling criteria for overload i.e. according to U and *shedbound* the identified set of schedulable tasks is scheduled, and the task a with the smallest absolute deadline is selected from the set of tasks.
Step 4: For the prioritization of *shedbound* of the first task in the task queue b to determine whether it is in the schedulable tasks and the scope, such as in the scope of the implementation of step 5, not in the set of schedulable tasks to step 6.
Step 5: Place the task a and b its absolute deadlines are compared and the smallest task is selected for scheduling by the main scheduler.
Step 6: Submit the previously selected minimum end-time task a to the main scheduler for scheduling.

5 Experimental Analysis

5.1 Experimental Preparation

The real-time performance of the embedded EDM WEDM CNC system is analyzed by applying the above proposed method. Figure 5 shows the structure of the embedded EDM WEDM CNC system.

WEDM will not be limited by the strength and hardness of the workpiece material and other physical properties during machining. At the same time, because it is non-contact machining, it will not produce cutting force and other macro contact forces, so it theoretically has the characteristics of machining any hard and tough conductive materials. In this system, the slow wire walking machine tool usually moves in one direction at a speed of 0.2 m/s. The copper wire is used as the tool electrode. The working fluid is mostly deionized water and kerosene. Its machining precision and surface quality are high, and the precision and surface roughness can reach 1 μm, Ra0.8 μm. The processing

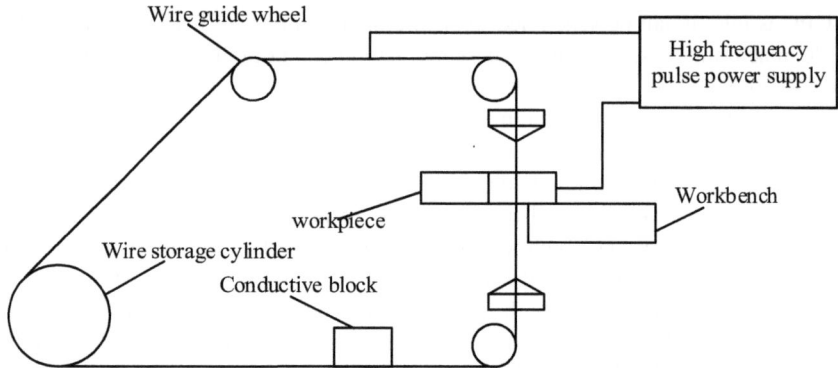

Fig. 5. Structure of the embedded EDM wire-cutting CNC system

speed of the fast wire walking machine tool is 6–12 m/s, and the materials of the working fluid are mainly emulsion, mineral oil, deionized water, etc. The electrode wire is made of wear-resistant molybdenum wire, and its fast speed will make the electrode wire wear less than the amount of workpiece erosion, so that the wear degree can be effectively reduced.

In the testing phase, the unimproved embedded wire electrical discharge machining numerical control system kernel and the kernel with the improved SPD scheduling strategy based on the EDF algorithm will be firstly programmed into the target development board. The real-time task scheduling delay will be analyzed and compared under two conditions: light load and overload. Subsequently, the context switching time, interrupt delay, and other indicators will be tested.

The hardware test environment is: the processor model is Exynos4412, the kernel is Linux4.4.184, the memory is 16 GB, and the operating frequency is 1.4 GHz–1.6 GHz. Exyos 4412 CPU is based on CortexA9 architecture, has 32 nm process, supports dual channel LPDDR21066, and can well meet the needs of intelligent terminal equipment, mobile communication, modern medical equipment, industrial control and other products. There are many testing methods for real-time operating systems. This paper selects corresponding testing tools according to the actual situation. The main tools used are as follows:

Cyclictest is a high-precision test program, which belongs to the test tool in the real-time tool set rt-tests of the system, and has a very wide range of applications. It can record the information about the kernel delay by executing a priority process circularly, and thus more accurately reflect the difference between the expected scheduling time of the process and the actual scheduling time. At the same time, it can accurately count the scheduling delay generated by the hardware and the system itself. This utility is employed for assessing the latency of the kernel's scheduling to ascertain its real-time capabilities. The scheduling capability of the system is reflected by evaluating the response time and bandwidth of the system. This article mainly uses this tool to measure the context switching delay of the improved kernel.

Gmuplot is an interactive drawing tool that uses the command line. During the test, the data obtained from the Cylictest test is visually displayed. Input relevant instructions in the Iinux terminal to set the style and style of the chart. This paper uses this tool to visually display the experimental measurement results for subsequent analysis.

5.2 Real-Time Task Scheduling Latency Test

The elapsed time from the commencement of a task in the ready queue to its actual invocation by the CPU represents the task's scheduling latency. The statistics of the scheduling delay of the standard version and the improved version of the kernel under light load and overload conditions can reflect the improvement effect of the kernel. The scheduling delay analysis of the two kernels is carried out under the following two conditions.

The scheduling delay analysis is tested with Cyclictest tool. First, the average delay scheduling of the standard kernel and the improved kernel is tested under light load. Two threads are created under the two kernels, and executed 1000000 times circularly to count their average delay scheduling time; In the overload environment, create processes on the two kernels, count their delay scheduling time, and use tools to draw according to the results.

When using the Cyclictest tool on the development board, it needs to be compiled and processed by the virtual machine on the PC side. Acquire the utility and designate the cross-compilation tool. Once compiled, it should be replicated to the root directory of the intended machine. Subsequently, access the utility directory on the target development board via the PC's super terminal (serial port interface) for utilization.

Based on the above experiments, the results were tabulated as shown in Table 3.

Table 3. Statistics of the results of the scheduling delay experiment

Delay time	Light load state		Overload state	
	Standard/us	After optimization/us	Standard/us	After optimization/us
Minimum delay time	9	6	23	15
Average delay time	17	12	33	26
Maximum delay time	757	32	986	56

As can be seen from Table 3, it can be seen that the average scheduling delay of the standard version kernel is further increased under the load condition, and the fluctuation is obvious. The average delay is 33us, and the highest delay is 986us, which is close to millisecond level. The average scheduling delay of the improved version of the kernel is about 26us, and the maximum is 56us. Although there are fluctuations, the fluctuation is small, which basically solves the problem that EDF algorithm can not be scheduled in the overload state and meets the hard real-time requirements.

5.3 Context switching tests

Context switching time refers to the time spent by two different processes on CPU usage right conversion, that is, the time interval from the former to the latter. It specifically includes three parts: the time to save the state, the time to allocate the scheduler to select the next task, and the time to restore the context of the task. The scheduling algorithm and related data structure selected by the process will affect the switching time, so context switching time is an important criterion to evaluate the real-time performance of the kernel.

This paper uses a test tool to compare and test the standard Linux kernel before and after optimization. The use method of this tool needs to be cross compiled, placed in the directory under the host, and used to mount. It enters the target machine through the serial port terminal for testing. After the test is completed, the test results can be generated. The test comparison is shown in Table 4.

Table 4. Context switching test statistics

Ctxsw Time/us Linux	Standard(us)	After optimization(us)
2p/0k	29.3	3.23
2p/16k	24.6	2.61
2p/64k	16.4	5.34
8p/16k	27.2	6.34
8p/64k	26.6	7.68
16p/16k	30.5	11.3
16p/64k	32.4	10.16

As can be seen from Table 4, 2p/0k and its corresponding numbers indicate the time consumed by context switching between two processes with size 0 (no task executed); The size of 2p/16K parallel processing of two processes is the time consumed by 16k context switching; 8p/64k processes eight tasks with the size of 64k in parallel, and the time consumed by context switching. Through vertical comparison, it can be seen that the optimized kernel context switching time is obviously reduced and the real-time performance is obviously enhanced.

6 Conclusion

In view of the development status of CNC equipment, it is of great practical significance to make full use of the advantages of embedded platform such as modularization and customization, and to develop a conforming system on it. Based on this, this paper puts forward the real-time analysis of embedded CNC system based on information fusion of Internet of Things. Through experiments, the real-time task scheduling delay test and context switching time test are adopted for the optimized embedded CNC system. The results show that the real-time performance of the optimized system has been greatly improved, which can meet the basic requirements of the CNC system. The software operating environment can work normally, and the main functions of the software are reading files, CNC code interpreter and trajectory interpolator.

When the system is overloaded, if all the scheduled real-time tasks are in the same priority list, the improved information fusion algorithm of the Internet of Things will still produce a domino chain reaction, and it can't handle the overload situation of the system either. Therefore, further analysis and exploration are needed to further improve the overload handling ability of the algorithm. At the same time, in future research, real-time analysis should be alert to potential security threats and data integrity challenges. Hardware is vulnerable to physical attacks, software may have loopholes, and network communication faces the risk of data being intercepted or tampered with. The problems of data synchronization and consistency, data tampering and forgery and possible loss and damage all pose challenges to the real-time performance and data integrity of the system. Therefore, it is necessary to strengthen security measures from various aspects to ensure the stable operation of the system and data security.

References

1. Heintzel, A.: Red Hat: IBM: GM: exida: safety certified linux operating system for autos coming soon. ATZ Electron. Worldwide **17**(10), 1–28 (2022)
2. Oliveira, D.B.D., Casini, D., Cucinotta, T.: Operating system noise in the Linux Kernel. IEEE Trans. Comput. **72**(1), 196–207 (2023)
3. An, Y.M., Kim, C.H., Chol-Jun, O.: A novel external modular hardware architecture for PC-based Soft-CNC system. Int. J. Adv. Manuf. Technol. **127**, 1–15 (2023)
4. Liu, S., Guo, C., Al-Turjman, F., et al.: Reliability of response region: a novel mechanism in visual tracking by edge computing for IIoT environments. Mech. Syst. Signal Process. **138**, 106537 (2020)
5. Belén, O.M., Stanislav, K.: Developing internet of things-related ISO 10001 hand hygiene privacy codes in healthcare. TQM J. **35**(5), 1194–1210 (2023)
6. Zhang, Q., Zhu, L., Li, Y., et al.: A group key agreement protocol for intelligent internet of things system. Int. J. Intell. Syst. **37**(1), 699–722 (2022)
7. Wei, Y., Ren, P., Liu, C., et al.: High-sensitivity fiber SPR strain sensor based on n-type structure. Opt. Lett. **48**(19), 5057–5060 (2023)
8. Zhao, Y., Rao, H.L., Le, K., et al.: RCFS: rate and cost fair CPU scheduling strategy in edge nodes. J. Supercomput. **80**(10), 14000–14028 (2024)
9. Hameed, K., Garg, S., Amin, M.B., et al.: A context-aware information-based clone node attack detection scheme in Internet of Things. J. Netw. Comput. Appl. **197**(1), 1–27 (2022)
10. Sharma, R., Nitin, N., Dahiya, D.: Fault tolerance in the joint EDF-RMS algorithm: a comparative simulation study. Comput. Mater. Continua **72**(3), 5197–5213 (2022)

A Multi-dimensional Early Warning Method for Digital E-commerce Supply Chain Risk Based on IoT Information Fusion

Liwen Zuo(✉)

School of Business, Sichuan University Jinjiang College, Meishan 620860, China
zuoliwen2021@163.com

Abstract. Due to the low prediction accuracy of the existing early warning methods, a multi-dimensional early warning method of digital e-commerce supply chain risk based on Internet of Things information fusion is studied. Based on the idea of scene information and complex event processing, the event information structure and event processing method based on the fusion of human-object-field information are proposed. Through negotiation between suppliers and retailers, the proportion of wholesale price and profit sharing is determined. Following the principle of selecting early warning indicators, the early warning indicators of financing risk are selected from four aspects: macro and industry environment, online supply chain operation, financing enterprises and core enterprises, and a risk early warning indicator system is constructed. The random forest algorithm is used to calculate the importance of each index and complete the index ranking accordingly, and the key early warning indicators are selected to calculate the potential risk loss of financing business for early warning. The experimental results show that when using this method to identify financing risks, the model can be the most accurate when the number of risk early warning indicators is 40; The accuracy of model prediction using test sample data is 94.59%.

Keywords: IoT · Information fusion · E-commerce · Risk · Multidimensional · Early warning

1 Introduction

In the wave of digitalization sweeping the world today, the e-commerce industry is developing at an unprecedented speed, the supply chain as the core component of e-commerce business, its stability and security is directly related to the survival of the enterprise. However, as the complexity of the supply chain increases and the uncertainty of the external environment intensifies, the supply chain risk is becoming more and more prominent, which brings great challenges to e-commerce enterprises. Therefore, developing a method that can comprehensively and accurately warn supply chain risks is of great significance for the sound development of e-commerce enterprises [1]. A multi-faceted early warning approach for digital e-commerce supply chain risks has emerged, leveraging the benefits of contemporary information technology. This method integrates big data, cloud computing, the Internet of Things, and other technologies with supply chain risk management, and forms a new risk management paradigm. The method takes data as the core, and realizes real-time perception and accurate prediction of supply chain risk by collecting, integrating and analyzing data information of each link in the supply chain. The multi-dimensional early warning method of digital e-commerce supply chain risk is an innovative risk management tool, which can provide enterprises with comprehensive and accurate risk early warning services, and help them maintain sound development in the fierce market competition. As digital technology continues to evolve and its application scenarios broaden, it is believed that this method will play a more and more important role in e-commerce supply chain risk management and inject a new impetus for the sustainable development of e-commerce enterprises.

In terms of data collection, the use of Internet of Things (IoT) technology interconnects equipment, commodities, personnel and other elements in each link of the supply chain to realize real-time data collection and transmission. These data not only include traditional sales data, inventory data, etc., but also cover multi-dimensional information such as logistics information, supplier credit records, consumer behavior, etc., providing a rich data base for risk early warning. In terms of data analysis, advanced data mining and machine learning algorithms are used to deeply process and analyze the collected data [2]. By constructing a multi-dimensional risk indicator system and an early warning model, we can accurately identify potential risk points in the supply chain and predict the risk development trend. Concurrently, visualization techniques are employed to display the analysis outcomes in a visually appealing manner, facilitating quick comprehension of supply chain risk situations by enterprise decision-makers. The multi-faceted analysis-based digital e-commerce supply chain risk early warning approach not only enhances the precision and promptness of warnings but also offers enterprises a more comprehensive and profound risk management perspective. By adopting this method, enterprises can promptly identify and respond to various risks within the supply chain, ensuring its smooth operation and bolstering their market competitiveness. The innovative points of the research methodology are:

(1) Determining the wholesale price and profit-sharing ratio through negotiation between suppliers and retailers, providing a collaborative approach to pricing and profit allocation.
(2) Adhering to the principles of selecting early warning indicators, a comprehensive risk early warning indicator system is constructed by selecting indicators from four aspects: macro and industry environment, online supply chain operation status, financing enterprises, and core enterprises. This approach ensures a holistic view of risk factors.
(3) Utilizing the Random Forest algorithm to calculate the importance of each indicator and subsequently ranking them. Based on this analysis, key early warning indicators are selected to calculate the potential risk loss of financing operations, thereby enabling effective risk warning. This method combines advanced machine learning techniques with risk management practices.

2 Digital e-commerce Supply Chain Risk Interoperability

2.1 Risks of e-commerce Supply Chain Financing

E-commerce supply chain financing refers to the e-commerce enterprise as the core of the whole business, upstream and downstream enterprises within the supply chain seek financing from financial institutions or e-commerce enterprises, ultimately utilizing the revenue generated by the financed enterprise or the cash flow stemming from transactions with the e-commerce enterprise as the repayment source for a specific financing activity as shown in Fig. 1.

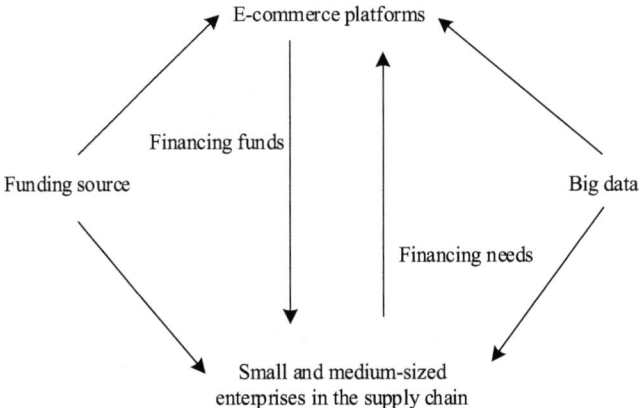

Fig. 1. Schematic diagram of e-commerce platform supply chain financing

The e-commerce platform records a large amount of data, including the authentication information of enterprises in the whole supply chain, records of transactions between enterprises and credit evaluation of enterprises, etc. [3]. The e-commerce company will sort out all the information, find out the key information needed, and then evaluate the creditworthiness of each enterprise, and ultimately use the rating results of the e-commerce platform to judge whether it can grant credit to the financing enterprise. For the lending enterprises, under the e-commerce supply chain financing mode, their own credit conditions and collaterals, such as orders signed with e-commerce companies, sales of goods to e-commerce companies to form the accounts receivable and so on, as a guarantee for their application for loans, which increases the chances of SMEs in the supply chain to obtain loans. For e-commerce enterprises, supply chain financing enables them to expand their business fields, make full use of idle funds, promote the establishment of long-term strategic partnership among supply chain enterprises, and increase the overall interests of enterprises in the supply chain [4]. Risk is the uncertainty of future loss or danger. Its main characteristics are objectivity, universality, inevitability, controllability and uncertainty. Supply chain financing risk refers to various hazards that arise during the financing process of the supply chain. These primarily encompass credit risk, market uncertainty, legal exposure, and operational vulnerabilities. All the above supply chain financing risks may bring the operation crisis to the enterprise, so the risk of supply chain financing should be emphasized. Supply chain financing under e-commerce platform optimizes and upgrades the traditional supply chain financing, and compares the two, mainly from the aspects of participating subjects, financing media, information collection and evaluator, credit granting method and risk control means, etc. In this context, three primary financing methods are chosen: accounts receivable financing, purchase order financing, and inventory financing based on warehouse receipts, because these three modes are the most common and most frequently used three modes to carry out supply chain financing, and the three modes are the most common and most frequently used three modes to carry out supply chain financing. As these three modes are the most common and most frequently used three modes for supply chain financing, through comparison, the benefits of financing within the e-commerce supply chain become evident, enabling a more effective utilization of financing outcomes. The three financing modes are listed for comparison, as shown in Table 1.

Table 1. Accounts receivable financing for e-commerce companies

Project	Traditional mode	Financing model for accounts receivable of e-commerce enterprises
Participants	Core enterprises	E-commerce enterprises and other financial institutions
Financing media	Offline	Internet
Information collection and evaluation	bank	E-commerce platform
Credit granting method	Direct bank credit	Credit evaluation system
Risk control methods	Pledge guarantee	

The table illustrates the disparities between e-commerce supply chain financing and traditional financing methods in aspects such as participants, credit extension mechanisms, and risk mitigation strategies. E-commerce supply chain financing has been optimized and upgraded in many aspects, such as: traditional supply chain financing has obvious weaknesses in information acquisition and big data analysis, while e-commerce enterprise supply chain financing has significantly improved in these aspects. When the e-commerce platform participates in the supply chain financing, banks and e-commerce enterprises can use the resources of the e-commerce platform to upgrade their credit granting methods, information collection and risk control management methods, so as to effectively prevent and manage various risks brought by supply chain financing, which is an unprecedented change for enterprises and banks in supply chain financing. Under the financing model of e-commerce enterprises in the supply chain, which involves entities such as warehouses, financial institutions, transportation companies, and regulatory organizations, loans can be tracked in real-time, enabling satisfactory financing solutions for upstream and downstream enterprises. This financing approach leverages an ERP integration system, utilizing big data analysis to offer data insights such as capital flows, logistics, and information, crucial for the advancement of supply chain financing.

It thereby enhances financing and support services for the entire supply chain's upstream and downstream enterprises. However, in the development of supply chain financing, all significant e-commerce platforms encounter certain challenges in risk management and control. Only by constantly revising and improving the construction of risk management and control, can e-commerce platforms give better play to the effect of supply chain financing and better serve enterprises.

2.2 IoT Information Fusion

In the field of digital e-commerce, based on the Internet of Things, big data and cloud manufacturing technology, facing the new trend of manufacturing development and the new characteristics of market demand, a new manufacturing model aims at comprehensive perception, real-time decision-making, the ability to use social resources and the ability to meet customers' personalized needs [5]. The underlying manufacturing process of intelligent manufacturing relies on the information physics fusion system. Its manufacturing process is digital. The enterprise can collect and perceive multi-source information in the manufacturing process in real time, especially the object information and environmental information in the manufacturing process, to form the management context information of "human-object-field" integration. Based on the management responsibility and management demand required by "human" and the real situation represented by "field", the management meaning of "object" information can be explained. At the same time, the information services at the lower level in the information physics system and the independent decisions of the higher level decision making units are all realized in a certain management context and management requirements. As a result, modeling the management context through "human-object-environment" information integration bridges the gap between the information physical system and management decision-making, serving as the foundation for describing the manufacturing process information within the intelligent manufacturing system. It also forms the basis for the information exchange among virtual manufacturing resources. And the environment on which the decision-making unit relies for independent decision-making under the multi-level decision-making model. The method based on event discovery is used to realize management context awareness. Events, as a high collection of management constraint information, can well summarize the human object field context semantically. At the same time, the production, manufacturing, management, service and other systems of enterprises are typical discrete event systems. From the point of view of discrete manufacturing system, events refer to the changes in the state of the manufacturing system that can be perceived.

Setting up the information fusion model, different levels have different processing functions, according to the degree of information abstraction, the first level of fusion is target identification, the second level of fusion is situational assessment, the third level of fusion is threat estimation, and the fourth level of fusion is process optimization. The specific information fusion model is shown in Fig. 2.

The data preprocessing function refers to the cleaning and filtering of the initial data source data. The obtained data are sorted and merged according to the observed time, location, data type, information attributes and characteristics, so that useful data can be introduced into the data fusion center. The data about target location, status,

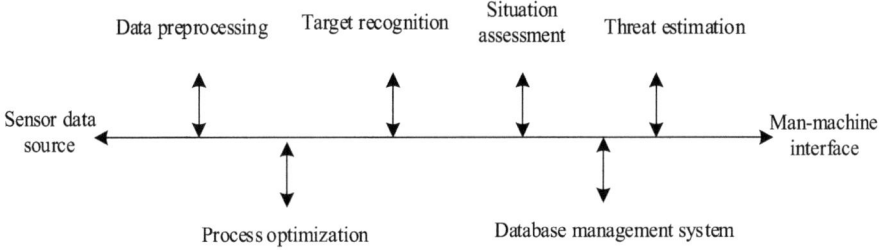

Fig. 2. Information fusion model

characteristics, attributes and identity collected by multi-sensor are combined, correlated and combined by clustering, neural network, template method, D-S evidence theory and Basys reasoning, so as to identify a single target entity. This part of the integration mainly includes: data calibration, interconnection, tracking and identity integration. The general situation representation is constructed according to the incomplete data set of the target, giving a reasonable explanation of the entity distribution, and deriving the results, so as to judge the intention, action plan and results of both sides, and build the situation map. According to the situation of both sides and the possible actions taken by the enemy, determine our weak links and possible events. Threat assessment is a very complicated process, because the factors of assessment not only include the assessment of possible results, but also consider the changes in the enemy's fighting strength, training level and political situation. Process optimization runs through the whole fusion system, and its main role is to complete the performance evaluation of the system, distinguish the value of data information, and optimize sensors and systems. The Internet of Things technology can perceive entities and entity attribute information in real time [6]. Under the constraint of scene state definition, the value of entity attribute is matched to state, so that the system can perceive state and state transition in real time. The transition of the state triggers the event. The occurrence of the event is combined with the scene rules to judge its meaning and then make decisions. The entity is represented by 0, and the A represents a set of entity attributes. Entities and entity attributes are generally in one-to-one or one-to-many relationships. Then for entity 0, its attribute set, the $A = \{a1, ..., an\}$, of which a represents an instance of the attribute, the n represents the number of entity attributes. The structure of an attribute instance is represented as:

$$a = < key, type, value > \tag{1}$$

In the formula: *key* indicates the name of the attribute, which is uniquely identifiable. *type* indicates the value type, including numeric type and text type; Value indicates the attribute value. In a certain period of time, the state of an entity is limited s represents a state instance, and a change in the value of an entity attribute affects a change in the state instance. The definition of state is divided according to the system and entity characteristics, and the state instance formula associates the entity attributes with the state definition, thereby enabling the assessment of the state's worth. Decision-making information mirrors the preferences of the decision-maker, and in this paper, the decision-maker's preferences towards distinct decision-making options are captured through the method of assigning weights. For an event, the e, its preference programs is:

$$p = \{action, \theta 1 | \sum_{k=1}^{k=i} \theta k = 1\} \quad (2)$$

In the formula: *action* is the i th decision-making scenarios of an event. $\theta 1$ represents the preference weight of the program *action*. Considering the characteristics of human, object and field information, under the general framework of information fusion, the event information structure and event processing method based on human-object-field information fusion are proposed by taking the research of context information as the basis for the scenario information of this paper, and by taking the idea of complex event processing as the basis for the establishment of the information processing logic and processing rules of this paper.

2.3 Supply Chain Entities Exchange Information to Aid in Decision-Making

The neurons within a neural network are interconnected and collaborate with each other to carry out information processing and storage [7]. Artificial neural networks mimic the functionalities of the human brain's nervous system, form a network of artificial neurons according to certain connection rules, and allow the connection weights of each neuron to change according to specific rules in order to realize the recognition and learning of input patterns. According to the interconnection structure, it can be categorized into no interconnection hierarchy within the layer without feedback, no interconnection hierarchy within the layer with feedback, no interconnection hierarchy within the layer without feedback, and interconnection non-hierarchical structure with feedback. In decentralized decision-making, each member of the multichannel supply chain, as an independent economic entity, aims to maximize its own interests, and therefore ignores the collective interests in price setting, which is in conflict with the centralized decision-making goal consistency. In the supply chain dominated by e-commerce platform, the game sequence of suppliers and e-commerce platform can be described as follows: (1) e-commerce platform obtains the market demand information; (2) e-commerce platform decides the online selling price r of the products; (3) the supplier determines the price r at which the product will be sold offline. In a decentralized supply chain, the transactional relationship between an e-commerce platform and a supplier is established through a revenue-sharing contract, and the revenue-sharing factor is ε. ie, when a supplier sells a product through an e-commerce platform, the e-commerce platform receives revenue for each unit of product sold ε, suppliers receive benefits $(1 - \varepsilon)r$. Consider the demand information symmetry situation, that is, both the e-commerce platform and the supplier know the actual situation of market demand. The competitive equilibrium solution of supply chain members can be analyzed by using the inverse induction method. At this time, given the online selling price r of the e-commerce platform, the supplier's profit function is:

$$\pi(r) = (1 - \varepsilon)rD \quad (3)$$

In the formula, D is a concave function; based on this, an e-commerce platform based on user privacy is proposed. In the process of the game, because the e-commerce platform decides the online selling price of its goods after obtaining the demand information r, therefore r in a sense, this is a manifestation of the private demand information held by the e-commerce platform. At the same time, given the online selling price of an e-commerce platform, the r. The suppliers may then be able to infer private demand information held by the e-commerce platforms. But in fact, e-commerce platforms sometimes do not want suppliers to infer their own market demand information through their own online selling prices, thus distorting their own online selling prices. From the game process between e-commerce platforms and suppliers, it can be seen that this is a typical signaling model, and it is also pivotal in achieving a balanced sharing of benefits among supply chain members. In pursuit of this objective, we delve into whether the e-commerce platform has any incentive to misrepresent market demand information. The e-commerce platform, based on the demand information obtained $ai(i \in \{H, L\})$, selecting the appropriate online selling price L. It is assumed that the supplier can, from the e-commerce platform's L inferring demand information from the $ai(i \in \{H, L\})$. It is clear that e-commerce platforms, in anticipation of high market demand, will engage in higher pricing online. Therefore, the following trust structure is given: the price threshold for online sales, the L, if the online selling price of the e-commerce platform reaches $L > Le$, then the supplier believes that the market demand is high; if the online selling price of the e-commerce platform satisfies $L < Le$. If the supplier believes that the market demand is low, then the supplier believes that the market demand is low. Guided by the aforementioned beliefs, the supplier is apprised of the online selling price on the e-commerce platform. Subsequently, it evaluates the demand information possessed by the platform to ascertain the offline selling price of the goods. Based on the inverse induction method, the online selling price of a given e-commerce platform can be obtained L and the offline selling price of the goods selected by the supplier.

$$L = \frac{b(2-\theta)pr + (1-\beta)a + c}{2} \tag{4}$$

In the formula: β is pricing the network. c is a stochastic parameter. Under the consideration of the supplier response function, the e-commerce platform will make network pricing in the network environment to maximize its own revenue. In this regard, a revenue function on e-commerce can be obtained as follows.

$$\pi(r) = \varepsilon r \frac{[b(2-\theta) - 2]pr + bc}{2} \tag{5}$$

In the formula: $\pi(r)$ is obtaining information on actual demand for e-commerce platforms. b is profit for the information delivered. Under the e-commerce platform-dominated supply chain, because of the competitive relationship between online and offline channels, the e-commerce platform can weaken the competitiveness of suppliers' online and offline channels by lowering the online sales price to increase the sales volume of online channels. Consider first the low type e-commerce platform decision-making problem, if the low type e-commerce platform follows the high type e-commerce platform and chooses a higher online sales price, the supplier then deduces that the e-commerce platform belongs to the high type, and chooses a higher offline sales price, and the supplier can choose a higher price by lowering the online sales price, and then the supplier can choose a higher price pr. The supplier has chosen a higher offline selling price in the process. In the process, the supplier chose a relatively high offline selling price pr for e-commerce platforms, it is advantageous to increase sales in the online channel. However, when a low-level e-commerce platform does not emulate a high-level platform, but chooses a lower online selling price, suppliers infer that the e-commerce platform is low-level and will choose a lower online selling price pr. In this process, due to the supplier offline sales price is low, competitive, e-commerce platform online channel sales did not increase. Comprehensive low type of e-commerce platform will tend to follow the high type of e-commerce platform. The more types of e-commerce platforms always choose a higher online sales price will be the actual demand for information conveyed to the supplier. Due to the influence of information asymmetry, double marginal effect and long whip effect and other factors, conflicts arise between multi-channel supply chains, thus causing system failure [8]. Multi-channel supply chain coordination can be regarded as cooperation and competition under the cooperative game of supply chain members, and the result of coordination makes the members within the multi-channel supply chain coordinate and present a harmonious state. Multi-channel supply chain coordination is essentially the same as two-channel supply chain coordination, and the primary distinction between them lies in their reliance on supply chain contract coordination. Its objective is to safeguard the interests of each supply chain member, optimize commodity pricing and resource allocation, and thereby maximize the overall benefits of the multi-channel supply chain. This is closely intertwined with the seamless cooperation among its members. In a multi-channel setup, decision-making variables often conflict, and maximizing individual interests does not necessarily align with optimal choices for the entire chain. Hence, the utilization of supply chain contracts becomes crucial for coordination. Among the various contract types, revenue-sharing contracts are particularly common. These contracts involve negotiation between suppliers and retailers to determine wholesale prices and profit-sharing ratios, and the a priori profit of the integrated supply chain is:

$$\pi c = \frac{1}{2}\pi(pc, ph) \qquad (6)$$

In the formula: pc is offline sales prices. ph is the online selling price. Typically, the wholesale price is driven down to incentivize the retailer to buy, but only if the retailer shares a portion of the sales revenue with the supplier, who then drives down the wholesale price to the benefit of both companies. The revenue-sharing contract essentially provides the retailer with a measure of control and relieves it of the financial pressures of advance shipments. The contract is designed to enhance the relationship between the companies through benefit sharing and risk sharing, thereby improving the overall interests of the companies, and the core issue is the determination of benefit sharing and risk sharing. In this case, when the retailer's order quantity is high and its wholesale price is low, more orders will be generated. Although the lower wholesale price will reduce the supplier's yield per unit of production, when the overall yield increase caused by the increase in orders and demand is positive, the manufacturer is willing to fulfill the contract, thus realizing cooperation. By setting two combinations of prices, small-demand users will not imitate high-demand users in pursuit of small marginal costs, while high-demand users will not imitate low-demand users in pursuit of fixed marginal costs, so as to realize effective differentiation between the two types of users. On the other hand, since specialized value-added services are mostly provided by high-demand users and such services can be completed with only one-time investment, high fixed fees can be levied on high-demand users to recover the costs, while low-demand users are mostly used to satisfy the purpose of information collection on the platform, so it is sufficient to levy fewer fixed fees, but it has to pay more marginal fees in order to achieve cross-subsidization of different users, which is the purpose of cross-subsidization of different types of users. However, it has to pay more marginal fees for value-added services in order to achieve the purpose of cross-subsidization for different users, which not only obtains benefits, but also takes into account the requirements of information collection and screening.

3 Risk Multidimensional Early Warning

3.1 Types of Supply Chain Risks in e-commerce Companies

The supply chain characteristics of import e-commerce are generally reflected in the procurement link. There are two main procurement methods: first, direct purchase overseas, and establish overseas warehouses and logistics in Europe, North America and other places. If an order is generated, choose international logistics for direct shipment; The second is bonded stock. Combining the current big data technology and using data mining to discover users' preferences and needs, a large number of required goods are directly purchased and stored in the warehouse in the bonded area in advance, and orders are directly delivered from the bonded area to consumers [9]. In terms of supply sources, the import e-commerce supply chain mainly includes four modes: foreign supply sources, intermediary trade or agents, e-commerce supply chain companies and domestic bonded zone supply chain companies. No matter which mode, subject to the double restrictions of international procurement and logistics, there are problems of duplication of modes and long chain. The mode emphasis is mainly reflected in the difference between domestic and overseas stock demand, and the chain length is mainly reflected in the circulation channel. Therefore, the resources required by the supply chain of import e-commerce enterprises include stock funds, customs clearance contacts, teams at home and abroad, and efficient ERP systems, all of which are indispensable. Many import e-commerce enterprises choose to set up overseas warehouses around the world in order to ensure the advantages of direct sourcing from the source of international brands and ensure that consumers can buy authentic products at the lowest possible price. However, these all need the strong supply chain support of import e-commerce enterprises. The supplier source is shown in Fig. 3.

Although some cross-border import e-commerce platforms have overseas warehouses and domestic bonded area, although the cost is higher, but the logistics process is difficult to do the whole process of monitoring, there are too many uncertainties, consumers can not fully believe the authenticity of the source. According to the results of the survey, domestic consumers are less satisfied with the goods purchased overseas, mainly because of the long logistics time, high freight costs and the return and exchange process is too cumbersome. Imported e-commerce enterprises should pay more attention to the speed of goods flow and customs clearance, and improve consumer experience

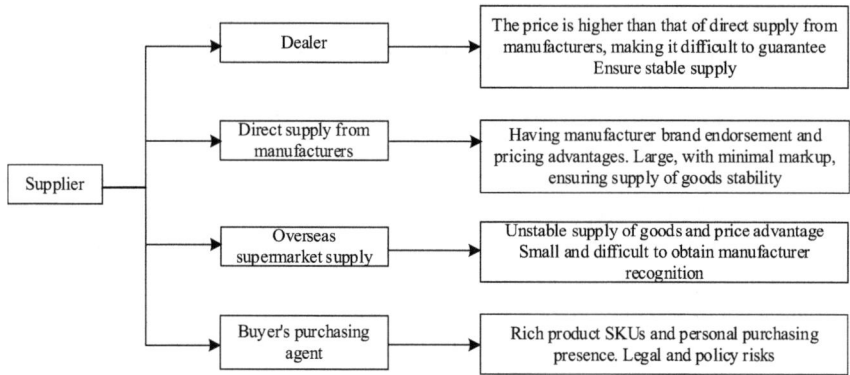

Fig. 3. E-commerce supplier sources

in logistics operation. Regarding logistics and overseas warehouse and bonded warehouse, in the international logistics section, although the cost of shipping is lower, but the transportation time is too long, the timeliness is poor, so companies generally choose air transport; in the airport transit section and customs clearance speed, each bonded area slightly different, generally speaking, will not spend too much time; finally, the domestic logistics section, according to the geographical differences, the goods eventually reach the hands of consumers. These several logistics segments contain uncertainty factors, are closely related to the supply chain effectiveness. The priori profits of the supply chain is:

$$\pi t = \pi r + \pi s \qquad (7)$$

In the formula, πr is the optimal profit; and πs is profit for the supplier. Look at the overseas warehouse and bonded area, the cost of overseas warehousing is very high, generally only large cross-border import e-commerce enterprises will choose to establish overseas warehouses in the origin of the goods, small and medium-sized enterprises can not afford the high cost of warehousing and operating costs. Domestic bonded warehouse resources are limited, far less than the enterprise demand, and the bonded area policy is not the same everywhere, to understand the policy and customs clearance process and other content in advance, in order to prevent the delivery of unstable and other issues. On the issue of time and process required for consumers to place orders for goods delivery, first of all, take the bonded zone model as an example, foreign commodities into China, after bonded customs clearance and storage in China's bonded zone, when our consumers place orders, through the domestic logistics system from the bonded zone for direct delivery. Cross-border import e-commerce business involves a very complex bonded warehouse docking and registration work, first of all, the enterprise should be completed before the sale of goods for the record of merchants and commodities for the record, after the order is generated, but also need to be three in one, that is, the purchase and sale of information, payment information and logistics information to the Customs system to report for the record, the Customs and Excise Department verifies the accuracy of all information before granting permission for shipment.

3.2 Establishment of a System of Risk Early Warning Indicators

Selection of risk warning indicators for core enterprises. Through the analysis of the above financing mode process, it can be seen that the core enterprise has a pivotal position in the financing business. Therefore, this paper comprehensively considers the financing risk brought by the core enterprise from three aspects: creditworthiness, profitability and solvency. Creditworthiness is an important condition for financial institutions to examine whether the core enterprise can meet the requirements of being a credit guarantor for SMEs, and its creditworthiness is directly proportional to the financing risk, and the better the creditworthiness, the smaller the risk borne by financial institutions. In this paper, the credit rating and operational stability of the core enterprise are used as two indicators to present its creditworthiness; profitability refers to the ability of the core enterprise to operate profitably, and the stronger the profitability, the stronger its ability to help financial institutions bear the risk. In this paper, operating profit margin and return on net assets are used to reflect this ability. Usually, the higher the value of these two indicators, the lower the probability of risk; solvency. Core enterprise solvency is the key to the financial institutions can recover the loan, and therefore should be included in the core enterprise related risk early warning indicators into account. In this paper, this ability is reflected by two indicators, one is the quick ratio, the other is the gearing ratio. On the basis of identifying the sources of financing risk, following the principle of selecting early warning indicators, we select the financing risk early warning indicators and construct the risk early warning indicator system from four aspects, namely, macro and industry environment, online supply chain operation status, financing enterprises and core enterprises, as shown in Table 2.

In this model, the financial institution extracts the financing business data from the database and adopts the random forest algorithm to screen the key early warning indicators of financing risk. First, the Random Forest algorithm is used to calculate the importance of each indicator and complete the ranking of indicators accordingly. Then the key warning indicators are selected based on the minimum value of the error rate of the out-of-bag data, and in this step, the addition of noise to the influencing factors will make the accuracy of the prediction samples decrease. This change indicates that the influence factor has an important role in sample classification. Under the conditional assumption, this paper takes the collected information of SMEs as the training sample dataset $S = \{(x1, y1), ..., (xn, yn)\}$, $X \subseteq R^n$ is the sample input space, the y is the category output space that represents whether SMEs are risky enterprises or not, based on the above assumptions, the process of constructing a random forest is summarized as follows: there exists a sample set W, first select from sample set W based on self-service sampling method k the next time it occurs k self service sample set B and k out of bag sample set (OOB), then use k self help sample set B, construct meta classifier C (x), so the results of the classification for any sample x combined on classifier $C(x)$ is:

$$\widehat{y} = C(xi) = \arg\max \sum_i \delta(C(xi) = y) \tag{8}$$

In the formula, $\delta()$ is a schematic function, and when the argument is real, the $\delta = 1$; $C(xi)$ is a random forest. In order to find out the key influencing factors of financing risk, it is necessary to judge through various rules. Among them, the random forest

Table 2. Supply chain risk early warning indicator system

Primary indicators	Secondary indicators	Third level indicators
Macro and industry Environmental risk	Macro environment	Macroeconomic situation
	Industry environment	Government support
	Supply chain relationships	Industry Outlook
	Supply chain relationships	Close cooperation between upstream and downstream enterprises
Online supply chain operation risks	Financing enterprise status	Dependency of upstream and downstream products
	Electronic Silver Relationship	The degree of sharing of transaction information
	Logistics capability	Logistics enterprise level
	–	Enterprise scale
	Basic conditions	Enterprise management level
	–	Quality of financial information disclosure
Financing enterprise risk	Debt paying ability	–
	Profitability	–
	Operational capability	–
	Development capability	–
Core enterprise risk	Credit status	Credit rating
	–	Business stability
	Profitability	–

algorithm uses the importance value of the influencing factors to the risk calculated by the algorithm to find the key influencing factors. The following details the basic idea of using random forests to select the importance of key financing risk influencing factors: as mentioned above, if some noise is added to an influencing factor in the sample data, the classification accuracy of the random forest model will decline. Subsequently, adhering to this principle, the particular procedures for employing the random forest approach to determine the significance value of an individual early warning indicator are as follows: calculate the accuracy of the decision tree according to the OOB data corresponding to each decision tree generated in the random forest. By randomly selecting an indicator $x_i^{(j)}$ from the OOB dataset to make changes or add noise for interference to correct the accuracy of that decision tree. Find the importance for metrics $x_i^{(j)}$ and is calculated by the formula:

$$impx_i^{(j)} \sum_{k=1}^{K} \frac{(acc - acck)}{k} \tag{9}$$

In the formula, *acck* is the accuracy value. This value is obtained by calculating the difference between the original OOB data accuracy and the modified OOB data accuracy. Repeat until the remaining two early warning indicators, and determine the final risk early warning indicator set based on the principle of minimum OOB error rate. The final feature subspace is determined according to the minimum OOB error rate criterion.

3.3 Multidimensional Early Warning

Risk early warning is an important stage in the process of financial risk management by financial institutions, as it is an ex-ante control of financing risks. In risk early warning, it is necessary to visualize the level of risk of financing business by classifying the level of risk. Risk leveling refers to the process of evaluating risks in a scientific and reasonable way and classifying risk levels according to the results, which can provide guidance for risk managers to take risk prevention and control measures according to the level of risks. Currently, the research on the classification of risk level mainly focuses on the fields of finance, production, transportation and dangerous goods, this primarily concentrates on assessing the potential for risk occurrence and the magnitude of its impact, and the classification standard of the risk level has a certain degree of flexibility [10]. The online supply chain financing risk is the focus of this paper. In the actual financing business, the impact of financing risk mainly has three aspects: First, financial institutions, financial institutions are the capital source of the financing business. Once the financing risk occurs in a certain link, the capital provider will directly suffer capital losses; Second is the core enterprise, as the financing business's credit guarantor, the core enterprise is susceptible to the financing risk, which can influence its credit standing. Thirdly, since the financing business is carried out as a whole by the supply chain, the occurrence of financing risk will have different degrees of influence on all the financing participants in the supply chain. Because in general, the adverse effects caused by the risk has similarity, the impact caused by the occurrence of financing risk as a fixed variable, the probability of the occurrence of risk output model as the independent variable, divided into the risk warning level, and reference to the traffic lights set the color of the risk warning lights to visually display the degree of financing risk. As shown in Table 3.

Table 3. Risk Early Warning Level Classification

Probability of risk occurrence	Risk level	Alarm light setting
0–20%	Level I	green light
20%–40%	Level II	blue light
40%–60%	Level III	yellow light
60%–80%	Level IV	orange light
80%–100%	Level V	red light

An early warning indicator of risk, namely, potential risk loss, is introduced to quantify potential financing risks. The so-called potential risk loss of financing business refers to the loss of the capital provider due to the default of the financing enterprise caused by the uncertainty of risk. Calculating this indicator not only helps financial institutions to estimate the potential loss of the financing business, but also helps the capital provider to make a decision on whether to continue the financing business. In this section, the potential risk of loss in a financing operation is defined as follows:

$$P_L = P \times F_R \tag{10}$$

In the formula, P_L represents the potential peril of financial losses in financing operations. P represents the likelihood of risk occurrence, specifically the probability of default by the financing enterprise, derived from the financing risk analysis phase. F_R represents the total amount of enterprise financing is the total amount of financing that the financing enterprise applies to the financial institution. Due to the unavailability of the total amount of enterprise financing, in order to assess the potential risk loss of the financing operations, this paper utilizes the enterprise's total assets as a proxy for the overall financing amount. In practice, through the synergistic effect of the main parties involved in the financing business, the financial institutions can collect the data related to the financing business in real time, derive the likelihood of risk associated with the financing enterprise via a multi-stage financing risk analysis, and at the same time measure the potential risk loss of the financing business, and early warning of the level of financing risk.

4 Example Analysis

4.1 Experimental Preparation

In view of the fact that there is no perfect financing transaction information database in China, and the small scale of SMEs and incomplete disclosure of financial statements make it difficult to obtain their data, this paper selects the data of 54 listed SMEs in the automotive supply chain as the research sample data, a total of 162 samples. The data comes from the annual financial reports of listed SMEs, CSMAR database, Sina Finance website, National Bureau of Statistics and other multi-channel information. Furthermore, this paper categorizes risk enterprises based on the metric of interest-bearing debt ratio index from the latest Standard Valuation Criteria for Enterprise Performance Assessment, formulated by the State Administration of Assets Supervision and Administration. The enterprises whose interest bearing debt ratio in the sample is 53.7% higher than the lower value of the industry are identified as risky enterprises and assigned a value of 1; The remaining samples are divided into risk-free enterprises and assigned a value of 0. According to the above classification criteria, the sample enterprises are classified into 29 risky sample enterprises and 133 risk-free sample enterprises. The equipment used in the experiment is shown in Table 4.

In this stage, random forest algorithm is used to screen indicators with high information content, so as to help financial institutions identify key early warning points of financing risks.

Table 4. Experimental equipment

Project	Device	Parameter
1	Hardware system	RGB54
2	Operating system	Windows11
3	RAM	8 GB
4	CPU	Intel Core i7
5	Running memory	16G
6	Running Platform	Matlab2020
8	Hard disk	1.3G

4.2 Experimental Hypothesis

Based on the recorded data of an enterprise's daily digital e-commerce supply chain, this paper simulates the data in a period by experiments, and after eliminating the influence of interference factors, uses this method to analyze the risk of the experimental data and make an early warning. In the course of the experiment, in order to ensure the reliability of the experimental results and avoid the influence of random errors on the experimental results, unified risk attribute parameters are set. Suppose the enterprise is in a multi-level supply chain, and there are many different risk events in its supply chain in one cycle. These risk events are simulated by experiments and the corresponding occurrence coefficients are set. Specific parameter settings are shown in Table 5.

Table 5. Table of Risk Events and Occurrence Coefficient

Risk type		Occurrence coefficient	Non-occurrence coefficient
Intrinsic risk	Information transmission risk	20%	10%
	Logistics operation risk	15%	7.5%
	Production organization risk	5%	2.5%
External risk	Market demand risk	20%	10%
	Economic cycle risk	10%	5%
	Environmental risk	30%	15%

After determining the risk coefficient, in order to ensure the fairness and justice of the experiment, it is necessary to assume the occurrence attribute and risk level of risk events. Therefore, in the experimental hypothesis, starting from the simulation requirements, the attribute of the risk event that occurred is recorded as 1, and the attribute of the risk event that did not occur is recorded as 0. According to the above contents, the occurrence of risk events in the digital e-commerce supply chain in the experiment is assumed.

In the whole experiment, the state function of risk is obtained, which is expressed as:

$$f(y) = \begin{cases} W_a, 0 < f(x) < 1000 \\ W_b, 1000 < f(x) < 2000 \\ W_c, 2000 < f(x) \end{cases} \tag{11}$$

In the formula, W_b represents low-risk state; W_b represents medium risk status; W_c represents high-risk state; $f(x)$ represents the cost function caused when a risk event occurs. According to the state function of the above risks, experiments are carried out by using this method, and a large number of experimental data are generated for subsequent experimental research, so as to ensure the fairness and justice of the experiments.

Calculate the value of the state function of the risk by Formula (11) when the number of risks is uncertain, and the results are shown in Table 6.

Table 6. Experimental results of state function value of this method

Risk quantity	State function value
20	99.76%
40	98.58%
60	98.33%
80	97.66%
100	96.48%

As can be seen from Table 6, with the increase of the number of risks, the value of the state function of this method changes little and tends to be stable, and the overall level is above 95%, which can meet the demand of risk early warning.

4.3 Results and Conclusions

Use the recursive feature deletion method of the Rcaret package to screen the financing risk early warning indicators, and the screening results are shown in Fig. 4.

The results in Fig. 4 show that when the random forest algorithm is used to identify financing risks, when the number of risk early warning indicators is 40, the accuracy of the model can be the highest, showing the ability to maintain high accuracy in both the training set and the test set, which proves the effectiveness and stability of the feature set in capturing financing risk characteristics. Based on the algorithm of the financing warning indicators of the importance of the measure is shown in Fig. 5.

Figure 5 shows the ranking of the importance degree of risk early warning indicators under two different standards. No matter which standard, the importance degree of the upper indicators shown in Fig. 5 is higher than that of the lower indicators. According to the prediction accuracy of the importance, it can be seen that among the 162 sample data, the prediction accuracy of the training sample data is 84%, and the prediction accuracy

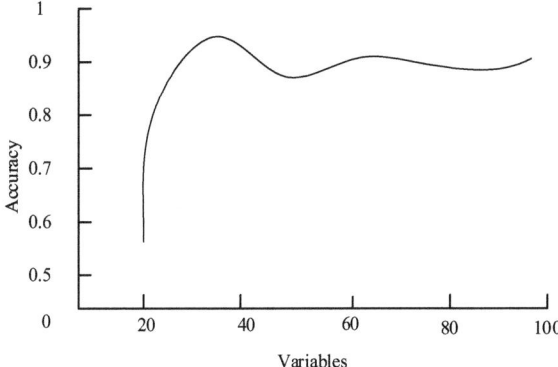

Fig. 4. Financing risk early warning indicator screening results

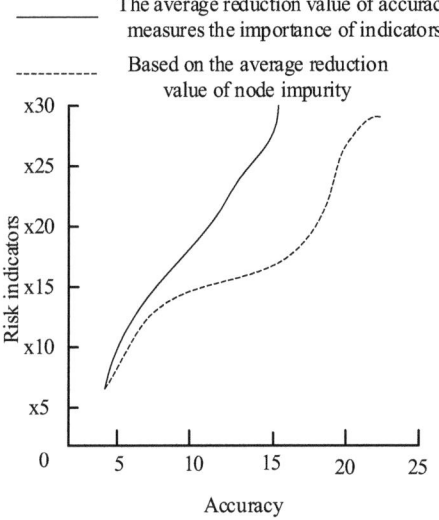

Fig. 5. Comparison of the importance of early warning indicators of financing risk

of the test sample data is 94.59%, which shows that the prediction effect of the early warning method of supply chain financial risk constructed in this paper is good, and it can effectively measure the online supply chain financial financing risk, which has certain practical significance.

5 Conclusion

The multi-dimensional risk early warning method of digital e-commerce supply chain based on information fusion of Internet of Things not only improves the intelligence level of e-commerce supply chain, but also provides comprehensive and accurate risk management means for enterprises. By collecting, processing and analyzing IOT data in real time, potential risks in the supply chain can be found in time, and targeted risk coping strategies can be provided for enterprises through multi-dimensional early warning mechanism. The following conclusions are obtained through experiments:

(1) In the face of different risk quantities, the state function value of this method changes little, tends to be stable, and the overall level is above 95%, which can meet the demand of risk early warning.
(2) The effectiveness and stability of this method in capturing the characteristics of financing risk;
(3) The prediction effect of this paper is good, and it can effectively measure the financing risk of online supply chain.

In the next step, we will continue to explore and innovate, continuously improve the accuracy and timeliness of the early warning system, and provide more reliable risk management solutions for the e-commerce industry. In the future, we will also face more new challenges and opportunities, and we need to constantly innovate and optimize the early warning model to adapt to the rapidly changing market environment.

References

1. Chen, G.: The development dilemma and countermeasures of Chinese cross-border e-commerce enterprises under the background of big data. J. Comput. Meth. Sci. Eng. **23**(2), 1087–1099 (2023)
2. Sharma, A., Sharma, M., Dwivedi, R.K.: Exploratory data analysis and deception detection in news articles on social media using machine learning classifiers. Ain Shams Eng. J. **1**(1), 1–6 (2023)
3. Roy, P.K.: Enriching the green economy through sustainable investments: an ESG-based credit rating model for green financing. J. Clean. Prod. **420**(9), 1–13 (2023)
4. Zhang, H., Chen, Y., Yan, X., et al.: Research on credit risk evaluation of B2B platform supply chain financing enterprises based on improved TOPSIS. Int. J. Comput. Sci. Eng. **26**(2), 220–230 (2023)
5. Liu, S., Wang, S., Liu, X., et al.: Human inertial thinking strategy: a novel fuzzy reasoning mechanism for IoT-assisted visual monitoring. IEEE Internet Things J. **10**(5), 3735–3748 (2023)
6. Computing, C.M.: Retracted: the combination of internet of things technology based on probability model network and mass education. Wirel. Commun. Mob. Comput. **12**(7), 1–6 (2023)
7. Xie, W., Kimura, M., Takaki, K., et al.: Interpretable framework of physics-guided neural network with attention mechanism: simulating paddy field water temperature variations. Water Resour. Res. **58**(5), 1–10 (2022)
8. Chen, Z.S., Wu, S., Govindan, K., et al.: Optimal pricing decision in a multi-channel supply chain with a revenue-sharing contract. Ann. Oper. Res. **318**(1), 67–102 (2022)
9. Balgomera, K., Cruz, A.E.D., Santiago, J.E.G., et al.: Consumer trust in mobile phone industry: comparative study on traditional commerce & E-commerce. J. Bus. Manage. Stud. **26**(3), 1–10 (2022)
10. Qiao, W.: E-commerce across boarder logistics risk evaluation model based on improved neural network. J. Funct. Spaces **2022**(5), 1–10 (2022)

Cloud Monitoring Technology of Fish Pond Water Quality Based on Internet of Things Multi-source Data Collection

Shaoyong Cao[1], Huanqing Han[1(✉)], and Cao Song[2]

[1] School of Industrial Automation, Zhuhai College of Bejing Institute of Technology, Zhuhai 519085, China
`suinkln45222@126.com`
[2] Dalian University of Science and Technology, Dalian 116000, China

Abstract. In order to achieve precise monitoring of key indicators of fish pond water quality and optimize water quality management, a cloud based monitoring technology for fish pond water quality based on multi-source data collection of the Internet of Things is proposed. Using the AS950 water quality automatic sampler as the data collection device for monitoring technology in this article, the reasonable layout of the AS950 water quality automatic sampler can achieve multi-source collection of fish pond water quality. Using STRIDE of the Internet of Things to build a lightweight cloud transmission protocol that is connected and provides reliable transmission, the cloud transmits multi-source fish pond water quality data collected by the AS950 water quality automatic sampler. At the same time, on the basis of improving the empirical mode decomposition method, analyze the cloud transmission of multi-source fish pond water quality data, and obtain accurate monitoring results for various indicators of fish pond water quality. The experimental results show that the cloud based monitoring technology for fish pond water quality based on multi-source data collection of the Internet of Things designed in this paper demonstrates excellent performance in comparison, with extremely high fit to the actual situation and minimal deviation. Specifically, the pH monitoring error is less than 0.01, the total nitrogen monitoring deviation is controlled within 0.05mg/L, the total phosphorus monitoring deviation is only 0.03mg/L, and the permanganate index monitoring results are basically consistent with the actual values. These data fully validate the high precision and reliability of this technology in water quality monitoring.

Keywords: Internet of Things Multi-Source Data Collection · Water Quality of Fish Ponds · Cloud Monitoring Technology · As950 Water Quality Automatic Sampler · Stride · Improved Empirical Mode Decomposition Method

1 Introduction

With the continuous development of the social economy, the aquaculture industry is facing increasingly severe environmental pollution challenges while thriving, among which the pollution of toxic heavy metals arsenic (As) and mercury (Hg) is particularly concerning [1–3]. Once these pollutants enter the water body, they not only directly harm the ecological environment of fish ponds, but may also accumulate through the food chain, posing a significant threat to the safety of aquaculture products and affecting consumer health. Pond aquaculture, as the core mode of freshwater aquaculture, directly affects the sustainable development of aquaculture and public health and safety in terms of water quality. Currently, although studies have revealed the presence of As and Hg pollution in pond aquaculture in some areas, and the degree of pollution varies depending on geographical location and aquaculture mode, systematic monitoring and research on As and Hg pollution in fish ponds are still insufficient.

Reference [4] analyzed the diversity of true green alkene compounds in drinking water sources and explored their impact on sustainability and water quality. However, water quality is influenced by various environmental factors, including natural factors such as climate and seasonal changes, as well as human factors such as pollution emissions. When analyzing the impact of true green olefin compounds on water quality, it is necessary to fully consider these complex factors. Reference [5] analyzed the pathway between the abundance and biomass of planktonic animals and water quality parameters in water source reservoirs in Zhejiang Province. In the analysis process, it is necessary to accurately identify and eliminate the impact of external interference factors (such as human activities, natural disasters, etc.) on the results. Reference [6] studied real-time water quality monitoring and technological development cooperation. Real time monitoring requires data to be transmitted to the data center in a timely and accurate manner for analysis and processing. However, in outdoor environments, data transmission may be affected by factors such as network stability and equipment failures. At the same time, data security is also a concern that needs to be addressed.

Therefore, this paper proposes a cloud monitoring technology for fish pond water quality based on multi-source data collection of the Internet of Things. On the basis of multi-source collection of fish pond water quality data, cloud transmission of fish pond water quality data is achieved using Internet of Things technology [7, 8], in order to achieve cloud monitoring of fish pond water quality. This technology combines sensors and devices with cloud computing to achieve comprehensive monitoring and management of fish pond water quality. Compared to traditional manual sampling and analysis methods, this technology can improve monitoring frequency and accuracy, timely identify potential water quality issues, and take corresponding measures to adjust and improve, thereby improving aquaculture efficiency and environmental protection level.

2 Research on Cloud Monitoring Technology of Fish Pond Water Quality

2.1 Fish Pond Water Quality Data Multi-source Collection

In order to realize cloud monitoring of fish pond water quality, it is extremely necessary to accurately collect specific fish pond water quality data. Therefore, this paper takes Hash American Sigma 950 series water quality sampler as the specific selection range [9]. American Sigma950 series water quality sampler includes three designs: portable automatic water quality sampler, refrigerated sampler and all-weather sampler, which are mainly used in wastewater sampling, specimen collection, industrial pretreatment sampling, environmental monitoring, rainwater sampling and other industries. It has the characteristics of accuracy, simplicity, flexibility, reliability and economy, and is suitable for equal volume sampling at regular intervals and flow proportional sampling [10]. AS950 series automatic water quality samplers also meet higher sampling requirements, such as monitoring the overflow of sewers where rainwater and sewage meet, monitoring storm drains, biological monitoring, or water quality research. Among them, Fig. 1 shows the sampling structure of the AS950 water quality automatic sampler.

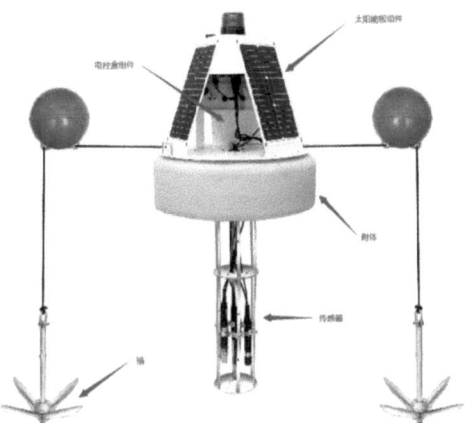

Fig. 1. Sampling structure of AS950 water quality automatic sampler

Combined with the specific requirements of cloud monitoring of fish pond water quality, this paper takes AS950 water quality automatic sampler as the final collection device. In the specific operation process [11], the AS950 water quality automatic sampler is seamlessly connected with the automatic sampling system CYQ-310H and communicates through the digital interface, and the water samples exceeding the standard are stored in the sample retention unit (conforming to HJ/T 372–2007) [12].

On this basis, this paper sets the operation mode of AS950 water quality automatic sampler in the specific data acquisition stage. As a portable water quality automatic sampler, AS950 water quality automatic sampler has the characteristics of light weight and easy to move with lifting handle. According to the actual situation, this paper configures it as a single bottle or multiple bottles for sampling. AS950 automatic water quality sampler is equipped with compact and standard base, the former can make the sampler used in sewer manhole. To some extent, this paper sets it as a dual-program mode, and at most two sampling programs can be run sequentially, in parallel or according to the weekly daily worksheet, so that a single sampler has the function of multiple sampling. In the aspect of sampling mode, the specific sampling methods are time weighting, flow weighting, timetable, flow table and event triggering. In terms of distribution, various adaptive modes are set up, such as single bottle mixed sample, multi-bottle mixed sample, multi-bottle discrete sample, single bottle multi-bottle, single bottle diversified or single bottle multi-bottle and single bottle diversified combination. The corresponding operation modes can be divided into two types: continuous or intermittent.

Among them, the settings of relevant realistic configurations are shown in Table 1.

Table 1. Operation Settings of AS950 Water Quality Automatic Sampler

Number	Parameter	Configuration
1	STATUS SCREEN	Display program running steps, whether there are any missing samples, when the next sample will be sampled, number of remaining samples, number of recorded channels, etc
2	call the police	Alarm information is displayed on the status screen and can be simultaneously recorded in the diagnostic alarm log. The system diagnosis and recording can set alarms
3	Manual sampling	Start sampling independent of the ongoing sampling program
4	Automatic shutdown in multi bottle mode	After the distribution arm fully rotates (except for selecting continuous mode)
5	Hybrid mode automatic shutdown	After the preset sample quantity has been fully transferred to the mixing container, the sample quantity is 1 to 999, or the bottle is full
6	Sample Volume	Program in increments of 10 ml (0.34 oz), ranging from 10 to 10000 ml (3.38 oz to 2.6 gallons)
7	sampling interval	From 1 to 9999 flow pulses in single increments (momentary contact closure for about 25 ms or 5–12 V DC pulses; optional 4–20 milliampere interface), or from 1 to 9999 min in increments of about 1 min
8	Set point sampling trigger	If a flow sensor, pH/temperature sensor, or other external detection equipment is installed, sampling will be triggered after exceeding the set limit range
9	Sample Historical Data Record	Store up to 4000 pieces of various data
10	Measurement data recording	Can store up to 325000 selected measurement channel data, stored at set recording intervals
11	Event log	Store 2000 records

In order to ensure the reliability of the collection results, the layout of AS950 water quality automatic sampler is set in this paper. Among them, at the same horizontal position, AS950 automatic water quality samplers are set at a distance of 200.0m; at the horizontal position, AS950 automatic water quality samplers are set at a distance of 5.0cm from the water surface and 5.0cm from the water bottom, respectively, and an AS950 automatic water quality sampler is set at the middle of two AS950 automatic water quality samplers. In this way, the reliability of collected data information is ensured.

2.2 Cloud Transmission of Fish Pond Water Quality Data Based on Internet of Things

When transmitting the collected water quality data of multi-source fish ponds in the cloud, this paper fully considers that STRIDE is a lightweight transmission protocol that is connection-oriented and provides reliable transmission. Each connection can only be point-to-point, and does not support multicast and multicast. In addition, the connection is a virtual connection, not a physical connection. It only exists in two end systems, and the intermediate devices of the network are completely unaware of the characteristics of the existence of this connection and its reliability in ensuring data security, which is the basis of transmission.

STRIDE designed in this paper is a reliable transport protocol based on UDP, and its location in the protocol stack is shown in Fig. 2.

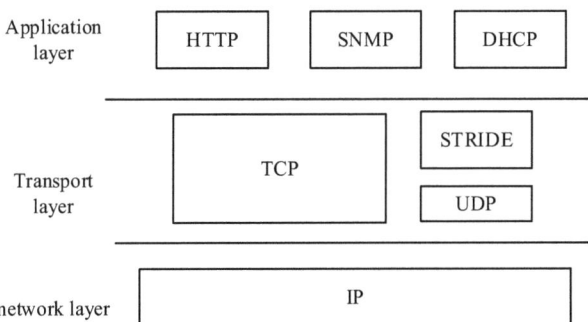

Fig. 2. STRIDE protocol stack

Combined with Fig. 2, it can be seen that using UDP as the bottom layer allows STRIDE to pass through intermediate devices, which is the premise of practical use of STRIDE.

In the specific data cloud transmission process, the data delivery process of STRIDE protocol consists of three stages, namely, connection establishment, data transmission and completion.

Connection establishment stage. Combining STRIDE is a connection-oriented lightweight transmission protocol. In this paper, establishing connection is an indispensable process in the whole data interaction process and an important step to ensure reliable transmission. The purpose of this paper is to solve the following three problems with the help of STRIDE.

1) Make sure that each part of the connection can know each other's existence.
2) Allow both parties to negotiate or collect some parameters (such as timestamp and receiving window).
3) Allocating and initializing transmission entity resources (such as cache size). Among them, the transmission entity resource encoding matrix model is:

$$Q = \begin{pmatrix} X_{11} & \cdots & X_{1b} \\ \vdots & \ddots & \vdots \\ X_{a1} & \cdots & X_{ab} \end{pmatrix} \quad (1)$$

Among them, X represents STRIDE design.

On this basis, this paper sets the establishment of STRIDE as the client-server mode. The STRIDE service process running at the receiving end first issues a passive open command. At this point, the service process is in a "listening" state, continuously monitoring the connection request of the waiting client process. Once the connection request is monitored, a response is generated immediately. The client process runs on the sending end. It first opens it actively, constructs the fish pond water quality data for connection request and sends it to the server, with the SYN chronization bit syn at the head set to 1, and selects a random serial number X at the same time. After receiving the fish pond water quality data of the connection request, STRIDE on the server returns the connection request to confirm the fish pond water quality data if it agrees. Set the SYN and ACK positions of the fish pond water quality data header to 1, set the confirmation number to x + 1, and add another randomly selected serial number Y to the header. After receiving the acknowledgement returned by the server, the client needs to give the acknowledgement to the server, with the ACK flag position 1 and the acknowledgement number y + 1, and its own serial number x + 1. After the server receives and analyzes the data sent by the client to confirm that the water quality of the fish pond is correct, it opens the connection to wait for the subsequent data from the client. The executable domain D where the server receives data is represented as:

$$D = \{i(x, y) | x \in I_l \cap I_s, y \in I\} \quad (2)$$

Among them, $i(x, y)$ represents the ACK position information in data transmission, I represents the set of STRIDE, I_l represents the set of surviving SYNs and ACKs, and I_s represents the set of idle SYNs and ACKs.

If this paper learn from the slow start algorithm of TCP, it will inevitably lead to a long convergence time, which is not ideal in a data center dominated by small streams. Therefore, this paper sets STRIDE to provide the startup rate by sharing the bandwidth among multiple streams, that is, the initial rate is the ratio of full bandwidth to the number of streams at the sender. The corresponding transmission logic can be represented as:

$$y - \lambda x = [\frac{x'V^{-\frac{1}{2}}}{y'}]_f \quad (3)$$

Among them, λ represents the initialization coefficient of STRIDE, (x', y') represents the position update result of i data within time f, and V represents the starting rate at the beginning of data transmission.

In this paper, the STRIDE message is composed of a header and a load. Its header length is fixed at 116 bits, including STT value (STT), sequence number (seq), acknowledgement number (ack-seq) and four flag bits (SYN,FIN,ACK,NACK). The meanings of the header fields are shown in Table 2.

Table 2. Composition of STRIDE message

Number	Constituent part	Functional effects
1	STT	If a data message is sent from the sending end, the STT records the sending time of the data message. If the data message is sent by the receiving end, the STT records the one-way transmission delay of the responded message
2	Seq	Seq is the sequence number of the first byte of the group
3	Ack seq	ack seq represents the sequence number of the first byte of the expected next packet to be received
4	SYN	SYN is used to identify the message as a connection request message when a connection is established
5	FIN	FIN is used to release the connection
6	ACK	The ACK message is used to provide feedback that the previous data segment has been successfully received
7	NACK	When the receiving end detects that the received message is out of order or out of time, it immediately constructs a NACK to notify the sending end to immediately retransmit the breakpoint and all subsequent messages

When transmitting water quality data for STRIDE fish ponds, the receiving end does not need to confirm the water quality data of each pond one by one, thereby reducing the number of confirmed data. When packet loss or disorder occurs, the receiving end immediately returns NACK to notify the sending end to resend the discarded or disordered water quality data. The receiving end determines whether the expected water quality data has been lost through a timeout timer, and determines whether it is out of order by maintaining the order of the received water quality data. Once a timeout or disorder occurs, the receiving end will no longer receive data other than the expected water quality data. Both the sending and receiving ends need to maintain a register that stores the length of accumulated water quality data. When the value of the register reaches the preset segment length, the receiving end will return confirmation of the last water quality data of each segment. Confirm that the data carries the timestamp of the last data, which is used to update the data transmission rate. The sending end adjusts the sending rate based on efficient congestion control algorithms by collecting timestamps of all confirmed data, ensuring a smooth and efficient network state.

End stage. The FIN flag of the last fish pond water quality data is set to 1 to inform the receiving end to release resources, and the receiving end returns the final fish pond water quality data confirmed by the release connection request and closes the connection. After receiving the confirmed fish pond water quality data, the sender also releases resources and closes the connection, thus completing the end of a connection.

2.3 Fish Pond Water Quality Cloud Monitoring

This paper utilizes the EEMD method for cloud monitoring of fish pond water quality. There are two important parameters used in the EEMD decomposition method, namely the number of times of repeated EMD decomposition M and the proportion of Gaussian white noise to the standard deviation of the amplitude of the original fish pond water quality data k. At present, there is no formula that can be applied to determine the values of M and k. Based on the suggestions of the algorithm proposer and the characteristics of the water quality data collected on site in this project, the M value is defined as 100, and k is taken as 0.05–0.5 times.

Among them, the realization steps of fish pond water quality cloud monitoring based on EEMD algorithm are as follows:

Initializing the number of repeated EMD decomposition m, and Gaussian white noise accounting for the standard deviation of the amplitude of the original fish pond water quality data.

The ratio k and the number of iterations m for EMD decomposition are set to 1.

Perform and record the process of EMD decomposition. The following is to perform the decomposition for the m time:

① According to the water quality data collected on site $x(t)$ and Gaussian white noise sequence $n_m(t)$, the expression of the m-th decomposition sequence is

$$x_m(t) = x(t) + k n_m(t) \tag{4}$$

② Decomposition $x_m(t)$, so that the initial value of the loop variable i is 1, then

$$x_{1,m}(t) = x_m(t) \tag{5}$$

Let the initial value of the loop variable j be 1, then

$$y_{1,m}(t) = x_m(t) \tag{6}$$

③ Find out among them. $y_{1,m}(t)$ all local maxima in a sequence $u_{j,m}(t)$ and local minima $v_{j,m}(t)$ envelope; The two upper and lower envelopes need to envelope all data points, and the average value is:

$$m_{j,m}(t) = \frac{u_{j,m}(t) + v_{j,m}(t)}{2} \tag{7}$$

Then the difference between the original fish pond water quality data and the envelope average value can be obtained as follows:

$$h_{j,m}(t) = y_{j,m}(t) - m_{j,m}(t) \tag{8}$$

④ Make a judgment to $h_{j,m}(t)$.. If the two conditions of IMF component are not met, then $j = j + 1$, and

$$y_{j,m}(t) = y_{j-1,m}(t) \tag{9}$$

Repeat ③; If yes, the I-th IMF component is obtained as follows

$$c_{i,m}(t) = h_{j,m}(t) \tag{10}$$

Then this paper can know that the residual component is:

$$r_{i,m}(t) = x_{i,m}(t) - c_{i,m}(t) \tag{11}$$

⑤ Judgment $r_{i,m}(t)$ whether the conditions for termination are met. If not, then

$$x_{i+1,m}(t) = r_{i,m}(t) \tag{12}$$

Repeat ②, ③, ④; Otherwise, decomposition to $x_m(t)$ is over.

Among them, for the end standard of the above process, NEHuang gives and defines the following standard deviation

$$S_D = \frac{1}{T}\int_0^T \frac{|h_k(t) - h_{k-1}(t)|^2}{|h_{k-1}(t)|2} dt \tag{13}$$

In general, setting the value of S_D is between 0.2 and 0.3. Its physical meaning is interpreted as: let $h(t)$ can meet the requirements of IMF as much as possible, and it needs to be able to control its screening times, so that the obtained IMF component can retain the amplitude modulation information of the original fish pond water quality data.

According to the above-mentioned way, the cloud monitoring of fish pond water quality data is realized.

3 Test Analysis

3.1 Test Environment

In the specific testing stage, this paper investigates the years of fish pond culture, the source of aquaculture water, the farming mode, the basic situation around the fish pond and whether to put in feed. The total area of the aquaculture pond has reached 200 acres, of which 150 acres are occupied by the aquaculture area, while the rest are slope protection, roads, and facilities. The water quality of the fish pond is excellent, mainly sourced from groundwater and natural rainwater. In order to maintain clean water quality, the fish pond is equipped with advanced purification facilities, including a water filtration system and oxygen replenishment equipment. Fish ponds mainly cultivate grass carp, carp, mud carp, and other species, with grass carp and carp being the largest in scale. At present, the average weight of grass carp in the fish pond reaches 10 kg, while the average weight of carp is 8 kg. The farmers have adopted scientific farming techniques, and through reasonable feeding and strengthened water quality management, the fish have grown rapidly. See Table 3 for the details of the fish pond.

Table 3. Statistical Table of Basic Situation of Test Fish Pond

Allocation	Parameter
Number of years of fish pond aquaculture	12a
Source of aquaculture water	River water flow
Breeding mode	Mixed breeding
Adjacent environment	Near the river, roadside
feed	Formula feed

Based on Table 3, the water quality parameter measurement results of aquaculture ponds were collected from multiple sampling points and time periods within a year through on-site real-time collection, and an experimental dataset was generated. The main content of this dataset includes but is not limited to the following indicators: pH value, dissolved oxygen (DO), total nitrogen (TN), total phosphorus (TP), permanganate index (CODMn), etc. The measured values of each indicator at different time points and positions are recorded. In addition, it may also include additional data related to water quality, such as temperature, turbidity, etc.

3.2 Test Results and Analysis

After completing the above preparations, the AS950 water quality automatic sampler should be reasonably arranged in different areas of the fish pond (such as the inlet, outlet, central area, edge area, etc.) to ensure comprehensive coverage and reflect the overall condition of the fish pond water quality. Each AS950 water quality automatic sampler is capable of real-time measurement and recording of various water quality parameters, including pH value, dissolved oxygen (DO), total nitrogen (TN), total phosphorus (TP), and permanganate index (CODMn). Sampling data is transmitted in real-time to cloud servers through IoT technology, avoiding errors and delays that may occur during manual recording. After completing the sampling, a lightweight cloud transmission protocol is constructed using the STRIDE protocol of the Internet of Things to ensure stable and reliable transmission of water quality data to cloud servers. The data is stored in the cloud for subsequent analysis and processing. At the same time, the received raw data is cleaned to remove outliers and noise, ensuring the accuracy of the data. Standardize the data and conduct in-depth analysis of multi-source fish pond water quality data transmitted from the cloud based on improved empirical mode decomposition method. This method can effectively extract useful information from data and identify the patterns and trends of water quality changes.

In order to analyze the monitoring effect of the design method in this paper, the water quality monitoring method based on big data and the water quality monitoring method based on sensors are taken as the test control group respectively. Among them, within 10 days, the relationship between the test results of different methods and the actual situation is shown in Figs. 3, 4, 5 and 6.

Based on the test results shown in Figs. 3, 4, 5 and 6, it can be clearly observed that the cloud based monitoring technology for fish pond water quality designed in this paper,

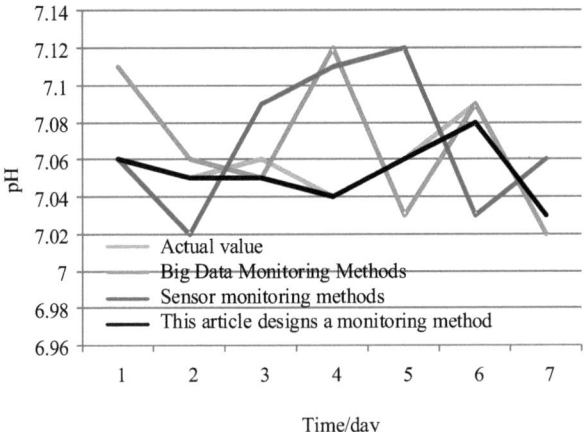

Fig. 3. Comparison chart of pH monitoring results

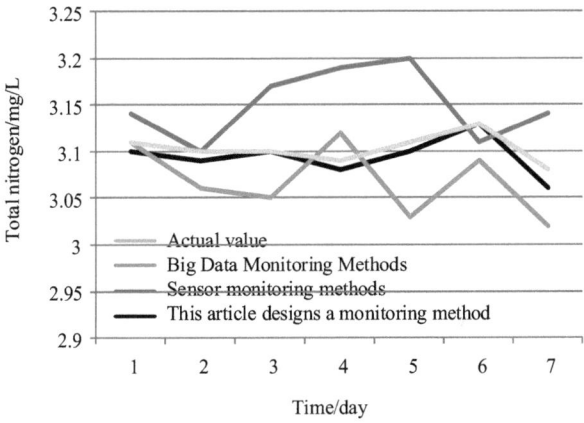

Fig. 4. Comparison chart of total nitrogen monitoring results

which is based on multi-source data collection of the Internet of Things, demonstrates excellent performance when compared with three different monitoring technologies. Specifically, this method has the highest fitting degree with the actual situation, while the corresponding deviation is also the smallest, fully demonstrating its accuracy and reliability. In the detailed data comparison, the error between the pH monitoring results of this method and the actual values is extremely low, only less than 0.01, which demonstrates the high accuracy of this technology in pH monitoring. For the monitoring of total nitrogen, the results of this method are basically consistent with the actual values, with a deviation controlled within 0.05mg/L, further verifying its accuracy in monitoring key water quality indicators. It is worth noting that although all three monitoring techniques have certain deviations in the monitoring of total phosphorus, the method proposed in this paper still performs well. The deviation between the total phosphorus monitoring results and the actual values is the smallest, only 0.03mg/L, which reflects the advantage

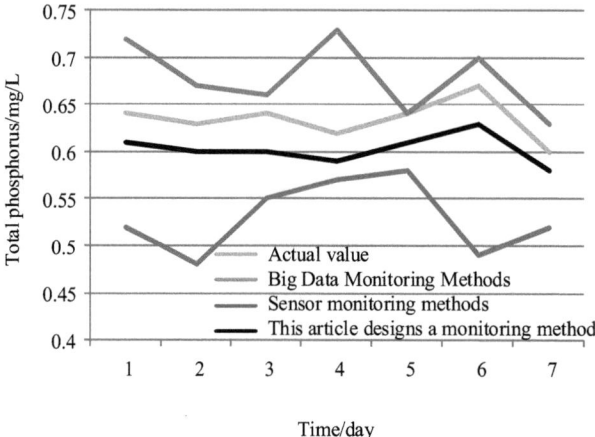

Fig. 5. Comparison chart of total phosphorus monitoring results

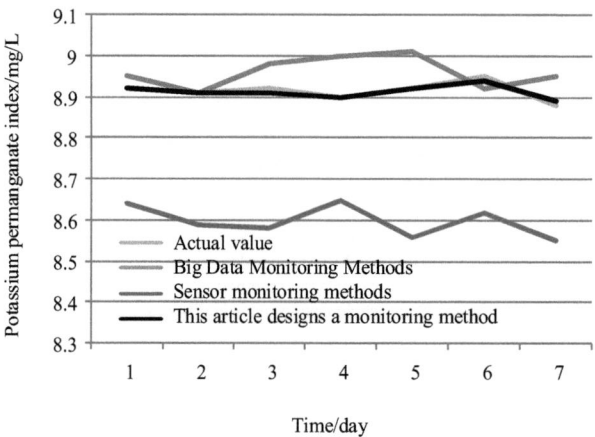

Fig. 6. Comparison of monitoring results of permanganate index

of this technology in dealing with complex water quality parameters. In addition, the monitoring results of the permanganate index using this method are basically consistent with the actual values, once again proving its comprehensive and accurate monitoring ability.

Through in-depth analysis of these results, it can be concluded that for different water quality factors, different outcomes are mainly due to the following reasons: firstly, technical factors are one of the key factors, including the accuracy of monitoring equipment, the stability of data transmission, and the accuracy of data analysis algorithms. These factors directly determine the reliability and accuracy of monitoring results. Secondly, environmental factors cannot be ignored, such as the complexity of water quality changes, spatial heterogeneity, and interference from natural and human factors, which may all have an impact on monitoring results. In addition, operational factors are equally important, such as the standardization of sampling and storage, and the timeliness of equipment maintenance, which may affect the quality of water samples and the accuracy of monitoring data.

4 Conclusion

The cloud based monitoring technology for fish pond water quality based on multi-source data collection of the Internet of Things proposed in this article demonstrates extremely high monitoring accuracy and reliability through precise data collection and efficient cloud analysis. Its core advantage lies in achieving accurate monitoring of key indicators of fish pond water quality (such as pH, total nitrogen, total phosphorus, etc.) with minimal deviation, providing strong support for water quality management. In the follow-up work, the investigation of plankton and microorganisms in the pond can be increased, and the migration characteristics of As and Hg in the fish pond can be deeply studied by isotope tracing method, so as to clarify the sources of heavy metals and other pollutants in the fish.

Acknowledgment. The Special projects in Key Areas of Guangdong Province (Grant 2021ZDZX4050).

References

1. Dilmi, S.: Calcium Soft Sensor Based on the Combination of Support Vector Regression and 1-D Digital Filter forWater Quality Monitoring. Arab. J. Sci. Eng. **48**(5), 6111–6136 (2023)
2. Ji, Y., Wei, Y., Liu, J., et al.: Design and Realization of a Novel Hybrid-Drive Robotic Fish for Aquaculture Water Quality Monitoring. J. Bionic Eng. **20**(2), 543–557 (2023)
3. Fan, L.X.: Research on Quality Control of Water Quality Sampling in Environmental Monitoring. Wiley InterScience **5**, 89–92 (2022)
4. Ouattara, .M , Zongo, F., Zongo, B.: Diversity of euglenoids in a drinking water source in Burkina Faso (West Africa): implications for sustainability and water quality. African J. Aquatic Sci. **48**(2), 178–188 (2023)
5. Gong-Guo, L.I., Hang-Ying, X.U., Hai-Yan, Y.U., et al.: Path Analysis Of Zooplankton Abundance And Biomass With Water Quality Parameters In Water Source Reservoirs. Zhejiang Province. Acta Hydrobiologica Sinica **43**(1), 165–172 (2022)
6. Fielke, S., Taylor, B., Coggan, A., et al.: Understanding power, social capital and trust alongside near real-time water quality monitoring and technological development collaboration. J. Rural. Stud. **2022**(92), 1320–2131 (2022)
7. Liu, S., Wang, S., Liu, X., et al.: Human Inertial Thinking Strategy: A Novel Fuzzy Reasoning Mechanism for IoT-Assisted Visual Monitoring. IEEE Internet Things J. **10**(5), 3735–3748 (2023)
8. Liu, S., Guo, C., Al-Turjman, F., et al.: Reliability of Response Region: A Novel Mechanism in Visual Tracking by Edge Computing for IIoT Environments. Mech. Syst. Signal Process. **138**, 106537 (2020)
9. Chen, Z., Jiang, P., Liu, J., et al.: An adaptive data cleaning framework: a case study of the water quality monitoring system in China. Hydrol. Sci. J. **67**(5–8), 1114–1129 (2022)

10. Tsai, K.L., Chen, L.W., Yang, L.J., et al.: IoT based Smart Aquaculture System with Automatic Aerating and Water Quality Monitoring. Journal of Internet Technology **23**(1), 177–184 (2022)
11. Chapman, D.V., Sullivan, T.: The role of water quality monitoring in the sustainable use of ambient waters. One Earth **5**(2), 132–137 (2022)
12. Zhan, L.J., Lu, D.Z., Gao, Y.Q., et al.: Design and simulation of remote sensing monitoring method for water pollution . Comput. Simulat. **39**(6), 289–292,312 (2022)

A Transient Response Control Method for Hybrid Wind and Solar Power Microgrids Based on Composite Information Fusion

Jingtao Zhao[1], Chao Song[1(✉)], and Shaoyong Song[2]

[1] Dalian University of Science and Technology, Dalian 116000, China
songchao1225@163.com
[2] School of Industrial Automation, Zhuhai College of Bejing Institute of Technology, Zhuhai 519085, China

Abstract. A hybrid wind and solar power microgrid is typically comprised of multiple distributed energy sources, such as wind turbines and photovoltaic systems. Each of these energy sources exhibits unique current and voltage response characteristics, contributing to the composite nature of the input power signal. This composite nature poses significant challenges to the transient response control of hybrid wind and solar microgrids, often leading to high voltage fluctuations. In order to improve the stability and reliability of hybrid microgrid operation, a transient response control method for microgrid based on composite information fusion is proposed. Firstly, an analysis was conducted on the topology structure of the wind solar hybrid power generation microgrid, clarifying the system composition and connection relationships, laying the foundation for subsequent transient response control. Secondly, sensors and communication modules are utilized to collect data from the hybrid wind and solar power microgrid. The composite information fusion technology is utilized to fuse the microgrid data. On this basis, the transient response control strategies for isolated and grid-connected operation are designed. Finally, the control commands are executed and feedback is provided. After experimental verification, it is known that this method significantly reduces the voltage and power fluctuations of microgrids in the transient stage, thereby effectively improving the quality of electrical energy and the stability of the entire system.

Keywords: Composite information fusion · Hybrid wind and solar power generation · Microgrids · Transient response · Power quality

1 Introduction

In the context of the global energy structure transformation today, the widespread application of renewable energy has become a key force in promoting green and low-carbon development. The wind solar hybrid power generation microgrid, as an important carrier for integrating clean energy such as wind and solar energy, is gradually becoming an important component of distributed energy systems. These types of microgrids often gather multiple distributed energy sources, such as wind turbines and photovoltaic power generation systems, which exhibit highly dynamic and differentiated current and voltage response characteristics based on changes in natural environmental conditions such as wind speed and light intensity. This composite power input signal not only enriches the energy diversity of microgrids, but also greatly increases the complexity of their operational control. Specifically, the intermittency of wind power generation and the daily periodicity of photovoltaic power generation result in significant fluctuations and uncertainties in their output power. This unstable power supply mode directly increases the difficulty of balancing power supply and demand within microgrids, leading to more frequent voltage fluctuations during transient processes. According to industry statistics, the voltage fluctuation rate of uncontrolled microgrids can reach over 10%, far exceeding the stable operation standards of traditional power grids, posing a serious challenge to the safe operation of power equipment and the quality of electricity consumption for users. Serving as a critical link between distributed power sources and loads, the transient response capabilities of microgrids have a direct bearing on the overall system operation [1]. Consequently, exploring transient response control strategies for hybrid wind and solar power microgrids holds significant value in enhancing microgrid performance, optimizing energy utilization, and fostering renewable energy consumption.

The hybrid wind and solar power microgrid is characterized by intermittency and fluctuation, which makes its transient process complex and variable. Traditional control methods often rely on a single source of information or a fixed control strategy. They lack comprehensive consideration of microgrid characteristics under different operating states [2]. The operation state of a wind and solar hybrid microgrid is affected by several factors. These include wind speed, light intensity, and load changes. These factors are random and fluctuating, leading to a complex and variable transient process in the microgrid. Traditional control methods struggle to accurately capture microgrid changes and adjust strategies in real-time. This results in limited transient response and stability [3, 4]. Additionally, traditional control methods often overlook the economy and reliability of the microgrid. In actual operation, microgrids need to meet stable operation requirements. They must also consider factors like operating costs and power supply reliability [5]. Traditional control methods often focus on a single performance indicator, such as reducing losses or improving voltage quality. They ignore the overall economy and reliability of the microgrid. This oversight can limit the practical application of traditional control methods. They may fail to meet the comprehensive needs of microgrids in real-world scenarios. Therefore, the challenge is to comprehensively utilize multiple information sources. Designing a composite information fusion control method can improve transient response speed and stability of microgrids. This is a hot and difficult issue in current research.

Composite information fusion technology involves the integration of data from various sensors or sources using specialized algorithms or models, resulting in more comprehensive and precise information [6–8]. In wind-solar hybrid power generation microgrids, this technology includes the fusion of diverse data like wind speed, light intensity, load variations, and energy storage status. By integrating these data, the operating state of the microgrid can be assessed more accurately, supporting the development of effective control strategies. This paper introduces a new transient response control method for hybrid wind and solar power generation microgrids, utilizing composite information fusion. The goal is to achieve fast response and stable operation during transient processes by leveraging multiple information sources and control strategies.

2 Design of Transient Response Control Methods for Hybrid Wind and Solar Power Microgrids

2.1 Topology Analysis of Hybrid Wind and Solar Power Microgrids

The microgrid configuration studied in this paper for wind-solar hybrid power generation comprises four components: the energy storage system, the AC power grid, the wind power generation unit, and the AC load [9–11]. As shown in Fig. 1.

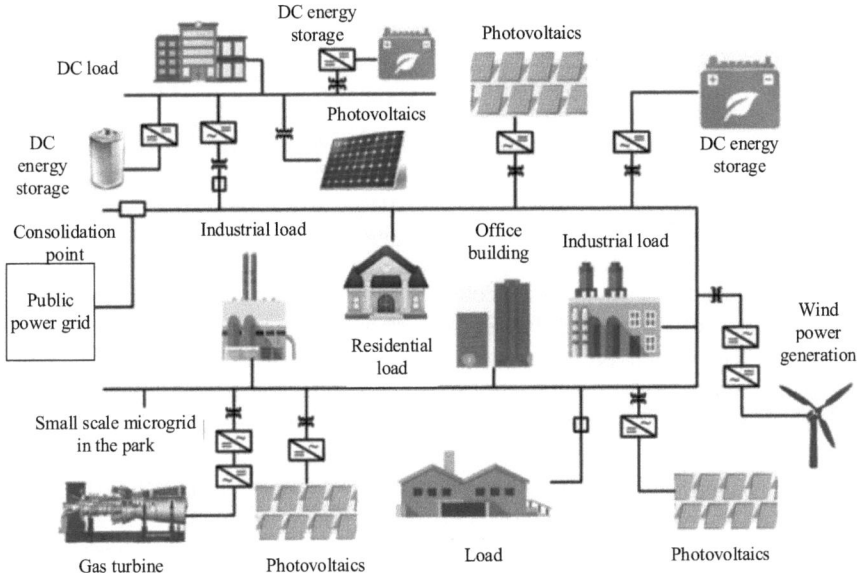

Fig. 1. Wind solar hybrid power generation microgrid structure

Among them, (1) The energy storage system uses a battery connected to the DC microgrid through a bidirectional converter (BDC). The BDC employs voltage droop control. (2) The DC microgrid connects to the AC power grid through a two-way converter (GVSC). The GVSC also uses voltage droop control to coordinate the DC bus voltage with the BDC [12]. (3) The wind turbine is connected to the DC microgrid via a converter (WVSC) using a permanent magnet synchronous generator. To maximize wind energy utilization, the WVSC operates in maximum power tracking mode. In order to make the best use of wind energy, WVSC operates in the state of maximum power tracking. As the driving energy of wind turbines, wind power P is usually used to quantify the intensity. It is defined as the kinetic energy of the airflow passing through the wind turbine in unit time, namely.

$$P_v = \frac{1}{2}\rho S v^3 = \frac{1}{2}\pi \rho R_w^2 v^3 \qquad (1)$$

Among them, ρ denotes the density of air. S denotes the swept area of the impeller. v indicates wind speed. R_w represents the blade radius. (4) AC load. AC load is powered by inverter LVSC. LVSC adopts constant AC voltage control, and the voltage characteristic of AC load is ignored. As power distribution between power sources can be realized without communication, droop control is widely used in power system [13]. In DC microgrid, the control method based on voltage signal is usually adopted. The voltage droop curve can be expressed as.

$$U = U_{ref} - \frac{1}{k}PP_v \qquad (2)$$

Among them, P indicates the output power. U indicates voltage. U_{ref} denotes the longitudinal intercept of the sag curve. $\frac{1}{k}$ indicates the sag factor.

2.2 Data Acquisition for Hybrid Wind and Solar Power Microgrids

Data collection for hybrid wind and solar microgrids is crucial for ensuring stable and efficient operation. These microgrids often include multiple distributed energy sources, each with different data collection methods and periods. This can lead to timing and format discrepancies, causing data inconsistency issues. To enhance transient response control, this paper focuses on optimizing sensors and communication modules to gather high-quality data from hybrid solar power generation microgrids [14, 15].

First, the sensors and communication modules required for microgrid data acquisition were selected, and their parameter settings were shown in Tables 1 and 2, respectively.

Ensuring accurate measurements and reliable data transmission requires careful sensor calibration and stable data transmission processes. Sensors are placed in critical locations within wind turbines and photovoltaic power systems to capture real-time environmental and equipment status data. Similarly, sensors are installed at key junctions and load points in the microgrid to monitor electrical parameters and load conditions. These sensors transmit data promptly to the data center through efficient data collectors, ensuring data freshness and accuracy. The data center contains a robust database system that archives and analyzes this data, maintaining its integrity and traceability [16]. This comprehensive approach enhances our understanding of the microgrid's operational status, aiding system optimization and stability.

2.3 Microgrid Data Processing Based on Composite Information Fusion

He addition of optimized sensors and communication modules can lead to data redundancy and multiplicity. Different sensors may collect the same or similar information, and communication modules may produce large amounts of duplicate or redundant data.

Table 1. Sensor parameters

No	Sensor type	Detailed parameter
1	Wind speed sensor	Measuring range. 0–50 m/s Accuracy. ± 0.5m /s Resolution. 0.1m /s Output signal. 4–20 mA Operating temperature −40°C to +85 °C
2	Light intensity sensor	Measuring range. 0–200,000 lx Accuracy. ± 5% Resolution. 1 lx Output signal. 0–10 V Operating temperature −20°C to + 70 °C
3	Voltage transformer	Rated voltage. 10 kV Ratio. 10,000.100 Accuracy. 0.2 grade Rated output. 100 mA Insulation class. F class
4	Current transformer	Rated current. 1000 A The ratio is 1000.5 Accuracy. 0.5 grade Rated output. 5 A Insulation class. H class

Table 2. Parameters of the communication module

No	Item	Detailed parameter
1	Transmission mode	Wireless/Wired (optional)
2	Communication protocol	Modbus RTU/TCP, MQTT, etc. Communication distance. up to 10 km in wireless mode
3	Transmission rate	Up to 1 Mbps
4	Operating band	2.4 GHz (wireless mode)
5	Interface type	RJ45, USB
6	Operating voltage	5–30 VDC
7	Operating temperature	−20 °C ~ + 70 °C

These redundant and superfluous data can increase the burden on data processing and storage, affecting subsequent data analysis and application. To address this, we use composite information fusion technology to refine the collected wind and solar hybrid microgrid data. We employ the standard deviation method to detect outliers, ensuring the efficacy and precision of subsequent analysis. This method identifies data points that deviate significantly from the mean, setting a reasonable threshold based on the standard deviation [17, 18]. Detected outliers are either removed or replaced to ensure data accuracy. Exponential smoothing methods are then applied to minimize noise and variations, smoothing out the time series data. This method smoothes short-term fluctuations while retaining long-term trends, making the data smoother and easier for subsequent analysis and modeling. The formula is as follows.

$$S_t = \frac{\alpha x_t + (1-\alpha)S_{t-1}}{Z} \quad (3)$$

Among them, S_t denotes the smoothed value at moment t; the x_t denotes the moment t the actual data points. α denotes the smoothing coefficient; the S_{t-1} represents the smoothed value of the previous moment. On this basis, the data are normalized by scaling them to fall into a small specific interval, eliminating the difference in magnitude between different features and making them numerically comparable. The normalization formula is as follows.

$$x_n = \frac{x - \min(x)}{\max(x) - \min(x)} \quad (4)$$

Among them, x_n denotes the normalized data; the x represents the original data. $\max(x)$ and $\min(x)$ are the minimum and maximum values in the data set, respectively. The information from different sensors is fused using composite information fusion algorithm. The formula is as follows.

$$F = S_t \sum_{i=1}^{n} w_i \times x_i \quad (5)$$

Among them, x_i indicates a value from a different data source. w_i denotes the corresponding weight; the n indicates the number of data sources. According to the importance and relevance of different information, different weights or non-linear combinations are assigned. The fused data are statistically analyzed to preliminarily evaluate and predict the operation status and performance of the microgrid. Processed data and analysis results are visually displayed and applied to monitor, control, and optimize the microgrid, enhancing its operational efficiency and stability.

2.4 Designing Transient Response Control Strategies

During data normalization, data of different types and ranges may be distorted after standardization. This issue is particularly pronounced when data sampling frequency is inconsistent or data quality is unstable, leading to missing data or abnormal values that can affect processing results. To ensure the transient response control effect for wind-solar hybrid power generation microgrids, a transient response control strategy is designed. This complex process involves multiple layers and requires a comprehensive evaluation of dynamic behaviors within the wind and solar power generation system, strict power quality standards for the microgrid, and synchronized control among various power sources. Various events that trigger transient responses in the microgrid, such as sudden changes in wind speed, light intensity, and load switching, are identified [19]. Transient performance requirements for microgrids are set, and seven transient response control strategies for wind-solar hybrid power generation microgrids are proposed. Mode switching is managed through DC bus voltage, its rate of change, and battery state of charge (SoC) to ensure transient stability of DC microgrids [20].

The following are specific descriptions of each mode of operation.

(1) Orphaned operation. The mode switching conditions under orphaned network operation are shown in Fig. 2.

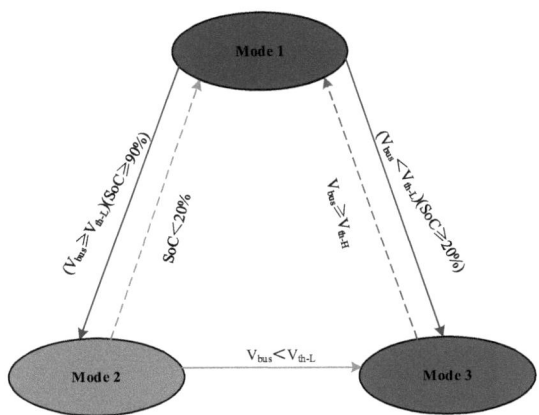

Fig. 2. Mode switching conditions for isolated network operation

In Fig. 2, the working states of the microsource unit are shown in three different modes, each with its specific operating characteristics and functions.

In Mode 1, the photovoltaic unit operates in maximum power point tracking mode, adapting to lighting conditions to maximize power output. It charges and stores excess battery power and stabilizes high-frequency fluctuations in the supercapacitor to ensure stable system operation.

In Mode 2, the primary task of the photovoltaic unit is to maintain a constant DC bus voltage. It adjusts its output power based on system requirements to ensure DC bus voltage stability. Both the battery and supercapacitor are in standby mode, ready to provide or absorb energy when needed.

In Mode 3, the photovoltaic unit returns to maximum power point tracking, similar to Mode 1, aiming for maximum power output. The battery operates in discharge mode, releasing stored energy to supplement the system. Supercapacitors continue to stabilize high-frequency power fluctuations, ensuring stable operation under such changes.

By switching between these three modes, the micro source unit can flexibly adjust its working state according to the actual needs and operating conditions of the system, in order to achieve optimal energy utilization and system stability. This flexibility and adjustability make microsource units play an important role in distributed energy systems.

It should be noted that the background of switching conditions from mode 2 to mode 1 in Fig. 1 is that the SoC of battery with sufficient power will naturally drop by less than 20% due to the passage of time during standby. This background is rarely seen in practice, so it is represented by dotted lines.

(2) Grid-connected operation [21]. When there is a shortage or surplus of energy in the microgrid, the microgrid can choose to connect to the AC grid for energy exchange as needed. The conditions for mode switching during grid-connected operation are depicted in Fig. 3. Notably, the dashed lines in this figure represent the switching conditions, and their corresponding backgrounds are identical to those in Fig. 2.

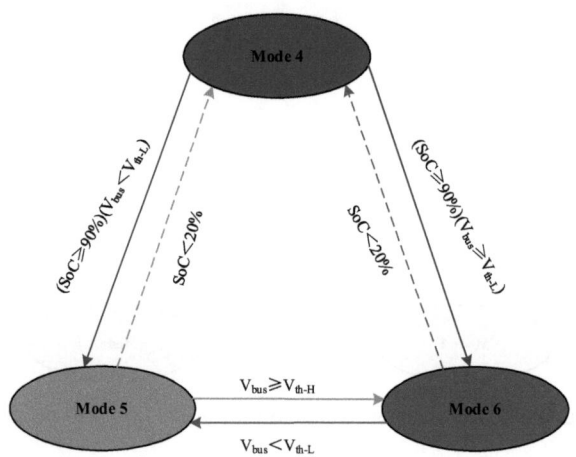

Fig. 3. Mode switching conditions for grid-connected operation

In Mode 4, the photovoltaic unit actively tracks the maximum power point to increase power generation. The battery charges and stores excess electricity, while the supercapacitor stabilizes high-frequency power fluctuations. The AC/DC converter rectifies AC power to DC power to ensure stable system operation.

In Mode 5, the photovoltaic unit maintains its maximum power point tracking status, effectively utilizing light resources for electricity generation. Both the battery and supercapacitor are in standby mode, ready to provide or absorb energy when needed. The AC/DC converter operates in rectifier mode, converting AC power into DC power to meet system demands.

In Mode 6, the photovoltaic unit continues in maximum power point tracking mode, ensuring optimal photovoltaic conversion efficiency. Batteries and supercapacitors remain as backup energy sources, ready to respond to system demand changes. The AC/DC converter switches to inverter mode, converting DC to AC to meet alternating current demands in the system.

Devise a synchronized control strategy tailored for wind and solar hybrid power generation systems, aiming to achieve smooth switching of generator units during transient processes and avoid adverse effects of power fluctuations on microgrids. At the same time, utilizing the fast response characteristics of energy storage systems, a charging and discharging control strategy is formulated to effectively suppress the output power fluctuations of wind and solar power generation systems, thereby improving the power quality of microgrids. In addition, based on changes in load demand, load management strategies are formulated to reduce the power supply pressure of microgrids during transient processes through demand side response or load scheduling, ensuring the stable operation of the system.

2.4.1 Control Instruction Execution and Feedback

After the design of the transient response control strategy is completed, on this basis, the control instructions are executed and feedback is provided. The control instruction execution and feedback flow is shown in Fig. 4.

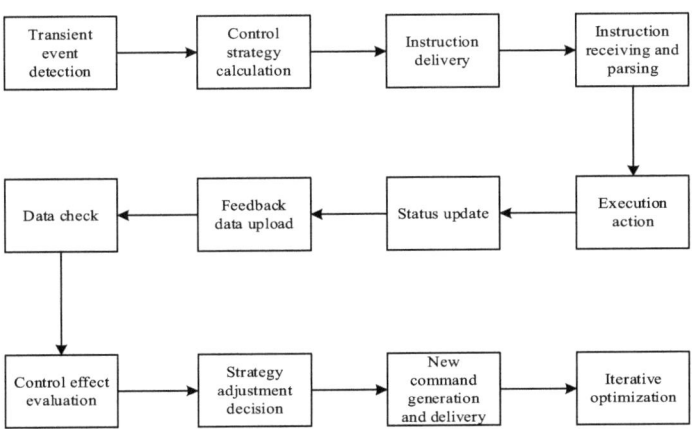

Fig. 4. Control instruction execution and feedback flow

Firstly, the microgrid's central controller monitors system status in real-time. Upon detecting transient events like sudden wind speed changes, light intensity variations, or load shifts, it calculates control commands based on the preset transient response strategy. These commands adjust wind power generation output, energy storage system charging and discharging power, and load scheduling. The central controller then sends these commands to corresponding execution units via the communication network. Each execution unit analyzes and confirms the instructions to ensure accurate understanding. They perform the corresponding operations based on the analyzed instructions. During execution, the units update their status information in real-time, such as output power, voltage, and current, allowing the central controller to monitor each unit's operation. Real-time status information and execution results are uploaded to the central controller through the communication network. Upon receiving the feedback data, the central controller verifies the data's accuracy and reliability. It then processes and analyzes the feedback data to evaluate the control instructions' execution effect. Based on the feedback data, the central controller assesses the microgrid's transient response effect, including voltage fluctuation, frequency deviation, response time, and other indicators to meet preset performance requirements. If the evaluation shows unsatisfactory control effects, the central controller adjusts and optimizes the control strategy according to the feedback data and the current system state. It generates new control instructions based on the adjusted strategy and sends them to each execution unit again. By repeating this process, the microgrid's transient response performance is gradually improved through iterative control strategy optimization.

The above process ensures the swift execution and feedback of transient response control commands, enabling the microgrid to promptly react and adapt its operational state during transient events. This enhances both system stability and power quality. However, in practical deployment, it's crucial to prioritize the reliability and real-time capabilities of the communication network, ensuring accurate command transmission and timely data feedback.

3 Test Analysis

3.1 Test Preparation

For this experiment, a representative wind-solar hybrid generation microgrid system is chosen as the experimental subject to validate the efficacy of the transient response control approach grounded in composite information fusion. This microgrid system specifically contains the components and detailed parameters shown in Tables 3, 4 and 5.

Table 3. Parameters of the wind power system

No	Item	Detailed parameter
1	Model number	WTG-2000
2	Quantity	2
3	Rated power	2MW
4	Cut-in wind speed	3m/s
5	Cutting wind speed	25m/s
6	Pitch Angle adjustment range	$0° \sim 90°$
7	Rotor diameter	2.5m

The micro grid load includes fixed load (office buildings, residential areas and other fixed electrical equipment), variable load (electric vehicle charging stations, industrial production lines and other electrical equipment that change with time and demand), and the total power demand range is 1MW to 3MW. Through the description of the above experimental sample objects, we can more comprehensively understand the composition and characteristics of the microgrid system, and provide accurate basic data for subsequent experimental design and research on transient response control methods. The experimental environment simulates a real wind solar hybrid micro grid operation environment, including wind speed, light intensity, load change and other factors. The specific configuration is as follows.

Table 4. Parameters of photovoltaic power generation system

No	Item	Detailed parameter
1	Model number	PV-1000
2	Quantity	3
3	Rated power	1MWp
4	Conversion efficiency	15%
5	Panel type	Polycrystalline silicon

Table 5. Parameters of the energy storage system

No	Item	Detailed parameter
1	Model number	BS-500
2	Type	Lead-acid battery
3	Capacity	500 kWh
4	Charging power	200 kW
5	Discharge power	300 kW
6	Charge and discharge efficiency	90%

(1) Utilize the wind speed simulation device to replicate the seamless fluctuations of wind velocity ranging from 3 m/s to 25 m/s, mirroring the authentic operational conditions encountered by a wind turbine in its natural environment.
(2) Light intensity simulation. by adjusting the light intensity simulation device, the simulated light intensity is between 0 and 1000 W/m^2 to simulate the actual operating environment of photovoltaic panels.
(3) Load simulation. The programmable load simulator is used to simulate the random fluctuation of the load between 1MW and 3 MW to simulate the actual load change of the microgrid.

In addition, the experiment is equipped with a data acquisition system for real-time collection of key information such as wind speed, light intensity, voltage, current and power in the microgrid for subsequent composite information fusion and transient response control.

In addition to the above settings, set the parameters for the execution of the proposed method as follows.

Wind turbine generator set. The rated power is 100 kW, the converter response time is 50 ms, and the sampling frequency is 1kHz.

Photovoltaic power generation system. The rated power is 50 kW, the maximum power point tracking (MPPT) response time is 100ms, and the sampling frequency is 500 Hz.

Energy storage device. Capacity is 50 kWh, charging and discharging efficiency is 95%, and response time is 20ms.

Load change. Simulate load sudden change, load fluctuation from 0 to 100 kW, frequency is 0.5 Hz.

Control strategy. PID controller is used to design the transient response control strategy for isolated network operation and grid connected operation, and the target adjustment time is set as 100 ms.

3.2 Experimental Process Design

Step 1. The experiment firstly starts the data acquisition system, which collects the key information such as wind speed, light intensity and load change in the microgrid in real time.

Step 2. To enhance the precision and dependability of the collected data, preprocessing steps including denoising, filtering, and standardization are conducted. These operations help refine the data for more accurate analysis.

Step 3. Utilizing the composite information fusion algorithm, the preprocessed data—including wind speed, light intensity, and load variations—is merged to efficiently integrate information from diverse sources. This allows us to acquire a more comprehensive and precise understanding of the microgrid's operational status.

Step 4. Considering the output characteristics of distributed power supply, charging and discharging capacity of energy storage system, load demand and other factors, formulate the corresponding transient response control strategy according to the fused microgrid operation state information.

Step 5. Apply the developed transient response control strategy to the actual microgrid, and observe and record the changes of voltage, current, power and other key indicators of the microgrid in the transient process.

3.3 Indicator Setting

Comparative experiments are conducted with two conventional transient response control methods for microgrids in reference [4] and reference [5] to verify the effectiveness and superiority of the transient response control method based on composite information fusion. The voltage fluctuation rate can reflect the stability of the microgrid voltage during the transient process. The power fluctuation rate can measure the stability of microgrid output power. Therefore, the voltage fluctuation rate and power fluctuation rate are chosen as the comparison indexes of this experiment, and the formulas are shown as follows.

$$V_{fr} = \frac{(V_{max} - V_{min})}{V_r} \times 100\% \tag{6}$$

$$P_{wr} = \frac{(P_{max} - P_{min})}{P_r} \times 100\% \tag{7}$$

Among them, V_{fr} denotes the voltage fluctuation rate. V_{max} and V_{min} are utmost and lowest voltage values. V_r indicates the rated voltage. P_{wr} denotes the power fluctuation rate. P_{max} and P_{min} denote the power maximum and power minimum, respectively. P_r indicates rated power.

3.4 Analysis of Results

Under the same microgrid system, two conventional control methods and a transient response control method based on composite information fusion are used for the experiments respectively. The environmental conditions, wind speed, light intensity, load change and other factors of the three experiments are ensured to be as consistent as possible. During the three experiments, key data such as voltage, current and power of the microgrid are collected in real time, and the start and end moments of the transient response are recorded. Based on the collected data, the specific values of the above comparative indexes were calculated. Using the statistical analysis method, the differences in voltage fluctuation rate and power fluctuation rate among the three control methods are compared, and the results are shown in Fig. 5 and Fig. 6, respectively.

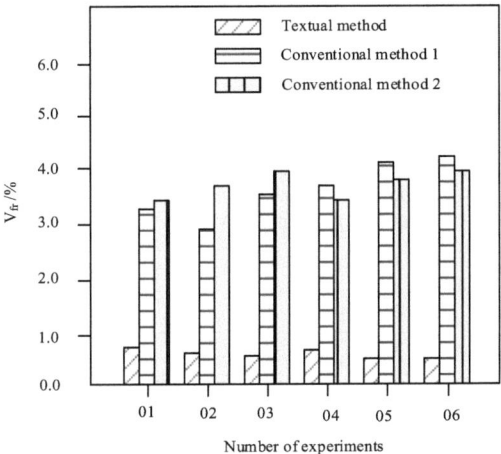

Fig. 5. Comparison results of voltage fluctuation rate of hybrid wind and solar power generation microgrids

It can be seen from the experimental comparison results that the hybrid wind solar power generation microgrid transient response control method proposed in this paper based on composite information fusion is significantly superior to the conventional control method in terms of voltage fluctuation rate and power fluctuation rate, and can effectively reduce the V_{fr} and P_{wr} improvement of power quality and system stability.

In order to enhance the reliability and universality of the experimental results, 30 additional experiments were conducted under each control method, and the average results of two experiments (i.e. Experiment 1–1 and Experiment 1–2) were taken. The experimental results after increasing the number of experiments using three methods are shown in Table 6.

According to Table 6, the transient response control method for wind solar hybrid power generation microgrids based on composite information fusion proposed in this paper is significantly lower in voltage fluctuation rate and power fluctuation rate than the two comparison methods. It has significant advantages in improving the transient response stability and power quality of microgrids. Therefore, the proposed transient

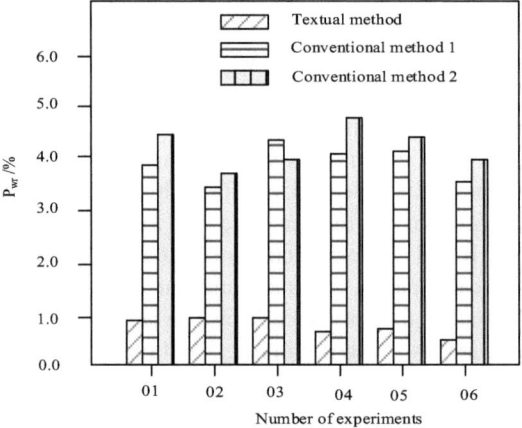

Fig. 6. Comparison results of power volatility of hybrid wind and solar generation microgrids

Table 6. Experimental results after increasing the number of experiments using three methods

Experiment number	Control method	Wind speed(m/s)	Light intensity(W/m^2)	Load variation(%)	Voltage fluctuation rate(%)	Power fluctuation rate(%)
Experiment 1–1	Conventional method 1	5.0 ± 0.5	800 ± 50	10	4.2	6.8
Experiment 1–2	Conventional method 1	5.1 ± 0.4	795 ± 45	10	4.1	6.7
Experiment 2–1	Conventional method 2	5.0 ± 0.5	800 ± 50	10	3.8	6.3
Experiment 2–2	Conventional method 2	5.1 ± 0.4	798 ± 48	10	3.7	6.2
Experiment 3–1	Textual method	5.0 ± 0.5	800 ± 50	10	2.5	4.5
Experiment 3–2	Textual method	5.1 ± 0.4	797 ± 47	10	2.4	4.4

response control method based on composite information fusion achieves finer and faster control of microgrids by comprehensively utilizing multi-source information, effectively reducing voltage and power fluctuations, and improving system stability and reliability.

4 Conclusion

The transient response control method for wind solar hybrid power generation microgrid based on composite information fusion proposed in this study successfully designs and implements transient response control strategies for isolated and grid connected operation by deeply analyzing the topology structure of the microgrid, accurately collecting multi-source data, and applying advanced composite information fusion technology. The test results show that this method significantly reduces the voltage and power fluctuations of microgrids during transient processes, effectively improving power quality and the stability and reliability of system operation. This achievement not only provides important technical support for the safe and efficient operation of wind solar hybrid power generation microgrids, but also opens up new ideas for the optimization control of distributed energy systems. The successful implementation of this study is of great significance for promoting the large-scale integration and efficient utilization of renewable energy. With the acceleration of global energy structure transformation, wind solar hybrid microgrids, as an important component of future smart grids, have a direct impact on the stability and reliability of energy supply security and sustainability. The method proposed in this study provides strong support for the reliable grid connection and flexible scheduling of renewable energy by enhancing the transient response control capability of microgrids, which helps promote the diversification and low-carbon transformation of energy structure.

Acknowledgment. Undergraduate Innovation and Entrepreneurship Training Program of Dalian University of Science and Technology: "Wind Dancing and Light Singing" Hybrid Intelligent Generator.

References

1. Wozniak, P.A.: Hybrid electric vehicle battery-ultracapacitor energy management system design and optimization. Elektronika ir Elektrotechnika **1**, 28 (2022)
2. Samia, B.: Frequency splitting approach using wavelet for energy management strategies in fuel cell ultra-capacitor hybrid system. J. Measurem. Eng. **10**(1), 12 (2022)
3. Abadi, S.A.G.K., Habibi, S.I., Khalili, T., et al.: a model predictive control strategy for performance improvement of hybrid energy storage systems in DC microgrids. IEEE Access 2022(10-), 10
4. Satheesan, J., Nair, R.: An adaptive two-level hierarchical controller for universal power sharing and performance enhancement of hybrid energy storage-supported AC/DC microgrids. Int. J. Circuit Theory Appl. **51**, 2122–2140 (2023)
5. Zhang, X., Pei, W., Mei, C., et al.: Transform from gasoline stations to electric-hydrogen hybrid refueling stations: An islanding DC microgrid with electric-hydrogen hybrid energy storage system and its control strategy. Inter. J. Electr. Power Energy Syst. **136** (2022)
6. Ding, J., Mao, Z.Y., Liu, W.X., et al.: Development and application of distribution network fault diagnosis system based on multi-dimensional information fusion. J. Electric Power Sci. Technol. **38**(3), 252–260 (2023)
7. Ning, N., Sun, R.Z., Pan, K.Y. et al.: Optimization of distribution network protection strategy based on the integration of operation mode and fault information. J. Electric Power Sci. Technol. **37**(2), 54–61 (2022)
8. Liu, X.F., Lin, Z.Z., Tian, Q.D., et al.: (2023) Research on the fusion technology of power multi-source heterogeneous information based on big data mining. Manufact. Autom. **45**(9), 75–78
9. Zhai, S.L., Zhang, X.Z., Duan, Y., et al.: Topology structure of smart energy station based on AC/DC microgrid. Smart Power **49**(5), 28–34 (2021)

10. Yin. X., Lai, J.M., Yin, X.G., et al.: Hybrid interlinking transformer and its fault blocking coordination control for microgrid. Electric Power Automation Equipment **43**(10), 145–151,207 (2023)
11. Li, H.T., Ma, Y.L.: Fault area determination technology of grid-connected photovoltaic microgrid in 4D coordinate system. Electric Mach. Control Appli. **50**(1), 83–89 (2023)
12. Abdullah, M.A., Al-Shetwi, A.Q., Mansor, M., et al.: Linear quadratic regulator controllers for regulation of the dc-bus voltage in a hybrid energy system: modeling, design and experimental validation. Sustainable Energy Technol. Assessments **50** (2022)
13. Lin, X., Zamora, R.: Controls of hybrid energy storage systems in microgrids: critical review, case study and future trends. J. Energy Storage **47** (2022)
14. Xiang, K.D., Wang, H., Peng, T.T. et al.: Optimal capacity allocation of wind-solar-storage system with hybrid energy storage. Sci. Technol. Eng. **23**(31), 13415–13422 (2023)
15. Zeng, Z.H., Liu, Y.P., Wei, Y.F., et al.: Optimal capacity allocation of hybrid energy storage system based on improved bat algorithm. J. Henan Polytechnic Univ. (Nat. Sci.) **42**(5), 130–136 (2023)
16. Zhang, J., Sun, Q., Zheng, Z.,et al.: Lightning surge analysis for hybrid wind turbine-photovoltaic-battery energy storage system. Electr. Power Syst. Res. **225** (2023)
17. Liu, .Y., Hao, T.N., Zhang, W.: Abnormal state detection of power system based on RMT eigenvalue fusion. J. Electr. Measurem. Instrument. **37**(12), 242–252 (2023)
18. Lin, Y.H., Hao, F.Z., Li. B.X. et al.: Method for identifying abnormal data in distribution network operation. Electric Power **56**(9), 134–139 (2023)
19. Rajput, A.K., Lather, J.S.: Energy management of a DCmicrogrid with hybrid energy storage system using PI and ANN based hybrid controller. Int. J. Ambient Energy **1**, 44 (2023)
20. Amine, H.M., Mouaz, A.K., Messaoud, H., et al.: The impacts of control systems on hybrid energy storage systems in remote DC-Microgrid system: a comparative study between PI and super twisting sliding mode controllers. J. Energy Storage **47** (2022)
21. Zhang, Y., Du, C., Chen, X.: Simulation of micro grid operation mode switching control with improved acceleration factor. Comput. Simulat. **40**(9) 76-80 (2023)

Data Intrusion Detection Method for Embedded Components of Power Terminal Based on Machine Learning

Hongwei Wang[1(✉)] and Xuedong Dai[2]

[1] Inspur Electronic Information Industry Co. Ltd., Jinan 250013, China
jhgaj84963@163.com
[2] Anhui Technical College of Industry and Economy, Hefei 230051, China

Abstract. In the past, the data intrusion detection methods of embedded components in power terminals only processed the data numerically, which led to low data quality and poor detection effect. Therefore, a data intrusion detection method for embedded components of power terminal based on machine learning is designed. The embedded components of power terminal are used to process and denoise the collected data, and their data characteristics are extracted and decomposed. By constructing an embedded component data intrusion detection system of power terminal and combining with neural network, intrusion detection is carried out. In the experimental test, the designed data intrusion detection method has better detection effect and can effectively improve the security protection ability of power system.

Keywords: Machine Learning · Power Terminal · Embedded Components · Data Intrusion · Detection Method · Method Design

1 Introduction

The integration of power grid systems and intelligent systems meets people's daily needs to a certain extent. And with the intelligence of power grid, its complexity is gradually increasing, and it is easy to be invaded by false data. In order to ensure the normal operation of the new power system and achieve the goal of "double carbon" as soon as possible, it is necessary to strengthen the research on embedded data intrusion detection of power terminals [1–3].

The traditional machine learning detection algorithm is based on feature extraction and feature separation, and it needs to learn each data attack type in advance by manual means. With the data intrusion becoming more and more complex and diverse, the limitations of this kind of detection algorithm are gradually enlarged, and it is often unable to achieve satisfactory results when dealing with new types of attacks.

On the basis of the above, this paper designs a data intrusion detection method for embedded components of power terminal based on machine learning. This method reduces data noise interference and improves data quality and efficiency by collecting embedded component data from power terminals and performing preprocessing. Then decompose it and construct a data intrusion detection model based on machine learning to achieve intrusion detection, effectively ensuring the effectiveness of embedded component data intrusion detection in power terminals.

2 Design of Data Intrusion Detection Method for Embedded Components of Power Terminal

Considering the data characteristics and intrusion detection requirements of embedded components in power terminals, the framework of data intrusion module proposed in this paper is shown in Fig. 1.

As shown in Fig. 1, the framework mainly includes three modules: data acquisition module, intrusion detection module and feedback response module. The data acquisition module is responsible for collecting the embedded component data of the power terminal, extracting relevant features from it and transmitting them to the intrusion detection module. The intrusion detection module deploys the classifier after training, processes and detects the data characteristics of the embedded components of the incoming power terminal in real time, identifies abnormal data and network attacks, and transmits the detection results to the feedback response module. The feedback response module firstly makes an autonomous response, checks the operation of the power terminal node, isolates abnormal data, suspends the business of the attacked node, and triggers an alarm to feed back to manual processing if it exceeds its processing capacity [4]. The detailed description is as follows.

2.1 Embedded Component Data Preprocessing and Data Feature Extraction

In this paper, a data set preprocessing method is proposed, which includes four steps: digitization, standardization, PCA dimension reduction and denoising. This method can sort out the original data set, extract the main features of the sample, and reduce the redundancy of features, subsequent calculation overhead and system operation cost.

1) Digitization. Intrusion detection data sets usually contain discrete features of string type, which cannot be directly used for subsequent processing [5]. Therefore, firstly, one-hot coding is used to convert the characteristic values of string types in each sample into numerical characteristic values, and the data set is expanded into a high-dimensional sparse characteristic matrix.
2) Standardization. The magnitude of features of different dimensions is inconsistent, which will cause the dimension reduction algorithm to deviate from the optimal value. Z-Score is standardized for each dimension after digitization, and the specific process of standardization is as follows:

$$X = \frac{x - p}{\sigma} \tag{1}$$

In the above formula, X represents the result of data standardization processing, x represents the eigenvalues of each dimension in the original data set, p represents the

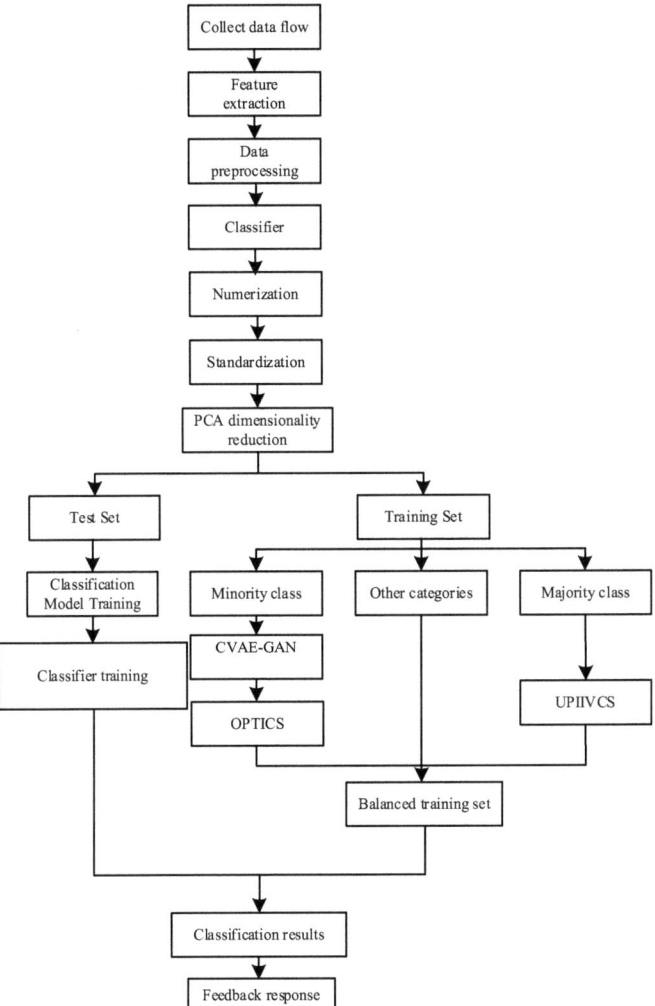

Fig. 1. Data Intrusion Detection Framework

average value of each dimension data feature, σ represents the standard deviation of data characteristics. Each dimension feature of the sample is transformed into a standard normal distribution with a mean value of 0 and a standard deviation of 1, so that the feature metrics of different dimensions are unified, the comparability is enhanced, and the convergence speed in operation is accelerated.

3) PCA dimension reduction. The high-dimensional sparse feature matrix obtained after processing is not conducive to deep network training. Therefore, the PCA algorithm is used to reduce the dimension of the feature matrix, extract the main feature components, and maximize the retention of the intrinsic information of the data [6, 7].

Mapping the obtained standardized data set to a k-dimensional space, wherein the covariance matrix of the feature calculation matrix is specifically described as follows:

$$\begin{cases} X = [x_1, x_2, x_3, \ldots, x_n]^T \\ x_i = [x_{1i}, x_{2i}, x_{3i}, \ldots, x_{mi}]^T \\ C = E[(x_i - \mu_i)(x_i - \mu_i)^T] \\ \mu_i = \frac{1}{m} \sum_{n=1}^{m} x_m \end{cases} \quad (2)$$

In the above formula, X represents a feature calculation matrix, x_i represents a standardized data set, C represents covariance matrix, m represents sample number, n represents feature dimension, μ_i represents mean vector representing all directions.

4) On this basis, the data is denoised, and the process of denoising is shown in Fig. 2.

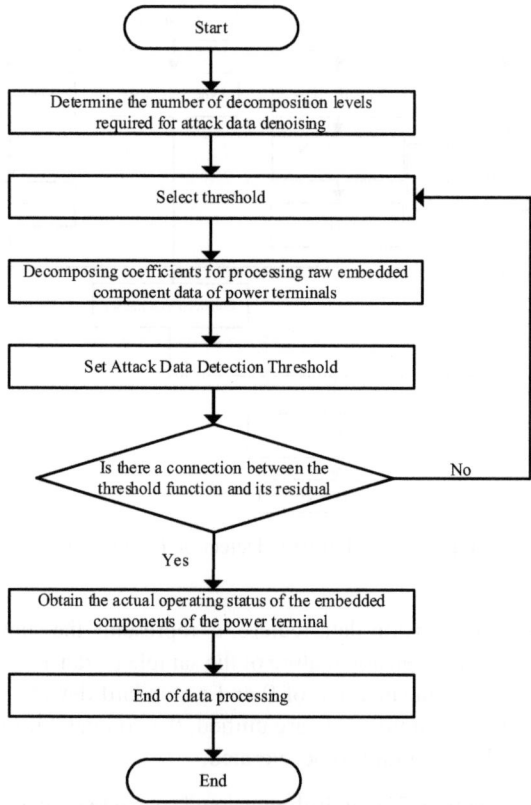

Fig. 2. Noise reduction processing flow

Taking the preprocessed data as the basis, data feature extraction is carried out. Firstly, the data set is cleaned up, and the lost data items, wrong data shifts and original samples with abnormal formats are processed. Then the original features are divided into more features for subsequent processing and analysis. Then convert hexadecimal data into decimal values. This step is for better numerical calculation and data analysis, because decimal values are easier to perform mathematical operations and statistical analysis. Normalize the converted decimal data. Normalization aims at scaling the data to a standard range (such as 0 to 1), which helps to eliminate the dimensional influence between different features and improve the efficiency and effect of model training. In the process of data processing, it is found that there will be dirty data in a large number of data streams, that is, data items are missing and data items are wrongly shifted. Considering that there may be some errors in the collected data in the real environment, it is necessary to delete or change the data in order not to affect the training and test results and prevent the data from being polluted. Firstly, the data set is cleaned. Return the original samples with abnormal data position and set all the missing data to 0, and delete the original samples with abnormal data format, thus obtaining the original features shown in Table 1.

Table 1. Original characteristics

	Timestamp	ID	DLC	Data	Label
Car-Hacking	1479121434.850202	0350	8	05	0
	1478191110.35041	0316	8	45	1
	1478193258.15421	043f	8	29	2
	1478198408.90276	0000	8	01	3
	1478195880.83481	062e	8	00	4
OTIDS	126.399740	0316	8	05	0
	130.276298	0164	8	90	1
	162.565819	0000	8	00	2
	1481193021.194441	04f0	8	00	3

As shown in Table 1, the data intrusion detection of embedded components in power terminals must be fast and effective. In view of the advantages of CNN model in the image field, we need to turn the intrusion detection problem into an image classification problem-we plan to convert the tabular data of network traffic data into images [8]. First, the original four feature data in Table 1 are split and converted into nine features, and the split results are shown in Tables 2 and 3. D is the CAN identifier, and the 8-bit Data of the CAN packet is divided into 8 columns D0-D7, each column contains two hexadecimal values. Convert the hexadecimal in D and DATA into decimal values, and then normalize the data. The split characteristics of Car-Hacking data set are shown in Table 2.

As shown in Table 2, on the basis of the above, the characteristics of OTIDS data set are split. The specific splitting results are shown in Table 3.

Table 2. Car-Hacking Data Set's Characteristics after Splitting

ID		0.013	0.012	0.017	0	0.024
DATA	D0	0.019	0.110	0.518	0	0.584
	D1	0.110	0.161	1	0	0.270
	D2	0.518	0.145	1	0	0.373
	D3	0.4	0.376	1	0	0.420
	D4	0.4	1	1	0	0.9
	D5	0.428	0.161	0	0	0.2
	D6	0	0	0	0.1	0
	D7	0.835	0	1	0	0.071
LABEL		0	1	2	3	4

Table 3. Characteristics of 3OTIDS data set after splitting

ID		0.013	0.012	0.017	0	0.024
DATA	D0	0.01	0	0.428	0	0.554
	D1	0.150	0	1	0	0.20
	D2	0.518	0	1	0	0
	D3	0.428	0.376	1	0	0
	D4	0	0	1	0	0.9
	D5	0.428	0	0	0	02
	D6	0	0	0	0.1	0
	D7	0.835	0	1	0	0
LABEL		0	1	2	3	4

As shown in Table 3, data feature extraction is carried out based on the above split features. At this point, the preprocessing and feature extraction of embedded component data of power terminal are completed.

2.2 Building a Data Intrusion Detection Model

Based on the above-mentioned extracted data characteristics, PMU can predict the system state of comprehensive energy such as new power system in real time, and the data in the analysis process is very reliable, so that the vector of PMU measurement data is:

$$Z = T[Z_1, Z_2, \ldots, Z_m]^T \quad (3)$$

In the above formula, T represents the measurement period. When measuring the attack data, there are often data such as node voltage and branch power of the embedded components of power terminal. The data concentrator is used to collect the attack data of various fields in the embedded components of power terminal, and the comprehensive energy data is transmitted to the control center to evaluate its state. The expression of the state evaluation model of the embedded components of power terminal is:

$$P = [G(Y)]^T + R^T \tag{4}$$

In the above formula, Y represents the energy data state vector of the embedded components of the power terminal; R represents the measurement error vector; G stands for Jacobian matrix [9, 10] of new power system. The state estimation of embedded components of power terminal needs to calculate the minimum state vector of embedded components of power terminal. Based on the weighted least square method, the formula for calculating the minimum state vector of embedded components of power terminal is as follows:

$$Y = \left(G^T TG\right)^{-1} G^T WP \tag{5}$$

In the above formula, W represents the weight value. Combined with the previous formula, the data residual H, that is, the difference between the existing and evaluated measurement vectors, can be obtained, and its expression is:

$$H = P^T - [G(Y)]^T \tag{6}$$

Through the above formula, the accuracy of measured data is obtained under the action of empirical threshold. If the matrix G is known to the false attack data of the embedded components of the power terminal, and the matrix can control most of the vectors, then the attack vector expression of the embedded component data is:

$$D = [D_1, D_2, D_3, \ldots, D_m]^T \tag{7}$$

Based on the above formula, the running state of the embedded components of the power terminal is predicted, and the measurement data is continuously added to obtain the corresponding intrusion detection model, which is described in detail as follows:

$$P_D = [G(Y)]^T + D^T \tag{8}$$

In the above formula, P_D represents the generated intrusion detection model. At this point, the design of data intrusion detection model for embedded components of power terminal is completed.

2.3 Based on Machine Learning to Achieve Data Intrusion Detection.

On the basis of the above design, the data intrusion detection of embedded components of power terminal is realized. In the design of this paper, the machine learning algorithm is used to detect data intrusion. Based on the above analysis, this paper introduces the neural network method into the data intrusion detection of embedded components in power terminals, which includes three parts: input layer, hidden layer and output layer. The overall model diagram of the training classifier is shown in Fig. 3.

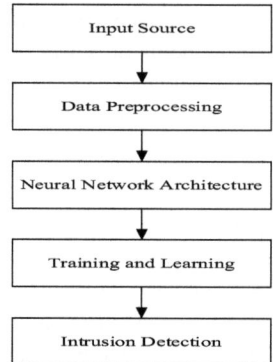

Fig. 3. The overall model diagram of the training classifier

As shown in Fig. 3, according to the algorithm structure shown in Fig. 2, wavelet denoising technology is combined with neural network. Firstly, the prediction value after wavelet denoising is input in the input layer, which is used as the training sample, and the wavelet basis function is used as the transfer function of the hidden layer node of the neural network, and with its help, the output result is generated, and whether there is false data in the output result is judged, so as to complete the detection.

Assuming that there are m detection samples and there are l hidden nodes in the samples, the calculation formula of the output value of the hidden layer in the neural network is:

$$K_j = h_j\left(\sum_{i=1}^{M} w_{ij}x_i + a_j\right) \qquad (9)$$

In the above formula, w_{ij} represents the weight between the input layer z and the hidden layer j, x_i represents the predicted value after wavelet denoising, that is, sample data, h_j is the wavelet basis function in wavelet denoising, a_j represents the offset between the input layer and the hidden layer. The calculation formula of the result in the output layer is:

$$Q_c = \sum_{j=1}^{N} K_j w_{jc} + b_c \qquad (10)$$

In the above formula, w_{jc} represents the weight between the output layer and the hidden layer, b_c represents the bias between the output layer and the hidden layer.

Machine learning is trained based on back propagation algorithm. The training process can be roughly divided into forward propagation of comprehensive energy information and back propagation of data errors. In this process, the weights and biases should be constantly adjusted according to the needs, the actual values and output values of the neural network are compared, and the values are updated in real time through the difference r between the two values, and the updated calculation formula of the weights is obtained as follows:

$$\begin{cases} w_{ij} = w_{ij} + \alpha K_j(1-K_j)x_i \sum_{c=1}^{M} w_{jc}e_c \\ w_{jc} = w_{jc} + \alpha K_j e_c \end{cases} \quad (11)$$

In the above formula, according to the trained network, the false intrusion data in the embedded components of the power terminal can be detected, and the false data can be detected when the deviation between the output value and the actual value is large. At this point, the design of data intrusion detection method for embedded components of power terminal based on machine learning is completed.

3 Experimental Test

In order to verify the overall effectiveness of the power terminal embedded component data intrusion detection method based on machine learning, the method in this paper (method 1), the power terminal embedded component data intrusion detection method based on improved generation countermeasure network (method 2) and the power terminal embedded component data intrusion detection method based on distributed attack (method 3) are tested experimentally. There are three network input nodes x (t), four hidden layer network nodes s (t) and three output layer network nodes Y}t}, with the dynamic parameter of 0.6, the allowable error of 1x10–4 and the number of iterations of 1,000 times. The specific test process is as follows.

3.1 Experimental Preparation

Car-Hacking data set and OTIDS data set are selected as data sets, which are widely used to simulate the data intrusion detection of embedded components of power terminals because they are collected from real environment.

In this paper, the corresponding data sets are obtained from relevant websites, and two data sets are obtained, one for information intrusion attack and the other for recording data traffic, and the captured communication data is taken as the data set. Car-Hacking data set contains four types of data attacks: Dos attack, Fuzzy attack, error deception and cheating attack per minute. OTIDS data set contains three types of attacks: Dos attack, Fuzzy attack and Impersonation attack. The number distribution of attacks in Car-Hacking dataset and OTIDS dataset is shown in Table 4.

As shown in Table 4, the above data set is taken as the data of this experiment. In the experiment, in order to ensure the smooth progress of the experiment, the corresponding experimental parameters are set. Among them, the experimental environment is Pytorch deep learning framework, NVIDIA2080Ti GPU} 8-core 16-process 32G CPU, and the development environment is Py _ Thon3.6. The optimizer adopts SGD, and the initial learning rate is 0. 000 1. Every 16 epoch, the learning rate is adjusted to the original 1/1/S}batch_size = 32. In addition, in order to prevent the model from over-fitting, dropout = 0. 5 is adopted in each layer. The word vector dimension is selected as 128,

Table 4. Data Set Composition

	Data attack types	total	Label
Car-Hacking	Normal	988872	0
	Spoofing flee RPM gauze	854365	1
	Spoofing flee drive gea	597521	2
	Dos attack	512477	3
	Fuzzy attack	122564	4
OTIDS	Normal	335648	0
	Fuzzy Attack	325647	1
	Dos attack	125471	2
	Impersonafion Attack	236454	3

and the multi-head attention head is set as 120. In the above experimental environment, the experimental test was carried out.

In the experiment, the types of samples in the above data set are divided. The specific division results are shown in Table 5.

Table 5. Number of sample divisions for each category of experimental data set

Type	Training Set	Test Set	Verification Set	Total sample
BENIGN	15914675	454620	227130	2273097
Dos Hulk	161751	46215	23107	231073
PortScan	111251	31786	15893	158930
DDos	89618	25606	12803	128027
Dos GoldenEye	7205	2059	1029	10293
FTP-Patator	5516	1588	794	7898
SSH-Patator	4128	1179	590	5796
Dos-slowloris	4057	1159	580	5796
Dos SlowHttptest	3849	1100	550	5499
Bot	3849	3930	197	1987
Web-Attack Brute Force	1055	301	151	1507
Web Attack XSS	457	130	65	652
Infiltration	26	7	65	652
Web-Attack-Sql Injection	15	4	2	21
Heartbleed	8	2	1	11
Total	1981519	566149	283075	2830743

As shown in Table 5, all the above samples are used in experiments, and the above data sets are divided according to a certain proportion and used for training, verification and testing.

3.2 Results and Conclusions

In order to compare the effects of the three methods in the practical application of China, in this experiment, three groups of data samples with 1 000 kb data volume are selected, in which the noise of one data set accounts for 2% of the whole, the other group accounts for 30%, and the last group accounts for 70%. Three methods are used to denoise the above three sample data, and the relationship between signal-to-noise ratio and root mean square error of each method is calculated respectively. The experimental results are shown in Figs. 4, 5 and 6.

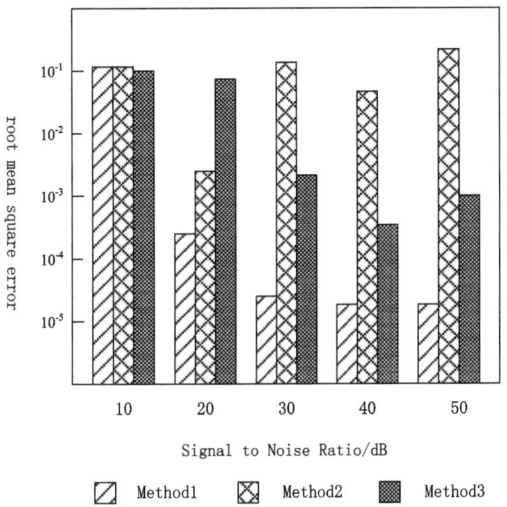

Fig. 4. Feasibility of three denoising methods for sample data with noise accounting for 2%

As shown in Figs. 4, 5 6, the error of method 1 is the smallest, which shows that the data denoising effect is the best. Compared with the contrast method, the root mean square error of the proposed method is smaller, and the denoising feasibility is higher. The main reason is that the proposed method uses wavelet threshold method to preprocess the power grid data by determining the decomposition levels and processing the decomposition coefficients. Therefore, the data intrusion detection method of embedded components of power terminal based on machine learning designed in this paper has a good effect in practical application.

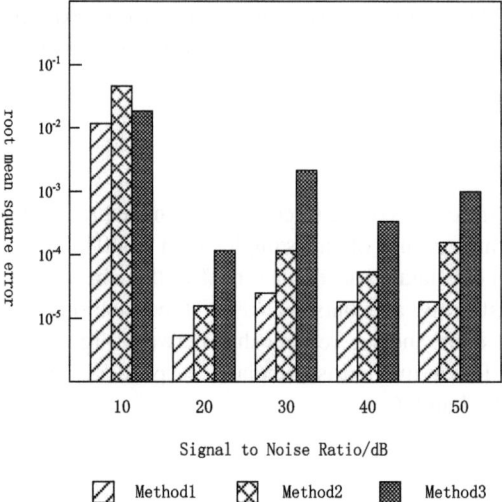

Fig. 5. Feasibility of three denoising methods for sample data with noise accounting for 30%

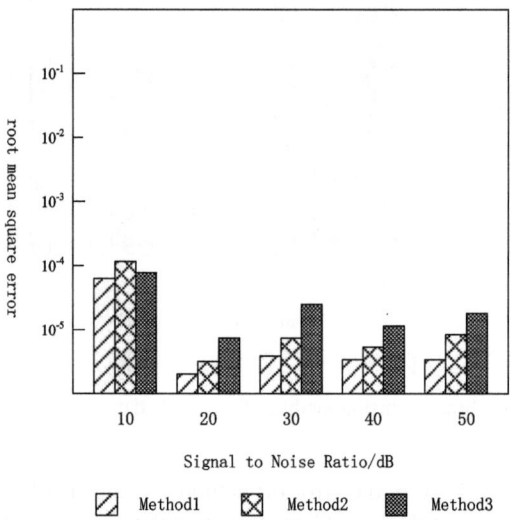

Fig. 6. Feasibility of three denoising methods for sample data with noise accounting for 70%

Further verify the detection time of the three methods and obtain comparative results as shown in Table 6.

According to Table 1, compared to Methods 2 and 3, Method 1 has the shortest detection time. Because this method collects embedded component data from power terminals and preprocesses it, it reduces data noise interference and improves data quality and efficiency.

Table 6. Detection time for three methods

Iterations/time	Method 1	Method 2	Method 3
200	4.3	6.8	8.8
400	5.8	7.5	9.7
600	6.9	8.2	11.2

4 Conclusion

False data intrusion easily leads to the sudden interruption of embedded components of power terminal, which will not only affect people's normal life but also bring serious economic losses. Therefore, a data intrusion detection method of embedded components of power terminal based on machine learning is proposed. By constructing the data intrusion detection model of power terminal embedded components, the current running state of power terminal embedded components is predicted, and the prediction results are generated and the data of power terminal embedded components are trained based on the combination of wavelet denoising and BP neural network, so as to detect false data and realize the false data intrusion detection of power terminal embedded components. The experimental results show that the proposed method has good false data intrusion detection effect, short detection time and high feasibility of denoising. In the era of rapid information development, the power quality and operation stability are improved, and the life efficiency is improved.

References

1. Rai, A., Shrivastava, A., Jana, K.C.: An empirical analysis of machine learning algorithms for solar power forecasting in a high dimensional uncertain environment. IETE Tech. Rev. **40**(4), 558–573 (2023)
2. Sclip, A., Bretas, V.P.G., Alon, I., et al.: Greenfield FDI attractiveness index: a machine learning approach. Competitiveness Rev. Inter. Bus. J. **32**(7), 85–108 (2022)
3. Fathollahi, R., Karimi, M., Khosravi, M., et al.: Determination of the heat capacity of cellulosic biosamples employing diverse machine learning approaches. Energy Sci. Eng. **10**(6), 1925–1939 (2022)
4. Víctor Rodriguez-Galiano, Guisado-Pintado E , Prieto-Campos A ,et al.A machine-learning hybrid-classification method for stratification of multidecadal beach dynamics. Geocarto International, 2022, 37(27): 16534–16558
5. Li, D., Dong, C., Chen, Z., et al.: A combinatorial machine-learning-driven approach for predicting glass transition temperature based on numerous molecular descriptors. Mol. Simul. **49**(6), 617–627 (2023)
6. Zhao, M., Tang, L., Chen. S., et al.: Machine learning based automatic foreshock catalog building for the 2019 MS6.0 Changning, Sichuan earthquake. Chin. J. Geophys.- Chinese Edition **64**(1), 54–66 (2021)
7. Sharma, A., Kumar, N., Kumar, A., et al.: Comparative investigation of machine learning algorithms for detection of epileptic seizures. Intell. Decision Technol. **15**(2), 269–279 (2021)
8. Liu, S., Chen, P., Woźniak, M.: Image enhancement-based detection with small infrared targets. Remote Sensing **14**, 3232 (2022)

9. Sridevi, M., Arun, K.B.R.: A framework for performance evaluation of machine learning techniques to predict the decision to choose palliative care in advanced stages of Alzheimer's disease. Indian J. Comput. Sci. Eng. **12**(1), 35–46 2021)
10. Jiang, S., Zhang, H.: Predictive control method of vehicle nonlinear suspension based on RBFNN observer. Comput. Simulat. **39**(01), 102–105+123 (2022)

The Intelligent Monitoring System of University Personnel File Falsification Data Based on Wireless Network

Hui Li[1(✉)] and Weiwei Zhang[2]

[1] Hunan Industry Polytechnic, Changsha 410000, China
m13874883043@163.com

[2] Computer Information and Engineering College, Guizhou University of Commerce, Guiyang 550014, China

Abstract. In the process of checking the fraudulent behavior data of university personnel files by monitoring system, the intelligent monitoring algorithm adopted has certain limitations, which can not cover all abnormal situations, resulting in poor detection effect of the system. Therefore, the intelligent monitoring system of university personnel file fraud data based on wireless network is studied. By initializing the wireless transmission module of the server, read the file information in XML format, and on this basis, define the server data transmission instructions to complete the server design of the intelligent monitoring system. Perfect the connection form of wireless network topology, determine the execution conditions of intelligent monitoring instructions, complete the definition of intelligent monitoring software, and cooperate with the server organization to realize the design of the intelligent monitoring system based on wireless network for college personnel file fraud data. The experimental results show that the application of the above monitoring system can increase the maximum real-time data transmission rate of university personnel file fraud to 9.5×10^8 bit/ms can effectively guarantee the network host's ability to accurately monitor information parameters.

Keywords: Wireless Network · Personnel Files · Fraud · Data Monitoring · Module Initialization · Xml Format File · Osal Mechanism

1 Introduction

Personnel file management is a concept of a larger category. In a broad sense, personnel file management is currently focused on the personal file management of social workers. Personnel files in colleges and universities are a special kind of files in personnel files, and are also the earliest, most complete, and most representative important part of personnel files. In recent years, there has been relatively more research on the management of personnel files of social floating talents, but there is relatively less research on the management of personnel files of colleges and universities, especially the research related to the falsification of personnel files, which is the top priority of student file work.

Relevant departments hope to find out the root cause of the falsification of personnel files in colleges and universities, find ways to avoid the falsification of students' personnel files, promote the digitization and informatization of students' personnel files, optimize the management of students' personnel files, improve the management level, and propose to establish a large university information database, strengthen the daily supervision of personnel work in colleges and universities. Improve the construction of the laws and regulations system of student personnel files, provide convenience for the organization and personnel departments to better select and employ people, help colleges and universities to reshape their image and rebuild their credibility.

Wu J studies the design of personnel file management system based on big data technology, analyzes the background of personnel file management system design, then expounds the main technologies of big data technology in system design, and finally puts forward detailed system design modules and implementation methods [1]. Wang H and others studied the personnel file management system based on blockchain, and adopted the improved PBFT consensus algorithm to store the file data safely and effectively in the personnel file blockchain system, thus ensuring the traceability and effectiveness of the file data. The local backup, transfer and sharing of personnel files are realized by using smart contract and IPFS [2]. However, in the design of the above-mentioned file management system, yes, it is easy to detect the falsification of personnel files.

The so-called wireless network refers to the network that can realize the interconnection of various communication devices without wiring. Wireless network technology covers a wide range, including global voice and data networks that allow users to establish long-distance wireless connections, as well as infrared and radio frequency technologies optimized for short-range wireless connections. Wireless networks are generally combined with telecommunications networks, and can be linked between nodes without cables. According to different network coverage, wireless networks can be divided into wireless wide area networks, wireless local area networks, wireless metropolitan area networks and wireless personal area networks. Wireless WAN is based on mobile communication infrastructure, and is operated by network operators. WLAN is a network responsible for wireless communication access function in a short distance range, and its network connectivity is very powerful. Wireless WAN and WLAN are not completely independent of each other. They can be combined to provide more powerful wireless network services. Wireless LAN can allow access users to share information within the local area, while wireless WAN can allow access users to share information outside the local area. Wireless MAN is a wireless network that allows access users to access fixed places. It connects multiple fixed places in a city or region. Wireless personal area network (WPLAN) is a wireless network in which users connect their portable devices through communication devices in a short distance.

In this paper, the intelligent monitoring system of University Personnel File Falsification Data Based on Wireless Network is studied. By initializing the wireless transmission module of the server, we can read the file information stored in XML format. Based on this information, we can define the server data transmission instruction, which can be used to build the server side of the intelligent monitoring system. At the same time, by optimizing the topology of wireless network, the execution conditions of intelligent monitoring instructions are defined, and the specific specifications of intelligent monitoring software are formulated. Finally, we cooperate closely with the server organization to realize the design of intelligent monitoring system for college personnel file fraud data based on wireless network.

2 Design of Intelligent Monitoring System Server

For the design of the server of the intelligent data monitoring system for the falsification of university personnel files, it is necessary to read the XML format file on the basis of initializing the wireless transmission module, and determine the data transmission behavior in the server system by combining relevant information parameters. This chapter conducts research on its specific design methods [3].

2.1 Initialization of Server Wireless Transmission Module

The intelligent monitoring server for the data of falsification of personnel files in colleges and universities uses the open source library Bluecover to program for the wireless network module. Bluetooth is an open source library developed for PC Bluetooth communication, which contains Bluetooth and other related class libraries developed by Bluetooth. You can program the server Bluetooth module by introducing this open source library on the PC side [4].

Get the local Bluetooth device through LocalDevice. GetLocalDevice() to determine whether the local Bluetooth function is supported. If not, prompt the user and exit the program. If it is supported, set the Bluetooth. After setting, start to monitor the falsification data of university personnel files and wait for the access of the client. CSAMT data collection station is set as the server side, and its discovery mode needs to be set to be discoverable. Call LocalDevice. SetDiscoverable() to set the discovery mode [5]. It supports two discovery modes. The first is the DiscoveryAgent. LIAC short-term discovery mode, which can only be discovered in a fixed short time. The second is the DiscoveryAgent. GIAC long-term discovery mode, in which the device is always discoverable [6].

The specific initialization process of the server wireless transmission module is shown in Fig. 1.

UUID is the universal unique identification code in the wireless transmission module. In Bluetooth programming, the service is identified by UUID. Only the identified service can be found and used. Each service has its own unique UUID to ensure the uniqueness of space-time. The UUIDs on both sides of the client and server in the same connection communication must be consistent [7–9].

The connection URL scheme of RFCOMMStreamConnection is adopted when Bluetooth serial port service is adopted. The URL connection string is url = "bt-spp://localhost:" + UUID + ";Name = RemoteServer ", after setting, you can use the acceptAndOpen() method to wait for the connection of the client. After the connection is successful, you can start a new thread to initialize the wireless transmission module of the intelligent monitoring system server [10].

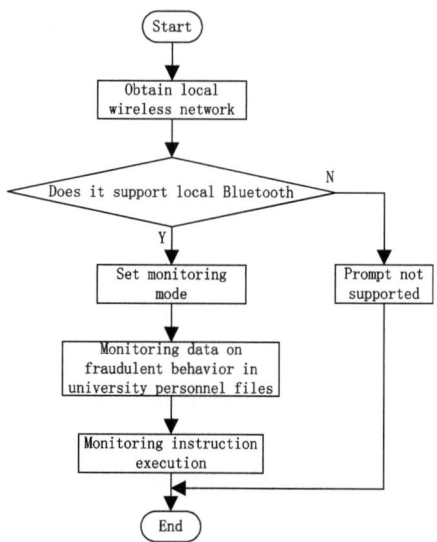

Fig. 1. Initialization flow chart of server wireless transmission module

2.2 Reading XML Format Files

The parsing method of XML files in different languages is basically the same, which is based on event flow or document structure. On this basis, the intelligent monitoring system for falsification data of university personnel files has developed four analytical methods: DOM, SAX, JAXB, and JDOM [11].

(1) DOM

DOM is the document object model. In the DOM mode, every element in the XML file can be treated as a node object. If you only look at the polymorphism of the data object of university personnel file fraud, it is very good, but it is not readable enough to be used in the server module [12, 13].

The definition formula of DOM parsing condition is:

$$w_1 = \frac{1}{\dot{\chi}} \times (\bar{q} + 1)^2 \tag{1}$$

$\dot{\chi}$ It represents the polymorphic expression characteristics of data objects of falsification of personnel files in colleges and universities, $\bar{q}$ It refers to the average transmission value of university personnel file fraud data in the unit monitoring cycle [14].

(2) SAX

SAX parses the falsification data of university personnel files by reading one line of files and parsing one line. This method is mainly based on events. Not all XML documents are read into memory, which is similar to the processing mechanism of tag library.

The definition formula of SAX resolution condition is:

$$w_2 = 1 - \sqrt{\frac{\hat{\alpha}}{\dot{\chi} \cdot \overline{q}}} \qquad (2)$$

$\hat{\alpha}$ It indicates the authority of the server wireless transmission module to analyze and process the data of falsification of university personnel files.

(3) JAXB

JAXB is an industry standard and a technology that can generate JAVA classes according to XML Schema. JAXB converts schema or DTD into JAVA objects, usually. Java files, through mapping, and then the server wireless transmission module parses the fraud data of the original university personnel file according to these. Java files.

JAXB parsing condition definition formula is:

$$w_3 = \beta \sqrt{\dot{\chi} \times \overline{q}} \qquad (3)$$

β It refers to the coding parameters of university personnel file fraud data based on JAVA protocol.

(4) JDOM

JDOM uses the JAVA data type to define each node of the operation data tree. Mainly through the use of pure JAVA technology to achieve the analysis, generation, serialization and other operations of university personnel file fraud data, JDOM also has excellent performance. Compared with the other three methods, JDOM parsing XML files is relatively simple, and is more suitable for processing files that are not very complex in structure [15].

The definition formula of JDOM parsing conditions is:

$$w_4 = 1 + \delta \sqrt{\frac{\overline{q}}{\dot{\chi}}} \qquad (4)$$

δ It refers to the definition item of the server wireless transmission module's ability to analyze the data of falsification of university personnel files.

Simultaneous Eqs. (1), (2), (3) and (4) can define the reading expression of XML format file in the intelligent monitoring system of university personnel file fraud data as:

$$Q = \frac{\hat{E}^2 \times (w_1 \cdot w_2 \cdot w_3 \cdot w_4)}{\sqrt{e_1^2 + e_2^2 + e_3^2 + e_4^2}} \qquad (5)$$

Among them, $\hat{E}$ Represent real-time query parameters of university personnel file fraud data based on XML format files, e_1 Represents the data reading vector matching the DOM file, e_2 Represents the data reading vector matching the SAX file, e_3 Represents the data reading vector matching the JAXB file, e_4 Represents the data reading vector that matches the JDOM file.

When the system host monitors university personnel file fraud data, it creates a SAXBuilder object, reads in the XML file to be parsed, and then uses getRootElement to obtain the root element. Next, it defines nine String strings, and uses root. GetChildTextTrim to obtain the sub elements under the root element and put them into the corresponding string. Then, nine strings are spliced into a String string, which is separated by a space to facilitate data restoration by the client [16].

2.3 Server Data Transmission

After the client of the intelligent monitoring system for falsification data of university personnel files is successfully connected to the server, the client can send various instructions to the server through the wireless network. The server responds differently to different instructions. When the server receives the university personnel file fraud data sent by the client, it automatically starts the Timer timer, reads and parses the XML data from the local disk, and then sends the data to the client after encapsulating it through a user-defined protocol [17].

The jdom. Rar package is introduced into the monitoring programming, a new thread is established to process the data, and Timer is used to realize the regular reading and transmission of XML files. After a client connects, wait for the sending command signal from the client, and then start the Timer after receiving it to read data at a fixed time interval for transmission. After the client connects to the server, the command sent by the client is read for judgment. Get the input stream through javax. Microedition. io. InputConnection. OpenInputStream(), get the data of the client, that is, Android, using the read() method, and enter different data processing through equals to determine the type of command. The data is transmitted in the form of output stream. Define a data output stream class and call its write method to send data to the client [18].

In the intelligent data monitoring system for falsification of university personnel files, the complete server data transmission principle is shown in Fig. 2.

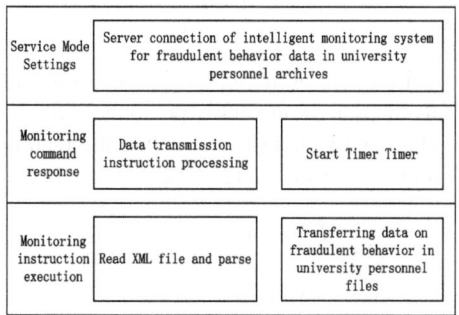

Fig. 2. Server Data Transmission Principle

The server side of the monitoring system contains three monitoring methods of data transmission modes. Click the client application icon, and after running the program, select the corresponding data monitoring mode according to personal needs to enter the drawing interface. Search the server in the drawing interface to connect. After the server is successfully connected, the client interface Toast prompts that the connection is successful. At this time, the client can send instructions to the server. When the client sends the command to obtain data to the server, the server sends data to the client. The client receives the data, parses and restores the data, and plots it on the screen in the form of logarithmic coordinates. If the user selects the data filtering function, the data will be smoothed based on the anisotropic diffusion algorithm before display. The client program can control the display and hiding of the selection curve by pressing the key.

3   Design of Intelligent Monitoring Software Based on Wireless Network

With the cooperation of the server system, in order to improve the design scheme of the intelligent monitoring system based on wireless network for university personnel file fraud data, a stable wireless network topology should also be established, and together with the OSAL monitoring mechanism, the intelligent monitoring implementation conditions should be defined.

3.1   Wireless Network Topology

In the intelligent data monitoring system for falsification of personnel files in colleges and universities, the wireless network always maintains a mesh connection, and its topology is shown in Fig. 3.

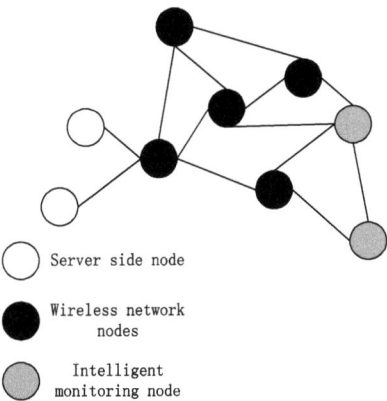

Fig. 3. Mesh topology of wireless network

In the wireless network, each university personnel file fraud data object has two addresses to distinguish each other, one is a 64 bit MAC address, and the other is a 16 bit network address. The 64 bit MAC address is the only global address fixed in the hardware device when it leaves the factory. The user cannot change it [19]. The 16 bit network address is the unique address allocated by the coordinator or router after the node joins the wireless network, which is mainly used to identify the node and transmit data in the network. The solution expression of wireless network topology for the data carrying capacity of university personnel file fraud is:

$$R = \frac{\sqrt[\varepsilon]{\gamma\left(\varphi \times \frac{\dot{y}}{|\Delta T|}\right)}}{Q^2 - 1} \tag{6}$$

Among them, ε It indicates the number of times the data of falsification of university personnel files is transmitted in the wireless network topology, γ Represents the topology connection parameters of the wireless network structure, φ It represents the real-time transfer coefficient of university personnel file fraud data in the wireless network topology, $\dot{y}$ It indicates the marking characteristics of wireless network topology for university personnel file fraud data, ΔT It refers to the unit transmission cycle of university personnel file fraud data in the wireless network topology.

The design of wireless network topology will affect the coverage range of the network. In the university environment, it is necessary to ensure that the wireless network can cover all areas related to Dang'an to ensure comprehensive data collection and monitoring. It is set that N wireless devices need to be installed in the university environment, and these devices need to be distributed in various areas involving Dang'an. Coordinates can be used to represent the position of each device, such as (x1, y1), (x2, y2), (xN, yN).

Establish a signal propagation model based on the free space path loss model, with the expression:

$$L = 20\log 10(d) + 20\log 10(f) - 147.55 \tag{7}$$

Among them, L represents the path loss of the signal during propagation, d represents the propagation distance, and f represents the frequency of the signal.

Based on the signal propagation model and device location, the signal strength or received power of each device can be calculated. Then determine the coverage range. According to the free space path loss model, it can be transformed into an expression for received power. Assuming the transmission power of the device is Pt, the received power Pr can be calculated using the following expression:

$$\text{Pr} = Pt - L \tag{8}$$

In order to more specifically describe the coverage range and convert the received power into signal strength, the above expression is rewritten as:

$$S = 10\log(\text{Pr}) + 30 \tag{9}$$

The child nodes in the wireless network topology will send their own device declaration to the nodes in the entire network after receiving the university personnel file fraud behavior data assigned by the parent node. After the parent node receives the declaration, it will modify the node association and binding table accordingly. In the Z-Stack protocol stack, the parameter MAX DEPTH determines the maximum depth of the entire network; The parameter MAX CHILDREN determines the maximum number of sub nodes that a router or coordinator can have; The parameter MAX ROUTER determines the maximum number of routing sub nodes that the coordinator and router can have.

3.2 OSAL Mechanism Monitoring

The OSAL mechanism includes the basic functions specified in the Zigbee protocol, and each function is realized through its corresponding functions. Later, in order to facilitate the management of these function sets, the protocol stack has added the OSAL real-time operating system to improve the functions of the protocol stack. Some software designs in wireless networks are based on Z-Stack, which can not only greatly improve the development efficiency, but also make the program more versatile and transplantable. The Z-Stack protocol stack and the OSAL operating system need to complete the initialization process when working.

The transmission process of wireless network communication is mainly divided into two parts, namely, the reporting process and the downloading process. First, the reporting process is described. When the monitoring host needs to report data, it will first send the encoded data to the NB IoT UE, and the terminal will send a request to the base station in a competitive manner. If the competition fails, the data will be delayed and then sent. When the competition succeeds, the data will be sent to the base station, and then transmitted to the host computer [20, 21] via the server. The upper computer decodes the monitoring instruction sent by the university personnel file fraud data, and returns the received response if the information is correct, otherwise it will be discarded. When the monitoring host does not receive a response for a period of time, it will directly retransmit. When the retransmission limit is exceeded, the information will be discarded.

The complete OSAL mechanism monitoring process is shown in Fig. 4.

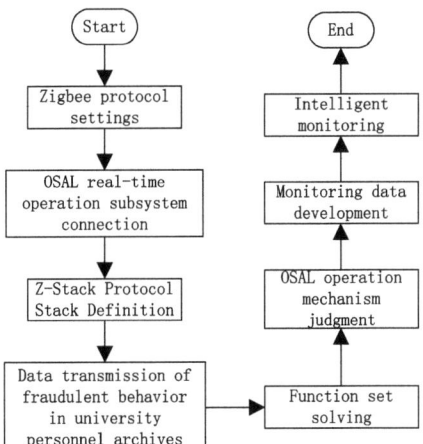

Fig. 4. OSAL Mechanism Monitoring Flow Chart

After completing the initialization process, Z-Stack will finally enter the OSAL scheduling management operating system, that is, OSAL_start_System(), which is the main loop function of the task event. In the protocol stack, OSAL is responsible for scheduling the operation of various tasks in the system through the main function of task processing. If an event is detected, it will select and call its corresponding event processing sub function for corresponding processing according to the data content of university personnel file fraud.

After receiving and analyzing data, the system needs to feed back the data information to the main interface for real-time display in order to enable users to intuitively monitor the data at the current time. The display and deletion of related nodes can be controlled by adding delete node keys. The main thread is responsible for the display and update of the OSAL mechanism, while the data receiving and parsing are in the child thread. Here, communication between threads is required.

set up i Indicates the minimum value of the scheduling parameter of the OSAL mechanism, o Represents the minimum value of the main function scheduling vector, ϕ It represents the code coefficient of the monitoring event of the falsification of university personnel files, u_1、u_2、$\cdots$、u_n express n Three randomly selected indicators for monitoring falsification of university personnel files. With the support of the above physical quantities, the simultaneous Eq. (6) can express the monitoring conditions of the OSAL mechanism as:

$$Y = \frac{\sum_{i=1}^{+\infty} (\phi - 1)^2 \cdot R}{\sum_{o=1}^{+\infty} \frac{1}{n!} \times |u_1 + u_2 + \cdots + u_n|^2} \tag{10}$$

Each event processing sub function will receive the task of the corresponding task_ID and event parameter representing the event content, and finally return the event processing result. In the OSAL mechanism, the main event processing function is designed for each type of college personnel file fraud data, and the various operations of the node are designed into corresponding events and event processing sub functions. These tasks will run orderly according to the requirements under the scheduling management of the OSAL operation mechanism to realize the functions undertaken by each node.

3.3 Intelligent Monitoring Execution Conditions

The wireless network needs to upload the university personnel file fraud data collected by the terminal node to the gateway through the network and then send it to the mobile monitoring terminal; At the same time, the data transmitted from the mobile monitoring end to the gateway needs to be distributed to achieve the control function of the terminal node. The sending and receiving operations of monitoring data obtained by various nodes are realized by calling the corresponding API functions in the OSAL mechanism, and the sending and receiving functions of data are realized by configuring reasonable parameters and establishing data buffers.

(1) Data transmission

For the data transmission function, the monitoring system defines an event SEND_DATA_EVENT, call AF in its event handler_DataRequest() function is used to send the data of falsification of personnel files in colleges and universities. This function has 8 parameters, of which buf and len are the more important parameters. Buf is the buffer area where data needs to be sent, and the data in the buffer area can be added through data receiving or data collection; Len is the length of the sent data.

The definition formula of data transmission function is:

$$p_1 = \frac{|\dot{d}|^2}{f\hat{s}^2 - (\lambda|\Delta A|)^2}\bigg|_{\lambda \neq 0} \quad (11)$$

$\dot{d}$ It indicates the sending characteristics of the data of falsification of university personnel files, $\hat{s}$ Represents the parameter of the data sending function definition item based on the OSAL mechanism, f It represents the data buffer coefficient of university personnel file fraud, ΔA It refers to the unit cumulative amount of falsification data of university personnel files, λ Indicates the real-time sending permission of the data sample object.

(2) Data reception

In order to realize the data communication of falsification of personnel files in colleges and universities, in addition to data transmission, it is also necessary to realize the function of data reception. An AF will be generated when the node antenna receives data_INCOMING_MSG_CMD message. The event processing function of the next task will execute the data receiving process [22]. The intelligent monitoring system defines relevant types of consortium data to store the received data. The data type here should be consistent with the data type when sending as much as possible to facilitate data storage. Then call osal_The memcpy() function copies the received data into the data, so that the data packets received by the node are stored in the data.

The definition formula of data receiving function is:

$$p_2 = \int_{\substack{k=1 \\ c=1}} \frac{\int_{l=1}^{+\infty} \vec{h} \times \tilde{g}}{(f-1) \times |\Delta A|^2} \Bigg|_{k \neq c}^{|\frac{c}{k}|^2} \quad (12)$$

k/c It represents two randomly selected instruction execution vectors for monitoring the falsification of university personnel files, and $k \neq c$ The inequality value condition of is always true, l Represents the data sampling parameters of university personnel file fraud based on the data union function, $\vec{h}$ Represents the data type definition vector of falsification of university personnel files, $\tilde{g}$ Represents the copy characteristics of data samples in wireless networks.

Simultaneous Eqs. (7), (8) and (9) derive the executive expression of the intelligent monitoring system for the data of falsification of university personnel files as follows:

$$F = \sqrt{Y \cdot \left(1 - \frac{1}{b^2}\right) \times \left(\left|\frac{p_1}{\hat{j}_1}\right|^2 \cdot \left|\frac{p_2}{\hat{j}_2}\right|^2\right)} \quad (13)$$

Among them, $\hat{j}_1$ Represents the monitoring instruction execution vector related to the data sending behavior of university personnel file fraud, $\hat{j}_2$ Represents the monitoring instruction execution vector related to the data receiving behavior of university personnel file fraud, b Represents the intelligent monitoring instruction sharing feature in the wireless network.

The implementation process of the intelligent data monitoring system for falsification of university personnel files includes the frequency synchronization and time synchronization of the wireless network. In order to achieve the final monitoring command synchronization, it is necessary to ensure that the data transmission frequency reaches the synchronization state before starting the corresponding time adjustment. The data transmission frequency is determined by the execution cycle of the monitoring command. Even if the data transmission and data reception commands use the same transmission frequency, there will be slight differences in their execution capabilities. Frequency adjustment is required to complete the synchronization process [23–25]. However, in practical applications, since the transmission frequency value level has remained fixed since its appearance, it cannot be adjusted from the hardware, so it can only be optimized through software algorithms. Therefore, in order to ensure the integrity of the monitoring execution program instructions, when defining the data sending function and data receiving function, it is necessary to ensure the integrity of the data transmission of university personnel file fraud. The overall design process is as follows (Fig. 5).

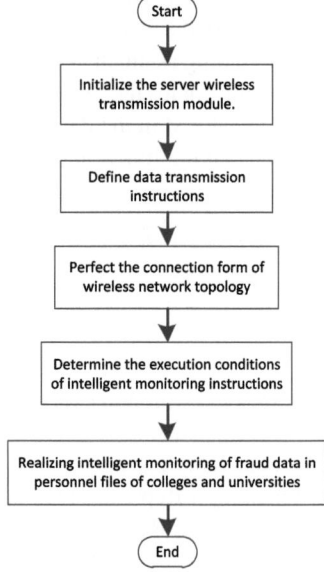

Fig. 5. Overall design process

4 Example Analysis

In order to verify the effectiveness of the intelligent monitoring system based on wireless network, the personnel file data of a university was collected as the benchmark data. These data include personal information, professional titles, academic qualifications, work experience and other information of faculty and staff. In order to highlight the practical differences between the intelligent monitoring system based on wireless network, the monitoring system based on Android and Zigbee network, and the monitoring system based on NB IoT, the following comparative experiments are designed.

4.1 Experiment Preparation

In the Internet environment, the data sample transmission rate affects the host components' ability to accurately monitor information parameters. If the university personnel management network is regarded as a local area network system, it can be considered that the faster the real-time transmission rate of university personnel file fraud data is, the stronger the host components' ability to accurately monitor information parameters will be.

Use the equipment components shown in Table 1 to build the Internet environment required for the experiment.

Table 1. Selection of Experimental Equipment

Item	Equipment components	Model and name
1	Data processor	EF860 M2 processor
2	Data Register	JK-TX-GST040/016
3	Internet sharing server	LD4112G-A 4U
4	Private cloud storage devices	QNAP TS-264C
5	Public cloud storage equipment	HDMI AP-S4-G2306
6	Monitoring server	NAS TS-416-4G
7	Data sharing device	TESA TWIN-T10
8	Transport and storage equipment	JK-TX-GST040/016

After data reception is completed, if the protocol frame is parsed to confirm that the information is correct, the successful reception response will be sent; otherwise, the failed transmission response request will be sent again. If you receive the order to report data, you will start to monitor the data of falsification of personnel files in colleges and universities after replying.

In order to ensure the accuracy of the test results, after the completion of the test data recording, it is necessary to ensure that the equipment components are restored to the initial connection state before the next test can be carried out.

4.2 3Steps and Processes

The specific implementation process of this experiment is as follows:

Connect each experimental equipment in Table 1 as required, close the control switch and start the experiment;

Input the executive program of the intelligent monitoring system for falsification of university personnel files based on wireless network into the processor host, record the specific value of the real-time data transmission rate under the effect of the system, and record the results as the experimental group variables.

Exit the experimental group execution program and disconnect the control switch.

Close the control switch again, input the execution program of the monitoring system based on Android and Zigbee network into the processor host, record the specific value of the real-time data transmission rate under the action of the system, and record the results as the control group 1 variable.

Exit the control group 1 execution procedure and disconnect the control switch again.

Close the control switch, input the execution program of the monitoring system based on NB IoT into the processor host, record the specific value of the real-time data transmission rate under the action of the system, and record the results as two groups of variables for comparison.

Statistic the variable data obtained and summarize the experimental rules. In order to avoid other interference conditions affecting the authenticity of the experimental results, the connection behavior between default equipment components will not change during the experiment.

4.3 Results and Conclusions

The following figure reflects the specific experimental values of the real-time data transmission rate of the falsification of personnel files in the experimental group and the control group (Fig. 6).

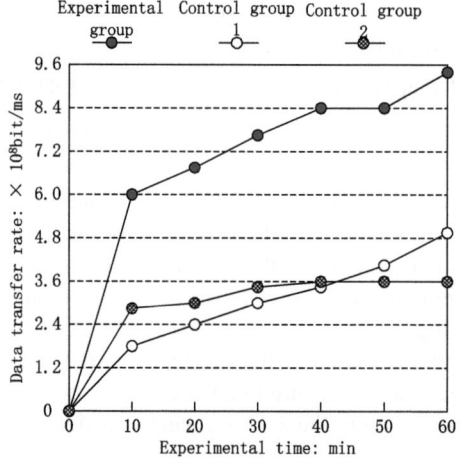

Fig. 6. Real time data transmission rate

Experimental group: the data transmission rate of falsification of university personnel files in the experimental group maintained a numerical trend of first increasing, then stabilizing, and finally continuing to increase. During the entire experiment, the maximum value reached 9.5×10^8 bit/ms.

Control group 1: the data transmission rate of falsification of university personnel files in control group 1 has always kept increasing, and the maximum value was 5. 0 at the end of the experiment $\times 10^8$ bit/ms, decreased by 4. 5 compared with the maximum value of the experimental group $\times 10^8$ bit/ms.

Control group 2: the data transmission rate of falsification of university personnel files in control group 2 kept a numerical trend of first increasing and then stabilizing, and the maximum value was 3. 6 during the whole experiment $\times 10^8$ bit/ms, decreased by 5. 9 compared with the maximum value of the experimental group $\times 10^8$ bit/ms.

To sum up, the conclusion of this experiment is:

(1) The monitoring system based on Android and Zigbee network has relatively weak application ability in improving the real-time data transmission rate, which does not meet the actual application requirements for accurate monitoring of information parameters.
(2) The application capabilities of the monitoring system based on NB IoT and the monitoring system based on Android and Zigbee networks are not significantly different, which is not enough to significantly improve the real-time transmission rate of the data of falsification of personnel files in colleges and universities. Therefore, its application capabilities in accurate monitoring of information parameters are relatively weak.
(3) The application of the intelligent data monitoring system for falsification of university personnel files based on wireless network can significantly improve the real-time transmission rate of data samples. For the network host, its application needs in accurate monitoring of information parameters can be better met.

In order to further verify the practicability of the method in this paper, the monitoring accuracy is used as the experimental index for comparative testing, and the test results are as follows (Fig. 7).

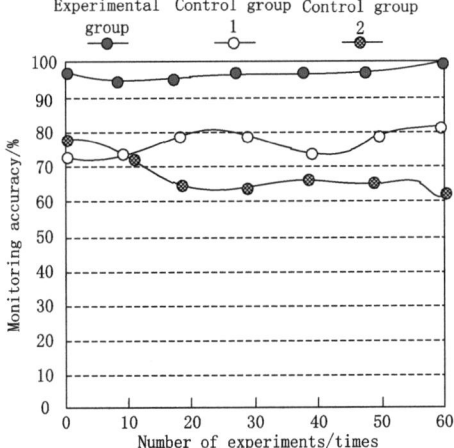

Fig. 7. Comparison chart of monitoring accuracy

As can be seen from the above figure, the monitoring accuracy of the experimental group is higher than 90%, and the highest is 98%, while the monitoring accuracy of the control group 1 is about 70% ~ 82%, and the monitoring accuracy of the control group 2 is about 63% ~ 79%. Therefore, the monitoring accuracy of this method is obviously higher than that of the comparison method, which shows that this method is practical.

5 Conclusion

The intelligent monitoring system of university personnel file fraud data based on wireless network is divided into server and client. On the server side, the scheme starts from the existing data monitoring system structure, and combines the analysis of the function and performance requirements of the data monitoring system in the current era of the Internet of Things and mobile Internet, and proposes the overall design scheme of the data monitoring system to meet the needs of the current era. Then, according to the content of the overall plan, the design and development of three structural parts, namely, the hardware platform of data acquisition and transmission end, the software of data acquisition and transmission end, and the software of mobile monitoring center, are completed.

With the support of wireless network, the application capability of intelligent monitoring system is mainly reflected in the following aspects:

(1) Aiming at the shortcomings of existing data monitoring system solutions, a wireless, mobile and low-power data monitoring system solution is proposed based on the selection of various schemes.
(2) The mobile monitoring software realizes the visual processing function of monitoring data by adding the OSAL mechanism. It mainly includes two main functions: real-time data dynamic curve viewing and historical data chart viewing.
(3) The delay part of the system is compensated through the reasonably designed system test link to improve the real-time performance.

References

1. Wu, J.: Design and implementation of Dang'an management system based on big data technology. J. Beijing Inst. Graph. Commun. **29**(1), 236–239 (2021)
2. Wang, H., Chen, B., Liu, Y.X.: Research on personnel file management system based on Blockchain. Comput. Sci. **48**(2), 713–718 (2021)
3. Songwu, L.I.: 6-DOF motion assessment of a hydrodynamic numerical simulation of a semisubmersible platform using prototype monitoring data. Chin. Ocean Eng. **36**(4), 575–587 (2022)
4. Ceresnak, R., Kvet, M., Matiasko, K.: Increasing security of database during car monitoring. Transp. Res. Procedia **55**(11), 118–125 (2021)
5. Zhao, S., Wang, H.: Enabling data-driven condition monitoring of power electronic systems with artificial intelligence: concepts, tools, and developments. IEEE Pow. Electron. Mag. **8**(1), 18–27 (2021)
6. Wang, C., Fan, J., Wang, Q., et al.: Use of L-band SAR data for monitoring glacier surging next to Aru Lake. Procedia Comput. Sci. **181**(8), 1131–1137 (2021)

7. Mabrouki, J., Azrour, M., Dhiba, D., et al.: IoT-based data logger for weather monitoring using Arduino-based wireless sensor networks with remote graphical application and alerts. Big Data Min. Anal. **4**(1), 25–32 (2021)
8. Zhang, J., Liu, H., Sun, X., et al.: Processing of building subsidence monitoring data based on fusion Kalman filtering algorithm. AEJ **60**(3), 3353–3360 (2021)
9. Nodeland, B., Belshaw, S.H.: Stolen at the pump: an empirical analysis of risk factors on gas pump skimmer fraud attacks. J. Fin. Crime **29**(3), 942–950 (2022)
10. Li, C., Ding, N., Zhai, Y., et al.: Comparative study on credit card fraud detection based on different support vector machines. Intell. Data Anal. **25**(1), 105–119 (2021)
11. Fang, K., Li, J., Ke, P., et al.: Energy consumption monitoring based on GM-BP neural network. Comput. Simul. **39**(10), 430–434 (2022)
12. Beery, S.: Scaling biodiversity monitoring for the data age. Crossroads **27**(4), 14–18 (2021)
13. Dang, W., Yu, T., Wang, H., et al.: PerfMon: measuring application-level performance in a large-scale campus wireless network. China Commun. **20**(3), 316–335 (2023)
14. Albogamy, F.R., Aiyashi, M.A., Hashim, F.H., et al.: Optimal resource allocation for NOMA wireless networks. Comput. Mater. Continua **74**(2), 3249–3261 (2023)
15. Injila, G., Begh, R.: In band full duplex (IBFD) technology for next generation wireless networks: a survey in cellular networks. China Commun. **20**(5), 20–39 (2023)
16. Chhea, K., Ron, D., Lee, J.R.: Weighted de-synchronization based resource allocation in wireless networks. Comput. Mater. Continua **75**(4), 1815–1826 (2023)
17. Wang, Y.J., Wen, D.Z., Mao, Y.J., et al.: RIS-assisted federated learning in multi-cell wireless networks. ZTE Commun. **21**(1), 25–37 (2023)
18. Cui, S., Yin, C., Zhu, G.: Special topic on federated learning over wireless networks. ZTE Commun. **21**(1), 1–2 (2023)
19. Almekhlafi, M.A., ElfadilEisa, T.A., Al-Wesabi, F.N., et al.: Efficiency effect of obstacle margin on line-of-sight in wireless networks. Comput. Mater. Continua **72**(7), 227–242 (2022)
20. Yu, J.: Research on the informatization value and optimization path of university Dang'an under the background of big data. Inside Outside Lantai **2020**(25), 1–3 (2020)
21. Xu, X.: The change of Dang'an in colleges and universities brought by the development of the network. Office Oper. **2020**(17), 132–133 (2020)
22. Liu, H.: The application of big data technology in the informatization management of Dang'an in colleges and universities. Sci. Technol. Vis. **0**(12), 157–159 (2020)
23. Yu, F.: Thoughts on the standardized management of Dang'an in colleges and universities. Inside Outside Lantai **2020**(25), 28–30 (2020)
24. Du, X.: Challenges and strategies of Dang'an management in colleges and universities in the era of Big data. J. Taiyuan Urban Vocat. Coll. **2020**(7), 58–59 (2020)
25. Zhao, C.: Research on Dang'an management in universities from the perspective of new public service theory. Sci. Technol. Inf. **18**(7), 246–247 (2020)

A Study of Resource Sharing Methods for Teaching English Reading Based on Blockchain Technology

Man Zhan(✉)

Sanya Aviation and Tourism College, Sanya 572000, China
zhanman0909ab@163.com

Abstract. The diversification and dispersion of English reading teaching resources lead to inefficient resource sharing and security and privacy problems. In response to this challenge, this paper proposes a method of sharing English reading teaching resources based on blockchain technology. Using homomorphic encryption technology to encrypt English reading teaching resources, using blockchain technology to re-encrypt English reading teaching resources and store them in blockchain, and using resource access control mechanism to authenticate the identity of resource sharers, so as to realize English reading teaching resources sharing. The experimental results show that this method significantly improves the security and efficiency of resource sharing, and maintains the integrity of resources, which provides a new solution for English reading teaching resource sharing.

Keywords: Blockchain technology · English reading · Teaching resources · Resource sharing · Homomorphic encryption technology

1 Introduction

With the advancement of educational modernization, traditional English reading teaching concepts and educational programs can no longer meet learners' learning needs. The educational idea of student-centered and teaching students in accordance with their aptitude has been recognized by most people as an educational method in the new era. In order to meet everyone's academic and career development, we must have a large number of high-quality English reading education resources as a support. However, there is no active and effective communication between universities in the co-construction and sharing of digital English reading education resources. Due to the regional differences and the differences between the comprehensive strengths of colleges and universities, compared with colleges and universities with abundant English reading teaching resources, they do not have high enthusiasm in sharing educational resources. In addition, at present, there is no professional institution or organization to manage and allocate resources among colleges and universities in a unified way, which is prone to fragmentation. Because there is less communication between colleges and universities, it leads to the repeated construction of large English reading teaching resources, resulting in greater cost losses.

With the development of society, the construction of "Internet + Education" has achieved rapid development [1]. The popularity and rapid development of computers and the Internet are constantly affecting the development of today's society and people's lifestyles and learning styles. Not only the application and popularization of the Internet is high, but also the technical level and attention of Internet research and development are quite high. Based on this, we should actively develop "Internet + Education", accelerate the modernization of education, and deeply study the integration of Internet and education under the background of "internet plus". The integration of Internet technology can create a free environment for students' study and provide rich learning resources. Mainly reflected in the Internet technology can be extended to every corner of the world through the network, breaking the constraints of time and region, realizing open and free teaching, and meeting the needs of lifelong education and national education for people of different ages and educational levels. It can provide an information platform for learners to communicate on the internet, share teaching resources, realize the redistribution of teaching resources in time and space to some extent, and greatly promote educational equity.

Blockchain technology has gradually entered people's sight with the development of Bitcoin. Through its continuous development, blockchain has gradually developed into a new technology. Blockchain technology is based on P2P network, which uses chain data structure to connect data blocks to achieve the function of storing data. Through the consensus mechanism, users in the blockchain can reach a consensus to update or store data, maintain the consistency of data, and form a decentralized account book. The development of blockchain technology provides technical support for the sharing of English reading teaching resources. Therefore, a method of sharing English reading teaching resources based on blockchain technology is proposed.

2 The Design of a Resource-Sharing Program for Teaching English Reading

The trusted data sharing framework designed in this paper is shown in Fig. 1.

The framework includes four types of entities: data owner (DO), data downloader (DD), blockchain and IPFS.

English reading teaching resources owner: the user who provides English reading teaching resources for sharing, mainly responsible for customizing the access policy, encrypting resources, publishing resources and distributing keys. The identity of the English reading teaching resources owner is not fixed, and he/she can also be the downloader of teaching resources to query and access the teaching resources.

English reading teaching resources downloaders: users who apply for access to teaching resources, mainly responsible for querying teaching resources and submitting information to apply for access to teaching resources. The identity of the downloader of teaching resources is not fixed, and he or she can also be the owner of the teaching resources to publish and share the data.

Blockchain: provide trusted infrastructure for untrusted parties, provide data publishing, data query, data verification and other services to DO and DD through smart contracts, reliably record all user operations on data, and ensure the security and traceability of data throughout its life cycle.

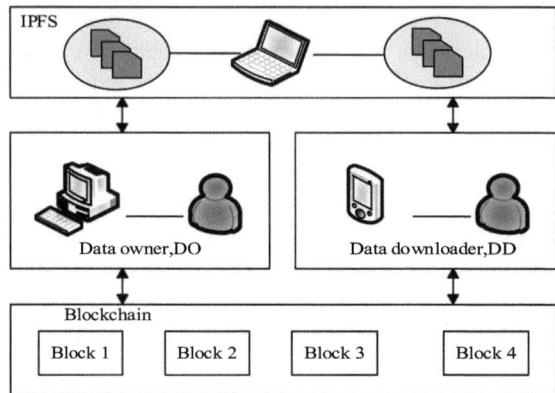

Fig. 1. A framework for sharing resources for teaching English reading

IPFS: It is responsible for storing the data uploaded by the owner of teaching resources, and returning the content id (CID) as the index for users to download corresponding teaching resources. It ensures the reliable storage of data through redundant backup technology.

3 A Resource Sharing Method for English Reading Teaching Based on Blockchain Technology

3.1 Homomorphic Encryption of English Reading Teaching Resources

The sharing of English reading teaching resources is actually the process of sending English reading teaching resources files from the place where the files belong to to the recipient, and ensuring the safety of English reading teaching resources in the process of file transmission. According to the actual needs, MapReduce technology is used to homomorphic encrypt English reading teaching resources, and the first security protection measure is set up. The schematic diagram of homomorphic encryption of English reading teaching resources is shown in Fig. 2.

As shown in Fig. 2, assuming that the English reading teaching resources to be shared are N, which contains k English reading teaching resource files were split into, according to the different categories of files n teaching resources of English reading are sorted and numbered according to the size of the fragments [2]. In MapReduce, Master is defined as the control node, MAP as the mapping node, and Reduce as the reduction node. Therefore, MapReduce consists of one control node, multiple mapping nodes, and reduction nodes. The master node, as the master node, sends the information mapping task to the mapping node and n English reading teaching resource slice is sent to the mapping node respectively, the mapping node encrypts the assigned slice, and its main role is to generate the ciphertext, the ciphertext generation rule is the first letter, number or character of the English reading teaching resource slice, and the number of ciphertext bits is 1 bit, the mapping function is utilized to extract the first data of the slice, which will be used as the ciphertext of the English reading teaching resource slice. The mapping

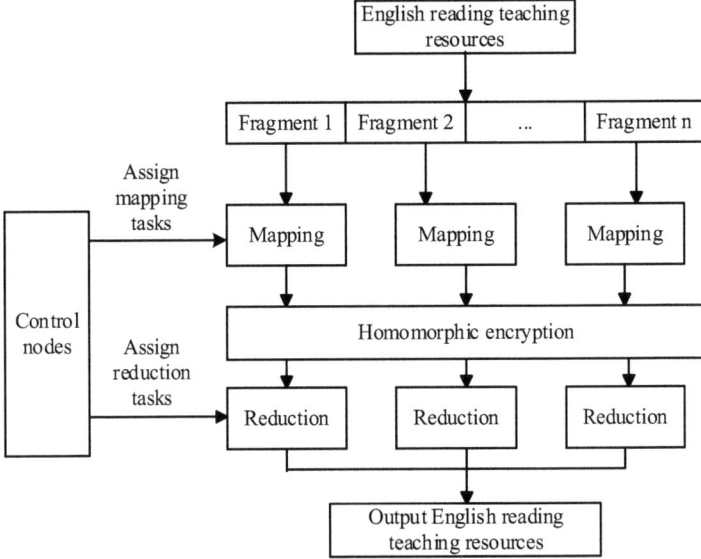

Fig. 2. Schematic diagram of homomorphic encryption of resources for teaching English reading

node sends the generated ciphertext to the attribution node, the control node sends the attribution task to the attribution node, and the attribution node summarizes the received ciphertext, generates the complete ciphertext of the English reading teaching resources file, and outputs the file along with it, thus realizing homomorphic encryption of the English reading teaching resources file.

3.2 Re-Encrypted Storage of Teaching Resources Based on Blockchain Technology

The blockchain is connected by blocks in chronological order in the form of chains, and the data structure of the blockchain is shown in Fig. 3.

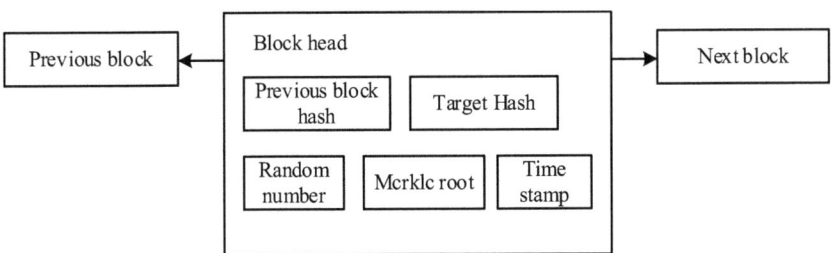

Fig. 3. Data structure of the blockchain

In Fig. 3, Mcrklc root is used for rapid location, search and integrity verification of data; target hash and random number are used for mining calculation and verification of workload proof consensus mechanism to ensure data consistency between nodes; the timestamp and the hash value of the previous block are used to determine the connection order of the blocks to ensure that the data cannot be tampered with and traceable. Smart contract and blockchain naturally fit together, giving blockchain programmability [3]. Use blockchain and smart contract to record data information, control data on the chain in the whole process, and promote data compliance sharing.

In this paper, blockchain technology is utilized to re-encrypt the storage of English reading teaching resources to ensure the security of teaching resources, and its encryption schematic is shown in Fig. 4.

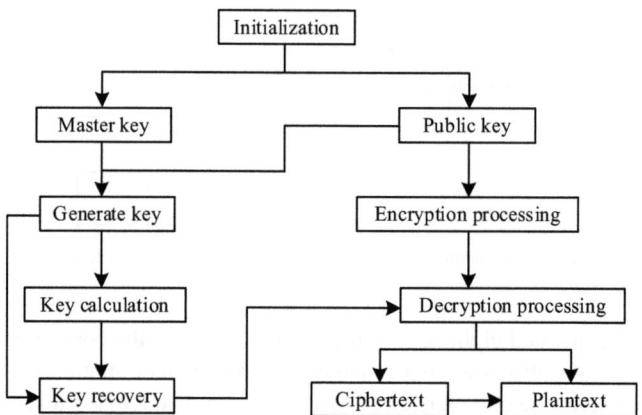

Fig. 4. Schematic diagram of encryption of teaching resources based on blockchain technology

As shown in Fig. 4, the encryption of English reading teaching resources using blockchain technology mainly includes five processes: data initialization, key generation (public key, private key), key calculation, key recovery and decryption.

3.3 Blockchain Cryptographic Key Generation

The English reading teaching resources to be stored and shared are mapped into the block, the English reading teaching resources are initialized in the block using the bilinear mapping principle, the utilization of the key matrix in generating the encryption key is pivotal, as the rationality of its design is intricately linked to the quality of encryption provided by blockchain technology for English reading teaching resources, in order to make the blockchain data parameter transmission rate can be greatly improved, in the process of encoding the key, the key template is used as a matrix to keep the state of the matrix. The key template is maintained in the state of matrix. The key matrix consists of horizontally defined coefficients and vertically defined coefficients, and the relationship between the two coefficients is always satisfied as follows:

$$\begin{cases} x = \dfrac{\lambda \cdot R}{E} \\ y = \dfrac{\sqrt{R \cdot E}}{\lambda^2} \end{cases} \qquad (1)$$

In Eq. (1), x denotes the key matrix transversal definition coefficient, which indicates the transversal transmission capability of the quantum-resistant blockchain data. λ denotes the query coefficient of the encoded information; the R denotes the data message encoding characteristics of any of the key metrics; the E denotes the average value of blockchain data transmission per unit; the y denotes the key matrix vertical definition coefficient, which indicates the vertical transmission capacity of blockchain data [4]. Assuming that the horizontal and vertical definition coefficients of the key matrix belong to the number greater than zero, the key matrix used to generate the key can be expressed as follows:

$$P = (s_{max} - s_{min})^2 \begin{bmatrix} S_{11} & S_{12} & \cdots & S_{i1} \\ S_{12} & S_{22} & \cdots & S_{i2} \\ \vdots & \vdots & \ddots & \vdots \\ S_{1y} & S_{2y} & \cdots & S_{xy} \end{bmatrix} \quad (2)$$

In Eq. (2), P denotes the key matrix used to generate the blockchain encryption key for teaching resources; the s_{max} denotes the maximum value of the data transmission vector that can be discriminated by the encrypted host of the teaching resource; the s_{min} denotes the minimum value of the data transmission vector [5] that the teaching resource encryption host can distinguish. Substituting the above generated key matrix into the encryption template, the teaching resource encryption key is generated as follows:

$$M = P \frac{\sqrt{g^2 - z^2}}{(\gamma \cdot |\Delta A|) - h^2} \quad (3)$$

In Eq. (3), M denotes the blockchain data encryption key. g denotes the text encoding information of the encrypted English reading teaching resources with arbitrary encoding precision. z denotes the initial encoding information of the text of the blockchain data encrypted English reading teaching resource; the γ denotes the ciphertext transfer coefficient. ΔA denotes the value of the amount of ciphertext transmitted per unit of time. h indicates the value of data encryption authority. The generated key is used to treat the shared part of the blockchain English reading teaching resources as decentralized, and the part of the English reading teaching resources is set as non-public information, and the rest of the data is public information.

3.4 Encrypted Storage of Teaching Resources

All English reading teaching resources deposited in the block are represented as a sequence of data, utilizing the generated key, which is encrypted in the block in order to prevent leakage of the generated key.

$$Rwkey(F, SK, BN, \varepsilon, SV, M) \to z \quad (4)$$

In Eq. (4), F denotes the block public parameter; the SK denotes the private key. BN denotes a column number of, and t of the matrix; the ε denotes the matrix that will be BN of columns mapped to attributes in the mapping function; the SV denotes the private key of the privacy information sharer. z denotes the encryption key [6]. And a random vector is chosen to partition the encryption key, and a key block is randomly chosen among the multiple key blocks obtained from the partition, which is used as the secret shared key share of English reading teaching resources and associated with the specified attributes of English reading teaching resources to generate the ciphertext, which is stored in the block, and is expressed by the formula as:

$$\zeta_i = \left(\frac{z}{\eta}, c, V\right)_i \tag{5}$$

In Eq. (5), ζ_i indicates the resources for teaching English as a foreign language stored in the i th blocks; the η denotes the number of key splits. c denotes a random vector; the V denotes the homomorphic encrypted English reading teaching resources file. Use the above program to re-encrypt all the English reading teaching resources for storage.

3.5 Sharing of Resources for Teaching English Reading

Considering that only the English reading teaching resources are partially encrypted, when the blockchain storing the English reading teaching resources is attacked, the attacker can use some decryption techniques to obtain the English reading teaching resources as well, therefore, on the basis of the above, an access mechanism is designed for the sharing of teaching resources [7]. Since only the basic description information of English reading teaching resources is recorded on the chain, the downloader of the resources who wants to obtain the complete teaching resources on the blockchain needs to verify the access to obtain the key of the ciphertext of the English reading teaching resources in order to successfully decrypt the original English reading teaching resources. In this paper, we utilize the lattice signature algorithm to treat the shared English reading teaching resources with linear signatures, which is used to authenticate the recipients of English reading teaching resources and ensure the privacy of blockchain data. The signer samples discrete vectors from a Gaussian discrete distribution, the lattice vectors, and the lattice vector values are denoted as:

$$e_L = ad_1 + d_2 \tag{6}$$

In Eq. (6), e_L denotes the value of the lattice vector. a represents the value chosen at random by the signatory, which is a positive integer. d_1. d_2 are two discrete vectors sampled from Gaussian discrete distribution. Together with the key as the input vector of the lattice signature algorithm, it generates the signature of the English reading teaching resources on the blockchain, which is expressed by:

$$U = (e_L, M) \tag{7}$$

In Eq. (7), U denotes the signature of English reading teaching resources on the blockchain [8]. In the signature, the e_L are vectors according to a Gaussian discrete distribution, so the generated signature needs to satisfy the following conditions:

$$\frac{1}{2}D_{er} = 1 \bigg/ \left(G \exp\left(-\frac{\|e_L\|^2}{2M}\right)\right) \tag{8}$$

In Eq. (8), D_{er} denotes a Gaussian discrete distribution; the G indicates the iterative expectation of the lattice signature algorithm. Using the above formula to test the generated data signature, if it does not meet the above conditions, then generate a new signature, and then test it again until it meets the conditions of Eq. (8); if it meets the above conditions, then the algorithm outputs the signature and realizes the signature of blockchain English reading teaching resources.

The above generated signature and the key on the blockchain are integrated into a digital certificate. The certificate format adopts the FG.151 international standard. The digital certificate contains the key, signature and ciphertext, which is expressed by the formula:

$$F = (M, U, Y, P) \tag{9}$$

In Eq. (9), F indicates a digital certificate. Y indicates the cipher text of the teaching resource to be shared. P represents the digital certificate sequence number, the sequence number for the digital certificate of all domains of the information hash summary.

When the data receiver receives the issued digital certificate, it can be said that the user legally owns the data key, but when the data receiver wants to obtain the plaintext in the certificate, it needs to extract the signature in the digital certificate to verify the identity, and its verification operation is:

$$o = F(\mathrm{mod}\, 2q, U) \tag{10}$$

In Eq. (10), o denotes the instructional resource hash value on the blockchain; the q denotes the verification key [9]. When the hash value of the teaching resource on the blockchain calculated by the above formula, the c equals to the hash value in the digital certificate, the verification is successful, the user identity is credible, and by recovering the key, pre-selected English reading teaching resources on the blockchain simulation color, according to the mode inverse element of the key is projected, obtained to the blockchain on the blockchain English reading teaching resources attribute key, the use of the recovered key decrypt the ciphertext processing, to get the plain text, which is expressed by the formula:

$$Z = \frac{C}{X(d,g)^{WQ}} \tag{11}$$

In Eq. (11), Z denotes the plaintext of the English reading teaching resources on the blockchain obtained after the decryption computation; the C denotes the secret shared key share. X indicates the recovery key. W denotes the identity random number key; the Q denotes the identity random number public key [10]. Using the above formula to decrypt the ciphertext, thus obtaining the English reading teaching resources on the blockchain, thus realizing the English reading teaching resources sharing based on blockchain technology.

4 Experimental Simulation and Analysis

4.1 Experimental Preparation and Environment

To assess the practicality and safety of the aforementioned proposal English reading teaching resource sharing scheme based on blockchain technology, the security analysis of the scheme is carried out from all stages of teaching resource sharing. The experimental environment is a CentOS Linux release 7 9. Server of 2009 (Core) system, processor is lntcl (R) Xcon (R) Silver X208 CPU@2 10 GHz. Establish a virtual machine server on the host host through VMware virtualization technology, and then virtualize multiple computers. Then manage the server through the client VMwarevSphere Client. Connect the server on the client, and then install multiple Linux operating systems on the server. Then install Hadoop, Spark, ElasticSearch and other software on the Linux operating system, and configure Hadoop configuration files, build a Hadoop cluster, implement MapReduce computing programming through the call of Hadoop's HDFS distributed storage system API interface, provide English reading teaching resource sharing services through WebServer, and upload and download shared curriculum teaching resource data. The experiment uses JKHFA software to simulate network attacks, invade the anti quantum blockchain database, the network attack frequency is 15.26 GHz, and generate anti quantum blockchain data encryption signatures according to the above process, as shown in Fig. 5.

46	52	45	46	52	45
23	12	46	23	12	46
79	35	26	79	35	26
85	08	35	85	08	35
46	52	45	46	52	45
23	12	46	23	12	46
79	35	26	79	35	26
85	08	35	85	08	35

Fig. 5. Cryptographic signature of blockchain teaching resources

The generated signature is verified against the verification conditions, and the signature is linearly calculated in order to pass the verification of the signature. The design approach outlined in this paper is capable of effectively fulfilling the objective of anti-quantum blockchain data encryption, and the specific encryption effect is evaluated in the following.

4.2 Experimental Results and Discussion

In order to test the integrity of data in the process of safe sharing of English reading teaching resources, the proposed method is evaluated in comparison with two existing methodologies, and the bit error rate (BER) is selected as the evaluation indicator. Bit

Error Ratio (BER) is a measure of the accuracy of data transmission within a specified time, which represents the probability of errors in binary data bit transmission. In data communication, if the transmitted signal corresponds to 1 while the received signal corresponds to 0, or vice versa, where the transmitted signal is 0 and the received signal is 1, an error occurs, that is, an error occurs. The generation of error code is due to the change of signal voltage due to interference and attack during signal transmission, resulting in signal destruction during transmission. As the most commonly used data communication transmission quality indicator, bit error rate is critical to evaluate the performance and reliability of data resource sharing methods. The higher the bit error rate, the worse the integrity of teaching resources. The calculation formula is:

$$h = \frac{\alpha + \varphi + \omega}{\sigma} \times 100\% \qquad (12)$$

In Eq. (12), h denotes the instructional resource sharing BER; the α denotes the number of samples with error codes in the shared teaching resources; the φ indicates the number of missing teaching resources. ω denotes the number of ineffective teaching resources. σ denotes the total amount of shared teaching resources. The BER curve of this paper's method in the application of English reading teaching resources sharing is shown in Fig. 6.

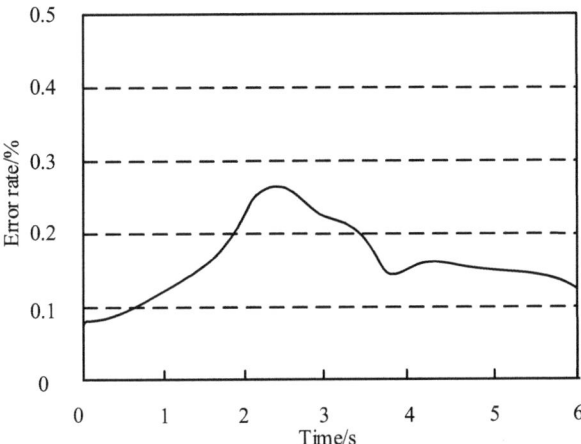

Fig. 6. BER graph of shared resources for teaching English reading

With 1000Byte as the benchmark, use the above formula to calculate the bit error rate after completing the sharing of 1000Byte teaching resources until the sharing of teaching resources reaches 8000 Byte. Use the spreadsheet to record the bit error rate. The specific data is shown in Table 1.

As shown in Table 1, in this experiment, the BER of the design method is within 1%, the average BER is 0.45%, the average BER of the control method 1 is 8.86%, and the average BER of the control method 2 is 5.68%, which are much higher than that of the design method, which is due to the use of the blockchain technology for sharing

Table 1. BER comparison of the three methods (%)

Resource quantity /Byte	Design method	Control method 1	Control method 2
1000	0.26	5.62	3.12
2000	0.32	6.14	3.56
3000	0.42	7.13	4.25
4000	0.44	7.86	4.96
5000	0.45	8.16	5.86
6000	0.52	8.83	6.75
7000	0.56	9.35	7.45
8000	0.59	10.42	8.15

teaching resources, the blockchain technology has the function of tamper-proof, which can guarantee the integrity of the data, and thus reduce the BER of sharing teaching resources. This is because the blockchain technology has the function of anti-tampering, which can guarantee the integrity of the data, thus reducing the BER of teaching resources sharing.

The data security bit can reflect the security of teaching resource sharing and is the main indicator of the reliability and sharing quality of data sharing methods, models or algorithms. It represents the proportion of successfully transmitted bits among the total number of transmitted bits (average value), i.e. the ratio of the number of erroneous bits in the digital signals received in a certain period of time to the total number of received digital signals at the same time. Data security bit is actually the average of many statistical results, is an average error bit rate. The higher the value of security bit, the higher the security of teaching resources sharing. The data security bit curve of this paper's method in the application of English reading teaching resources sharing is shown in Fig. 7.

In the experiment, 10 sub blocks are randomly selected in the blockchain. In the process of teaching resource sharing, IHFAR software is used to calculate the data security bits of each sub block, and spreadsheets are used to record the experimental data. The specific data is shown in Table 2.

From the data in Table 2, it can be seen that the three methods show obvious differences in data security bits. Under the application of the design method, the highest level of teaching resource security bits can reach 199.52 bits, the lowest level is 176.59 bits, and the average value of blockchain data security bits is 186.48 bits, indicating that the design method has good security for teaching resource sharing. Under the application of comparison method 1, the highest level of security bits of teaching resources is only 135.42 bits, the lowest level is 115.26 bits, and the average security bit value is 121.46 bits, nearly 61 bits lower than the design method. Under the application of comparison method 2, the highest level of security bits of teaching resources is only 125.32 bits, the lowest level is 101.23 bits, and the average security bit value is 112.41 bits, nearly 64 bits lower than the design method, therefore, the above analysis can prove that the design method in terms of security is better. Combined with the above comparison of

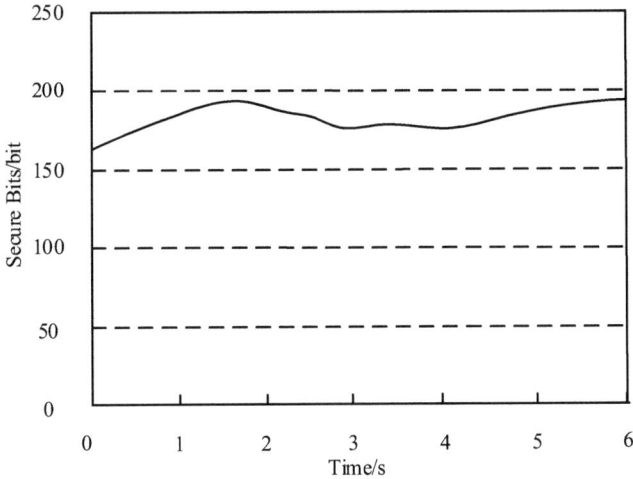

Fig. 7. Data security bit curves in the sharing of resources for teaching English reading

Table 2. Comparison of security bits of blockchain data under the application of three methods (bits)

Subblock number	Design method	Control method 1	Control method 2
1	186.15	124.16	124.15
2	176.59	135.42	111.22
3	185.42	124.17	125.32
4	186.16	112.56	100.24
5	194.25	125.36	102.31
6	195.63	132.12	104.26
7	199.52	133.33	112.32
8	187.52	121.41	112.42
9	186.32	115.26	111.48
10	188.52	123.41	101.23

method performance, it proves that the design method has good applicability in English teaching resource sharing. The English reading teaching resource sharing method based on blockchain technology stores English reading teaching resources through blockchain, and shares resources on this basis. Compared with traditional resource sharing methods, it has considerable advantages in resource security and integrity, and the security performance of English reading teaching resources has been effectively improved and guaranteed.

In order to verify the performance of the method designed in this paper when multi-users share teaching resources concurrently, simulate multi-users sharing teaching resources simultaneously, record the bit error rate, and analyze the influence of multi-users concurrent sharing on the bit error rate.

Table 3. Comparison of multi-user concurrent shared bit error rate (%)

subscriber number	Design method	Control method 1	Control method 2
10	0.50	9.20	6.50
20	0.70	11.00	8.00
30	0.90	13.50	10.00
50	1.20	15.00	12.00
100	2.00	18.00	15.00
200	3.50	22.00	18.00
300	5.00	25.00	21.00
400	6.50	28.00	24.00
500	8.00	30.00	26.00

As can be seen from Table 3, with the increase of the number of users, the bit error rate of the design method gradually increases, but it remains at a low level. In contrast, the error rate of comparison method 1 and comparison method 2 increased more significantly, especially when the number of users reached 500, the error rate of comparison method 1 and comparison method 2 was much higher than that of design method. It shows that the design method in this paper has better performance and scalability when dealing with a large number of users sharing concurrently.

In order to verify the performance of the method designed in this paper when sharing different types of teaching resources, the bit error rate and data security bits are recorded for the shared text, pictures, audio and video teaching resources, and the influence of different resource types on the bit error rate and data security bits is analyzed.

Table 4. Resource Type Diversity Test Data Security Bit Comparison (bits)

Resource type	Design method	Control method 1	Control method 2
text	186.48	121.46	112.41
picture	185.20	119.80	110.50
audio frequency	184.50	118.20	109.00
video	183.70	117.50	108.50

As can be seen from Table 4, the data security bit of the design method on text resources is 186.48 bits, which is 65.02 bits higher than that of the control method 1 and 74.07 bits higher than that of the control method 2. In terms of picture resources, the data security bit of the design method is 185.20 bits, which is 65.40 bits higher than that of the control method 1 and 74.70 bits higher than that of the control method 2. For audio resources, the data security bit of the design method is 184.50 bits, which is 66.30 bits higher than that of the control method 1 and 75.50 bits higher than that of the control method 2. Finally, in terms of video resources, the data security bit of the design method is 183.70 bits, which is 66.20 bits higher than that of the control method 1 and 75.20 bits higher than that of the control method 2. These results show that the design method has consistent and superior data security performance when dealing with different types of teaching resources. Whether it is static text and pictures or dynamic audio and video, the design method can provide higher data security and ensure the integrity and confidentiality of teaching resources in the sharing process.

5 Conclusion

In this paper, the application of blockchain technology in English reading teaching resource sharing is fully studied, which ensures the safe storage of data, improves the enthusiasm of users for sharing, and proves the intellectual property rights of users. It not only improves the integrity and confidentiality of teaching resource sharing, but also shows good performance in multi-user concurrency and different types of resource sharing. These achievements provide a new solution for the sharing of English reading teaching resources and help to promote the development and application of educational technology. Looking ahead, I firmly believe that blockchain technology will play an increasingly important role in the field of English reading teaching resources sharing. We will continue to pay attention to the development trend of technology, constantly optimize and improve the function and performance of the resource sharing platform, and provide more convenient, efficient and safe resource sharing services for English teachers and students. At the same time, we also look forward to cooperation and exchanges with more peers, experts and scholars, and jointly promote the research and application of English reading teaching resource sharing methods based on blockchain technology to achieve more fruitful results.

References

1. Liu, Y., Fang, W., Feng, T., Xi, M.: Blockchain technology adoption and supply chain resilience: exploring the role of transformational supply chain leadership. Suppl. Chain Manag. Int. J. **29**(2), 371–387 (2024)
2. Sun, D., Deng, L., Ying, W.: Can intermediary and disintermediary be compatible in coordination management: affordance, fusion and actualization of Blockchain and conventional systems. Int. J. Emerg. Mark. **19**(3), 582–604 (2024)
3. Jie, Y., Liu, C.Z., Choo, K.K.R., Guo, C.: An incentive compatible ZD strategy-based data sharing model for federated learning: a perspective of iterated prisoner's dilemma. Euro. J. Oper. Res. **315**(2), 764–776 (2024)
4. van Roode, M.Y., et al.: Six dilemmas for stakeholders inherently affecting data sharing during a zoonotic (re-)emerging infectious disease outbreak response. BMC Infect. Dis. **24**(1), 185–185 (2024)

5. Sreenivasan, A., Suresh, M.: Start-up sustainability: does blockchain adoption drives sustainability in start-ups? A systematic literature reviews. Manag. Res. Rev. **47**(3), 390–405 (2024)
6. Sadeghi Mahsa, Mahmoudi Amin, Deng Xiaopeng, Moslemi Naeni Leila.: Enterprise blockchain solutions for vibrant construction ecosystem: grey ordinal priority approach. Grey Syst. Theory Appl. 2024, 14(1), 115–143
7. Luo, Z., Shao, X., Ma, X.: Enhancing learners' performance in contest through knowledge mapping algorithm: the roles of artificial intelligence and Blockchain in scoring and data integrity. JOEUC **36**(1), 1–21 (2024)
8. Amir, F., et al.: A conceptual framework of decentralized Blockchain integrated system based on building information modeling to steering digital administration of disputes in the IPD contracts. Construct. Innov. **24**(1), 384–406 (2024)
9. Abdulrahman, S., Al-Tabtabai, H.: A paradigm shift toward the application of Blockchain in enhancing quality information management. Construct. Innov. **24**(1), 407–424 (2024)
10. Huang, W., Wu, Y.: Simulation of information public key searchable encryption based on cloud computing platform. Comput. Simul. **40**(11), 143–146+383 (2023)

A Hybrid Teaching Resource Sharing Method for Higher Vocational English Based on Cloud Storage

Man Zhan[✉]

Sanya Aviation and Tourism College, Sanya 572000, China
zhanman0909ab@163.com

Abstract. In order to optimize the sharing effect of blended teaching resources of higher vocational English and minimize the loss of such resources, a blended teaching resources sharing method of higher vocational English based on cloud storage is proposed. A blended teaching resources database for higher vocational English is established to collect higher vocational English blended teaching resources from multiple channels. Unified data resource conversion interface to integrate vocational English blended teaching resources in the resource database. Utilizing cloud storage, we constructed a model of sharing higher vocational English blended teaching resources, and integrated and shared higher vocational English blended teaching resources. The experimental results demonstrate that upon implementing this method, there is a notable reduction in the number of lost blended teaching resources for higher vocational English, even as the number of synchronous online students gradually increases, which can effectively solve the problem of lost teaching resources caused by the number of synchronous online students, improve the utilization rate of teaching resources, and ensure the completeness and stability of the sharing of teaching resources.

Keywords: Cloud storage · Higher vocational English · Blended teaching · Resource sharing · Cloud storage

1 Introduction

As a basic course in vocational education, the effectiveness and practicability of English education for higher vocational students hold paramount importance in nurturing their proficiency in English application. Presently, however, English instruction in higher vocational settings confronts numerous challenges. On the one hand, the traditional teaching mode often focuses on the inculcation of knowledge and ignores the subjectivity and practicability of students, resulting in students' low interest in learning and weak application ability. On the other hand, the learning characteristics and needs of higher vocational students are diversified, and it is difficult for a single teaching method to meet the needs of all students [1]. Therefore, higher vocational English teaching needs to explore new teaching modes to better meet the needs of students and the development of the times. Blending teaching, which integrates the strengths of both traditional classroom instruction and online learning, has emerged as a pivotal direction for reforming higher vocational English teaching practices[2]. By integrating online and offline teaching resources, blended teaching realizes the diversification of teaching content and the flexibility of teaching methods. Online, students can learn independently through online platforms and obtain rich teaching resources; offline, teachers can provide personalized guidance and real-time feedback through face-to-face teaching [3]. This teaching mode can not only play the guiding role of teachers, but also stimulate students' interest and initiative in learning.

Cloud storage not only provides huge storage space, but also realizes efficient data sharing and real-time updating, which provides a new path for the sharing of blended teaching resources in higher vocational English [4]. Cloud storage technology, with its high reliability, high scalability and high availability, provides powerful technical support for the sharing of hybrid teaching resources in higher vocational English [5, 6]. Through cloud storage, teaching resources can realize centralized storage, unified management and fast access, which reduces the management difficulty and cost of teaching resources. At the same time, cloud storage can also realize dynamic updating and real-time sharing of teaching resources, ensuring that teachers and students can obtain the latest teaching resources in time, and improving the timeliness and practicability of teaching.

Therefore, the study of blended teaching resources sharing method for higher vocational English, leveraging cloud storage, holds significant theoretical and practical importance for improving the quality of higher vocational English teaching, cultivate students' self-learning ability and recommend the process of education informatization. Through in-depth research and practice, a set of operable and popularized resource sharing scheme will be formed to provide powerful technical support for blended teaching in higher vocational English is crucial. At the same time, this study will also provide reference for resource sharing in other disciplines and promote the development and innovation of education informatization.

2 Design of Blended Teaching Resource Sharing Method for Higher Vocational English

2.1 Establishing a Blended Teaching Resource Base for Higher Vocational English

In order to meet the demand for integrating and sharing of blended teaching resources for higher vocational English, a teaching resource library is established. First of all, we communicate with higher vocational English teachers to understand their specific needs for teaching resources, including the type, quantity, and frequency of updating, etc. We conduct research on students to ensure that the resource library can meet their learning needs. Conduct research on students to understand their learning habits, preferences and needs, so as to ensure that the resource library can meet their learning needs. Collect higher vocational English teaching resources from multiple channels, such as textbooks, courseware, video tutorials, online courses, etc. The collected resources are screened and categorized to ensure the quality and applicability of the resources. Standardize the resources, such as unifying the file format and adjusting the resolution, to ensure the ease of use and compatibility of the resources. On this basis, a blended teaching resource base for higher vocational English is established as shown in Table 1.

Table 1. Higher level English blended teaching resource base

No	Fields	Type	Description
1	Quality education resource coding	Int (10)	Quality education resource coding
2	Resource acquisition channel	Int (10)	Resource acquisition channel
3	Bibliographic teaching resources	Char (20)	Bibliographic teaching resources
4	Courseware	Char (10)	Courseware
5	Real case	Char (20)	Real case
6	Online course	Char (50)	Online course
7	Frequently asked questions	Char (30)	Frequently asked questions
8	Resource directory index	Char (10)	Resource directory index
9	Other	Char (15)	Other

Upload the organized resources to the resource library and set appropriate titles, descriptions and labels for each resource. Establish resource updating mechanism, regularly update and safeguarding the contents in the resource library, thus ensuring the currency and precision of the resources. Establish resource audit mechanism to audit the uploaded resources to ensure the legality and compliance of the resources. Design resource access methods, such as online browsing, downloading, sharing, etc., to meet the different needs of users. Establish permission control mechanism to set different access rights for different users to ensure the security and controllability of resources [7]. Implement access logs to record users' access behavior and provide basis for resource management and optimization. Establish a user feedback mechanism to collect users' opinions and suggestions on the resource base and provide reference for the optimization of the resource base. Regularly evaluate and optimize the resource library, and adjust the structure and content of the resource library according to user feedback and teaching needs. Cooperate with other educational institutions or platforms to share and introduce high-quality resources, so as to continuously enrich and improve the resource library.

Through the above process, a comprehensive, systematic and efficient blended teaching resource base for higher vocational English can be established.

2.2 Teaching Resource Processing and Integration and Consolidation

Considering the diversity of teaching resources acquired through multiple channels, it is necessary to process and integrate the teaching resources in the resource library after the completion of the establishment of the English hybrid teaching resource library in higher vocational colleges, so as to lay a good foundation for efficient sharing of subsequent resources. First of all, we will evaluate the quality of the blended teaching resources of higher vocational English in the resource library to ensure that their content is accurate, the format is unified, and there is no copyright dispute. Transform the format of unqualified resources, such as converting PPT to PDF, to ensure that all resources can be opened successfully on multiple devices [8]. Adjust the resolution of video, pictures and other resources to ensure that they can be clearly displayed on different devices. Add descriptive metadata for each resource, such as title, author, creation date, etc., to facilitate subsequent management and search. According to the teaching characteristics and resource types of higher vocational English, a reasonable classification system should be established, such as classification by course chapters, resource types, difficulty levels, etc. Add multiple tags to each resource to facilitate users to search for related resources through keywords. On this basis, the labeled teaching resources are processed. In this process, it is necessary to do a good job in the unified processing of the data resource conversion interface, as shown in Fig. 1.

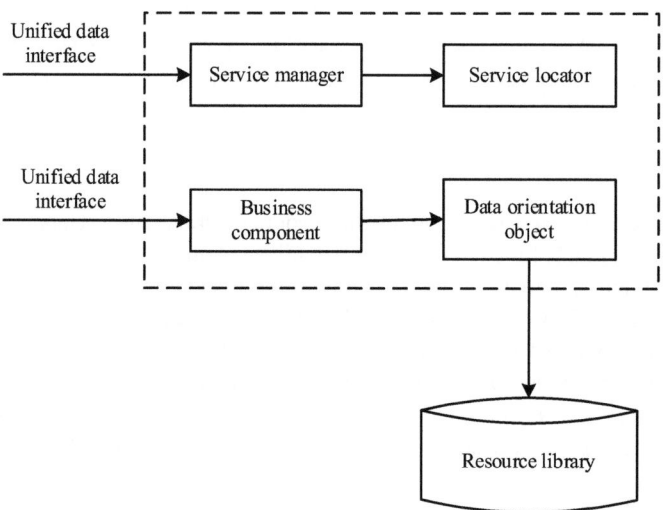

Fig. 1. Harmonized processing of data resource conversion interfaces

2.3 Building a Hybrid Teaching Resource Sharing Model Based on Cloud Storage

After the processing and integration of higher vocational English hybrid teaching resources, cloud storage technology is further utilized to build a teaching resource sharing model. This step is aimed at improving the efficiency of resource utilization and the convenience of teaching. Based on this, a cloud storage platform is used to achieve unified management and sharing of teaching resources. Before constructing the teaching resource sharing model, it is necessary to compare and analyze multiple cloud storage platforms. Stability is one of the most important indicators for evaluating cloud storage platforms, and it is necessary to examine the reliability of the platform in the process of data transmission, storage and access to ensure the integrity and availability of teaching resources. Security is also critical, cloud storage platform should have data encryption, access rights control and other security measures to protect teaching resources from illegal access or tampering. In addition, scalability is also a key factor in evaluating cloud storage platforms. With the continuous increase and updating of teaching resources, the cloud storage platform should be able to flexibly expand storage capacity and performance to meet the changes in teaching needs. Ease of use is related to the convenience of teachers and students in using the cloud storage platform, which should provide an intuitive and friendly interface and a simple and clear operation process to reduce the threshold of use. By comparing the performance of multiple cloud storage platforms in terms of stability, security, scalability and ease of use, we can select the cloud storage platform that best suits the needs of hybrid teaching resources sharing in higher vocational English. The logical architecture is shown in Fig. 2.

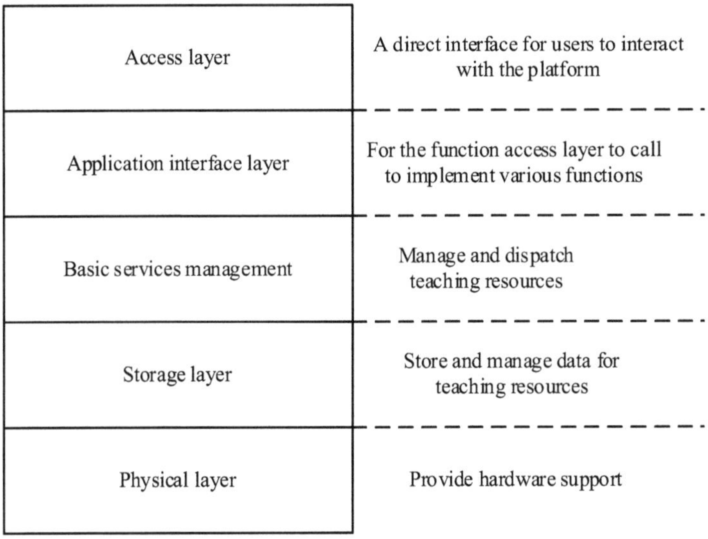

Access layer	A direct interface for users to interact with the platform
Application interface layer	For the function access layer to call to implement various functions
Basic services management	Manage and dispatch teaching resources
Storage layer	Store and manage data for teaching resources
Physical layer	Provide hardware support

Fig. 2. Logical architecture of cloud storage platform

As shown in Fig. 2, the cloud storage platform integrates data storage, basic service management, application program interface, multi-access and other functional modules. On this basis, the resource sharing model is designed, and its operation flow is shown in Fig. 3.

According to the process shown in Fig. 3, firstly, a resource sharing request is sent, and according to the resource type, subject area, difficulty level and other factors, an all-round search is carried out in the constructed higher vocational English blended teaching resource library, from which the required resource features are extracted and indexes are set up. Subsequently, the searched resources are processed and integrated. The similarity of teaching resources is calculated and the semantics of heterogeneous resources are identified to maximize the sharing of teaching resources [9]. Optimize the model according to the actual needs of the teaching process to improve the efficiency and

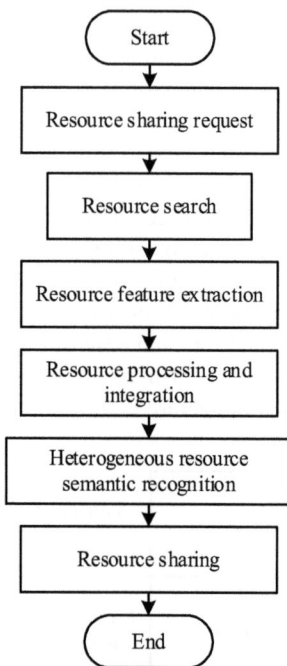

Fig. 3. Instructional Resource Sharing Model Operation Process

effectiveness of resource sharing. Organize the collected teaching resources, including format conversion and metadata addition. Upload the organized resources to the cloud storage platform to ensure the stability and security of the upload process. On this basis, set the sharing privileges of the resources, such as public, private, and specific user sharing. Implement access control mechanisms to ensure that only users with appropriate permissions can access and download resources. Regularly update teaching resources, including new resources and old resources. Regular maintenance of resources to ensure the integrity and availability of resources.

Through the above process, an efficient, convenient and safe hybrid teaching resource sharing model based on cloud storage can be constructed for higher vocational English teaching.

2.4 Heterogeneous Resource Semantic Recognition

From the construction of the above teaching resource sharing model, it can be seen that the semantic identification of heterogeneous resources is the core of sharing model realization. Therefore, the realization of teaching sharing requires heterogeneous semantic identification of resources, and the process of semantic identification, rules, and the calculation method of conceptual similarity of hybrid teaching resource sharing are studied separately. The process of heterogeneous resource semantic identification is as follows:

(1) By means of segmentation and other means, extract the characteristic items of knowledge points of two resource documents from the educational resource website,

and record them as resource document knowledge point 1 and resource document knowledge point 2.
(2) The recorded knowledge points will be categorized or not according to the classification criteria of the resource knowledge points and the shared knowledge base selection.
(3) Use the rules inherent in the inference to obtain heterogeneous results. If the relationship cannot be inferred at this point, it needs to be re-inferred using the customized rules of the knowledge point [10].
(4) The implied relationship diagram based on this design is shown in Fig. 4.

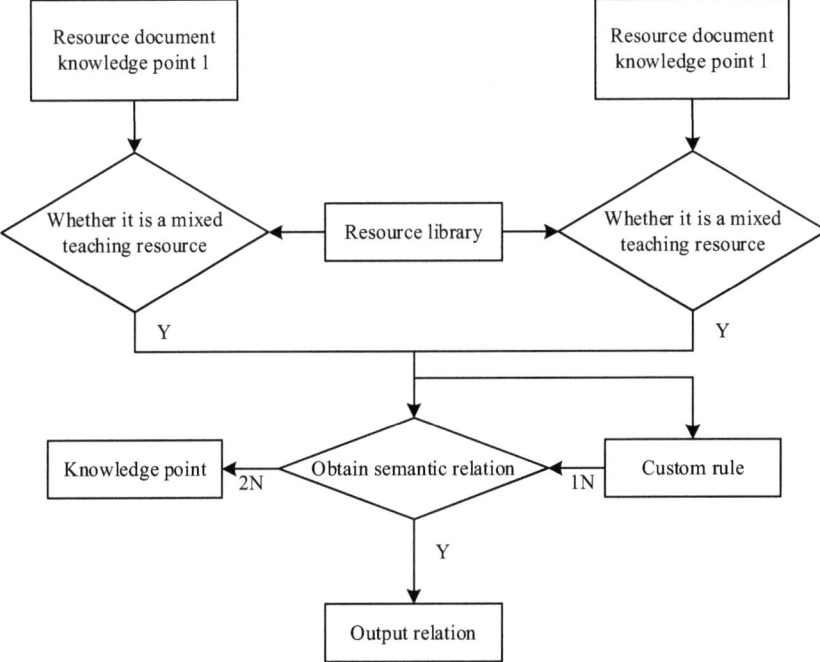

Fig. 4. Heterogeneous resource semantic identification implicit relationship map

As shown in Fig. 4, the semantics designed in this paper for recognizing heterogeneous resources is formulated according to the nature of resources, which can be realized by adding implicit relations. If it is still impossible to reason after resource heterogeneity, new conceptual knowledge points can be extended according to customized rules to get a set of related concepts.

The design of sharing technology implies that there is an inseparable connection between two knowledge points of feature items, that is to say, each knowledge point in the set of knowledge points is directly related to the feature items, and the incorporation of artificial intelligence will maximize the discrimination of similar features, and then calculate their semantic similarity to get the most similar semantics to achieve the purpose of resource sharing. The semantic similarity is calculated within a certain threshold in order to obtain the relationship between the feature items of the shared resources and the related knowledge points, and the formula is as follows:

$$Sim(C_1, C_2) = \frac{|Count[U(C_1) \cap U(C_2)]|}{|Count[U(C_1) \cup U(C_2)]|} \tag{1}$$

Among them, C_1, C_2 denote the related concept nodes respectively. Through this formula, the semantic similarity of higher vocational English blended teaching resources is derived to strengthen the interaction between semantics. Based on the similarity value, a word neighbor matrix of teaching resources is established, as shown in Table 2.

Table 2. Neighborhood matrix of teaching resource words

Word	W_1	W_2	...	W_i	...	W_n
W_1	1	-	-	-	-	-
W_2	-	1	-	-	-	-
...	-	-	-	-	-	-
W_i	-	-	-	$L(W_i,W_j)$	-	-
...	-	-	-	1	-	-
W_n	-	-	-		-	-

In Table 2, W_i, W_j denote the dynamic name and static name of teaching resources respectively. On the basis of satisfying the relationship of the adjacency matrix of teaching resource words, the characteristics of the vocabulary are set to be φ, when $L(W_i, W_j)$ has a value greater than or equal to the vocabulary's feature φ, it indicates that the candidate words play a major role in the characterization of teaching resources. All the condition sets that satisfy the $L(W_1, W_2) \geq \varphi$ of W_j being used as the next notice, the W_i. The primary content and terms expressed in the instructional resources are considered featured words F. The selection of the name of the feature was guided by two points. First, if there is a high-level connection between the pedagogical resource and the lexicon under discussion, it is considered a strong expressive force and can be selected as a featured sign. Second, in the case of a high degree of lexical characterization, the featured word F and the records related to teaching resources are also very high. Through the above process, the goal of semantic recognition of higher vocational English hybrid heterogeneous teaching resources is realized.

2.5 Sharing of Teaching and Learning Resources

The premise of resource sharing is to ensure the reliability of communication connection. In order to meet the demand in this regard, we should integrate and share high-quality teaching resources. First, analyze the resource demand in the teaching process, determine which teaching resources need to be shared, as well as the goal and scope of sharing. According to the results of demand analysis, specific goals of teaching resource sharing are set, such as improving resource utilization, promoting collaboration among teachers, and facilitating students' learning. Through the data analysis tool of the cloud storage platform, analyze the use of teaching resources, including the number of visits, downloads, frequency of use, etc. In order to improve the sharing effect of teaching resources, the distributed computing model MapReduce is introduced, as shown in Fig. 5.

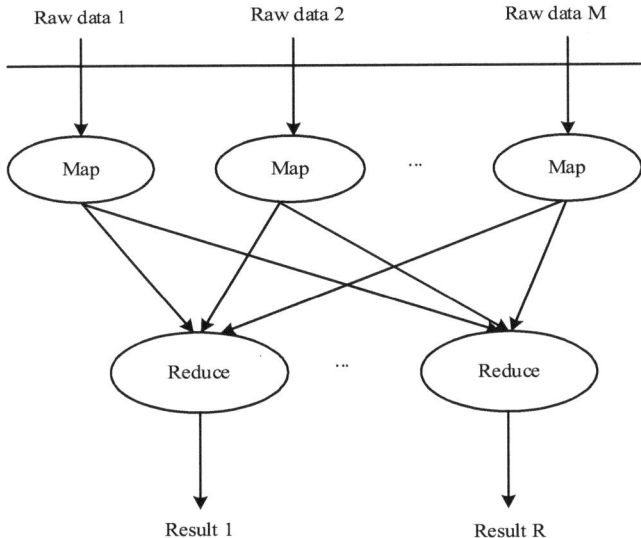

Fig. 5. MapReduce Model

MapReduce is a computing model that uses the underlying distributed computing environment for parallel processing. Its computing process is divided into two stages: Map and Reduce. Map decomposes large computing tasks into multiple smaller computing tasks, and then allocates multiple smaller computing tasks to multiple computer nodes in the cluster for parallel computing processing. Reduce then merges the completed multiple smaller computing tasks, and finally completes the computing tasks through multiple iterations of splitting, merging, and merging. The basic data unit for MapReduce to process data is Key Value (key value/value pair). Its data type can be a complex data structure, but the key value pair must be maintained. The key value/value pair is used as input and output in both Map and Reduce stages. The MapReduce framework process is a process of calling mapper function and reducer function. The mapper process involves dividing the input data of a task into smaller, fixed-size fragments, and then further decompose each task fragment into a batch of key value pairs $<$ Key1, Value1 $>$. The mapper process is to calculate the $<$ Key1, Value1 $>$ pairs in the fragment as input to obtain intermediate results $<$ Key2, Value2 $>$, sort the intermediate results, generate a new list, and then reduce. On this basis, teaching resources are integrated and shared. The expression is as follows:

$$F_i = \frac{1}{x^2}(1-S)D_iH_i \qquad (2)$$

Among them, F_i represents the integration and sharing of class i high-quality teaching resources. x indicates the transmission link. D indicates a shared computer instruction. H represents the bandwidth capacity for the transmission of network resources. In order to improve the adaptability of shared resources and students' needs. According to the analysis results, we adjust the resource classification, optimize the resource searching method, and increase the popular resources. According to the optimization strategy, the sharing platform is adjusted and improved to enhance the effectiveness and efficiency of resource sharing.

3 Test Analysis

3.1 Test Preparation

In the process of exploring the method of sharing English teaching resources in higher vocational colleges based on cloud storage, it is very important to plan and build an efficient and stable experimental environment to ensure that the feasibility and performance of this method can be comprehensively evaluated. These tools and environments not only need to support the effective management, storage and sharing of teaching resources, but also ensure the performance and stability of the system. The experimental development software and tool software needed for the sharing method of higher vocational English mixed teaching resources based on cloud storage are shown in Table 3. At the same time, in order to ensure the accuracy and reliability of the experimental results, a high-performance hardware environment is adopted as the support. The hardware environment required for the experimental test is: wave server, model NF5270M4, CPU E5-2650V3 (2.3 GHz/10c); Memory: DDR 4 1333 MHz 256 GB;; System CentOS6. It provides a solid hardware guarantee for data processing and concurrent access.

Table 3. Experiment development software and tool software models

No	Software	Type
1	Virtualization software	VMware ESXi5.5
2	Linux network operating system	CentOS 6.9
3	Virtualization software customer tool software	VMware vSphere client5.5
4	Terminal connection software tool	PieTTY 0.3.26
5	Real-time processing tool	Spark-1.6.3
6	Distributed coordinator	Zookeeper-3.4.6
7	Distributed database	Hbase-1.0.0
8	Compiler tool	Jdk-8u40-linux
9	Platform development tool	MyEclipse10.0

Select a stable and secure cloud storage platform and make necessary configuration and settings to ensure that teaching resources can be uploaded, stored and shared smoothly. The IP address division and resource allocation of the cloud storage platform server are shown in Table 4.

Table 4. Distribution of service IP addresses used by the system

No	Host name	Role in a Hadoop cluster	IP
1	ESXi	Provide web hosting	10.0.173.1/24
2	Namenode A	Master namenode node	10.0.173.9/24
3	Datanode A	Slaves datanode node hbase master	10.0.173.8/24
4	Datanode B	Slaves datanode node hbase slave	10.0.173.7/24
5	Datanode C	Slaves datanode node hbase slave	10.0.173.6/24
6	Datanode D	Slaves datanode node hbase slave	10.0.173.5/24
7	dataServer	Mysql server	10.0.173.4/24
8	WebServer	Web Server Mysql	10.0.173.10/24
9	Zookeeper A	Distributed application coordination service	10.0.173.11/24
10	Zookeeper B	Distributed application coordination service	10.0.173.12/24
11	Zookeeper C	Distributed application coordination service	10.0.173.13/24
12	Hive	Hive data warehouse server	10.0.173.14/24

The underlying platform of the hybrid teaching resource sharing system for higher vocational English based on cloud storage is deployed as a distributed system. Three higher vocational English classes are selected as experimental objects to ensure that the number of students, learning background and teaching progress of the two classes are basically the same. Prepare rich and diversified resources for teaching English at higher vocational level, including courseware, videos, audios, practice questions, etc., and ensure the quality and adaptability of these resources. Ensure that all three classes have the hardware and software environments for blended teaching, such as computers, projectors, and networks.

3.2 Analysis of the Effects of Sharing

The key to solving the problem of resource sharing in blended English teaching in vocational colleges is to ensure the integrity and stability of teaching resources. The loss of teaching resources not only affects students' learning experience, but also may have a negative impact on teaching quality. Therefore, this comparison process aims to evaluate the advantages of the hybrid teaching resource sharing method based on cloud storage in terms of stability and reliability of teaching resources by comparing the number of lost teaching resources in the experimental group and the control group. Prepared teaching resources are uploaded to the cloud storage platform, and appropriate permissions and access methods are set to ensure that students can easily access and obtain these resources. Collect students' learning data, including learning hours, learning resources access, learning performance and so on. In the experiment, control group 1 and control group 2 used the traditional teaching resource sharing method 1 and sharing method 2 respectively, while the experimental group utilized the teaching resource sharing method proposed in this paper. Three methods, we have successfully integrated and shared the mixed teaching resources for higher vocational English. Throughout the experimental phase, the loss of teaching resources within both the experimental and control groups was meticulously tracked and documented.

After completing the resource sharing, the background data in the experiment were called up in the computer terminal to analyze whether there is any shared data loss phenomenon during the application of the three methods. The number of online students in the test terminal was taken as a variable to count the number of lost higher vocational English blended teaching resources in the sharing, and the results are shown in Table 5.

Table 5. Comparative results of the number of teaching resources lost during the sharing process

Sync online student numbers	The method of this article loses quantity/item	Conventional method 1 number of lost items	Conventional method 2 number of lost items
20	0	1	2
40	0	3	7
60	2	24	26
80	4	25	26
100	5	27	29

From the comparison results in Table 5, it can be seen that, in the case of the gradual increase in the number of synchronous online students, after applying the sharing method proposed in this paper, the number of higher vocational English blended teaching resources lost is significantly smaller than that of the other two control groups, with a maximum of no more than five items. This experimental result shows that the designed sharing method can effectively reduce the number of lost teaching resources, can effectively solve the problem of lost teaching resources due to the excessive number of synchronous online students, to uphold the the integrity and stability of teaching resources, ensuring a more consistent and dependable learning experience for students.

To further validate the effectiveness of the proposed sharing method, a test was conducted to assess the utilization rate of blended teaching resources for higher vocational English. The formula for calculating the utilization rate of teaching resources is shown as follows:

$$Q = \frac{W_s}{W} \times 100\% \qquad (3)$$

Among them, Q indicates the resource utilization rate. W_s indicates the number of actual visits. W represents the total resources. This indicator can reflect the utilization of teaching resources by students. During the experiment, the resource utilization rates of the three methods were tracked and recorded in detail. First of all, we counted the number of visits to online teaching resources of students in the hybrid teaching group, including the number of videos viewed, the number of documents downloaded, and the number of online tests completed. At the same time, it also records the frequency of traditional teaching group students using textbooks, PPT and other resources. Sort out and analyze the collected data, calculate the specific value of resource utilization ratio, and compare the differences in resource utilization ratio of the three methods. The results are shown in Fig. 6.

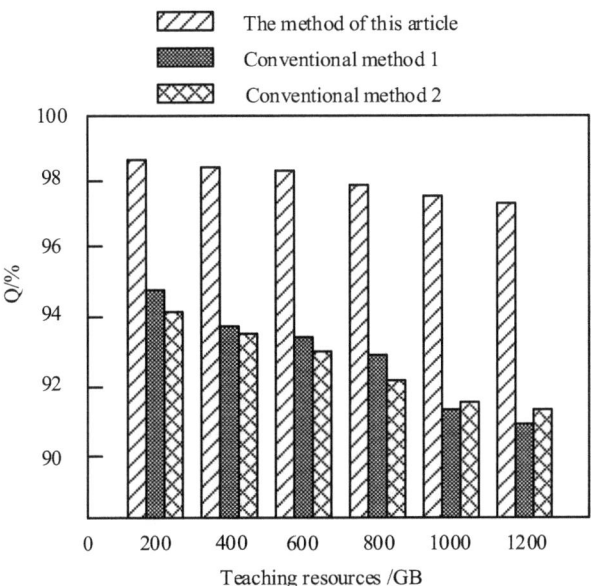

Fig. 6. Comparative results of the utilization of blended learning resources

It can be seen from the comparison results in Fig. 6 that the implementation of the sharing method introduced in this paper results in a notably higher resource utilization rate within the hybrid teaching group, in contrast to the traditional teaching group. This result further verifies the advantages of hybrid teaching in promoting the utilization of student resources, and can effectively improve the utilization of teaching resources. The richness and convenience of online teaching resources make it easier for students to access and use resources, thus improving their learning effects. Therefore, it is suggested to actively promote and apply the hybrid teaching resource sharing method in English teaching to optimize the allocation of teaching resources.

In order to further verify the stability of this method in application, the loading speed of teaching resources including videos and documents under different sharing methods is measured. Using automatic script timing method, the loading time of teaching resources in three different sharing methods under each network environment is recorded. In order to ensure the accuracy of the data, repeat the test at least three times and take the average as the final data. The loading time results of teaching resources in three different sharing methods are shown in Table 6 below.

Table 6. Loading times of teaching resources for different shared methods

network environment	Sharing method	video/s	document/ s
campus network	The method of this paper	2.8	1.1
	Sharing method 1	4.5	2.2
	Sharing method 2	4.0	1.8

As shown in Table 6, in the campus network environment, the method in this paper is superior to the sharing method 1 and method 2 in the loading time of videos and documents. The advantage of this method is more prominent. The loading time of video and document is 2.8 s and 1.1 s. In contrast, the loading time of sharing method 1 is 4.5 s and 2.2 s, while that of sharing method 2 is 4.0 s and 1.8 s. This shows that this method can provide faster and more stable resource sharing service in the campus network environment. The results show that the efficient cloud storage sharing method is more important, especially in the educational environment that needs fast access to teaching resources, that is, the mixed teaching scene of higher vocational English. This method can significantly improve the access speed of teaching resources, thus optimizing the learning experience and improving teaching efficiency.

In order to verify the performance of the teaching resource database in this method under the condition of high concurrency. Constructing a simulated platform for sharing English mixed teaching resources in higher vocational colleges. Simulate 100, 200, 500 and 1000 concurrent users, access the teaching resource database at the same time, and record the system response time. Monitor the system resource occupation of teaching resource pool under different concurrent users. The experimental results are shown in Table 7.

Table 7. Performance results of the teaching resource base

Number of concurrent users	Average response time / s	CPU occupancy rate /%	Memory footprint / MB	network /KB/s
100	1.2	45	600	500
200	2.0	60	800	900
500	3.5	75	1200	1500
1000	5.8	85	1800	2000

As shown in Table 7, the teaching resource database can effectively handle concurrent requests. However, with the increase of users, although the system response time is slightly prolonged and the CPU and memory resources are significantly increased, the system can still maintain stable operation and show good scalability. The experimental data show that the resource library has the basic processing ability and the ability to meet the teaching needs when dealing with high concurrent access.

4 Conclusion

With the rapid development of information technology and the deepening of education reform, the sharing method of mixed English teaching resources in higher vocational colleges based on cloud storage has become an important means to improve teaching quality and promote the cultivation of students' autonomous learning ability. After a period of research and practice, the advantages and application strategies of cloud storage in teaching resource sharing are deeply explored, and a series of positive results have been achieved. This paper provides an innovative solution for English teaching in higher vocational colleges by establishing an efficient teaching resource database, processing and integrating teaching resources, constructing a mixed teaching resource sharing model, realizing semantic recognition of heterogeneous resources and analyzing the sharing effect. This scheme not only improves the utilization rate and access speed of teaching resources, but also ensures the integrity and stability of resources, thus optimizing the learning experience and improving the teaching efficiency.

Looking ahead, the application of cloud storage technology in the field of education is expected to become more extensive and profound. We expect that through continuous research and practice, we can further strengthen and improve the sharing methods of blended teaching resources for vocational English by utilizing the capabilities of cloud storage, so that it can play a more important role in the field of education, with the goal of cultivating high-quality and highly skilled talents, we recommend the reform, innovation and development of English teaching in vocational colleges.

References

1. Zhu, X.: Study on regional digital teaching resource sharing platform based on internet of things and big data. Int. J. Ind. Syst. Eng. **44**(4), 458–474 (2023)
2. Wu, X.: Research on the reform of ideological and political teaching evaluation method of college English course based on online and offline teaching. J. High. Educ. Res. **3**(1), 87–90 (2022)
3. Pérez-Paredes, P.: Review of Viana (2022), teaching English with corpora: a resource book. Int. J. Corpus Linguist. **29**(1), 116–122 (2024)
4. Lu, J., Gao, H.: Online teaching wireless video stream resource dynamic allocation method considering node ability. Sci. Programm. **2022**, 1–8 (2022)
5. Yan, P., Zhou, L., Yan, H.: Research on privacy data protection model of internet of things under hybrid cloud storage. Comput. Simul. **40**(02), 530–534 (2023)
6. Liu, M., Li, Q.: Intelligent integration method of AI English teaching resource information under multi-agent collaboration. Adv. Multimedia **2022**(3), 1–11 (2022)
7. Gong, Y.: A transformation of oral English teaching method in the network environment. Sci. Program. **2022**(5), 1–10 (2022)
8. Li, B.: Design and research of computer-aided English teaching methods. Int. J. Hum. Rob. **20**(02n03), 2240004 (2023)
9. Yang, Y.: Machine learning for English teaching: a novel evaluation method. Int. J. Comput. Appl. Technol. **71**(3), 258–264 (2023)
10. Xia, T., Ahmad, M.T., Jan, N.: Method of ideological and political teaching resources in universities based on school-enterprise cooperation mode. Math. Probl. Eng. Theory Methods Appl. **2022**(8), 1–9 (2022)

A Personalized and Accurate Push Method for Online Teaching Resources Based on Social Media Information Integration

Jianhua Jiang[✉] and Ziyu Ai

Computer Engineering Technical College, Guangdong Polytechnic of Science and Technology, Zhuhai 519090, China
Jiangjh0034@163.com

Abstract. A study aimed at improving learners' efficiency, satisfaction, and the effective sharing of high-quality educational materials through personalized and precise delivery of online resources is underway. The research leverages social media data integration to achieve these goals. Initially, a learner profile is developed by amalgamating social media insights, which includes learners' fundamental details, areas of interest, and favored subjects. Subsequently, educational content is curated according to these preferences, ensuring the recommended resources are tailored to individual needs and interests. Lastly, a customized and precise recommendation system for online educational resources is devised by analyzing user behavior patterns, personal data, and social network attributes. The results demonstrate that this method achieves a personalized recommendation accuracy rate of over 98%. As a result, it effectively furnishes users with customized and precise educational resource suggestions based on their multifaceted attributes.

Keywords: Social media information convergence · Online teaching resources · Personalization · Precision push

1 Introduction

The proliferation of information technology and the ubiquitous use of social media have rendered online educational resources an indispensable element of contemporary education and self-directed learning. In the age of big data and artificial intelligence [1], pinpointing content that aligns with individual requirements and learning traits from a vast array of educational materials and ensuring personalized and precise delivery has emerged as a formidable and extensively studied issue within the educational domain. Social media information fusion technology, serving as a pivotal platform for information acquisition, exchange, and interaction, presents new opportunities for the personalized and precise delivery of online teaching resources. In recent years, there has been a significant enhancement in both the quantity and quality of online educational materials. Nevertheless, effectively harnessing these resources to cater to the individualized requirements of a diverse learner population continues to pose a challenge within the educational realm.

Currently, numerous scholars and research institutions both at home and abroad have focused on the issue of personalized and accurate recommendation of online teaching resources, and have put forward a variety of solutions. Among these, the recommendation method proposed in reference [2] which takes node capacity into account, the resource recommendation method based on the online teaching model proposed in reference [3], the resource recommendation method based on optimized allocation proposed in reference [4], and the resource recommendation method based on the school - enterprise cooperation model proposed in reference [5] are relatively common approaches. However, these methods still have certain limitations and challenges in practical application, such as data sparsity, cold-start problem, and the accuracy and interpretability of the recommendation results, which make it difficult to meet learners' personalized needs. Meanwhile, as the number of social media users keeps growing and their activities keep increasing, social media platforms have amassed a substantial amount of user - behavior data and preference information, offering a bountiful data source for personalized and accurate recommendation. With the continuous progress of artificial intelligence and big - data technology, research on personalized and accurate online teaching resources will exhibit the following trends: Firstly, data sources will become more abundant and diverse, encompassing user - behavior data, social - relationship data, sentiment - analysis data, etc. Secondly, the recommendation algorithm will be more intelligent and accurate, being able to better capture users' personalized needs, interests, and preferences. Thirdly, the recommendation results will place more emphasis on interpretability and user satisfaction.

Social media information fusion involves consolidating and integrating data from diverse social media platforms to create a more comprehensive and valuable information repository, which can improve the shortcomings of the traditional online teaching resources pushing method. Therefore, this paper proposes a personalized and accurate push method of online teaching resources based on social media information fusion, aiming to use social media information fusion technology to realize personalized and accurate push of online teaching resources by mining and analyzing users' behavioral data and preference information on social media. This endeavor not only aims to enhance the learning efficiency and satisfaction of learners but also seeks to foster the effective utilization and sharing of high-quality teaching resources. By delving into the fusion of social media information technology and conducting research and experimental validation of personalized and precise push algorithms, we anticipate offering an effective and feasible solution to the educational sector.

2 Online Teaching Resources Personalized and Accurate Push

2.1 The Construction of a Learner Model Based on the Integration of Social Media Information

The values of the learner model are usually obtained through the learner analysis module. In the past, the teacher could only obtain the learner's characteristics in the whole learning process by subjective judgment. However, in today's data explosion environment, the human-computer interaction between the learner and the system allows the computer to mine and process the learner's behavioral data only through the learner's learning process

[6–8]. Therefore, the analysis of the learner must adhere to established rules, guided by the principles of scientific data mining and behavioral analysis. This necessitates the development of a module designed for the effective storage and extraction of behavioral data information. However, the collection and processing of this information represent a substantial and comprehensive undertaking. It is only through the rigorous screening, processing, analysis, and summarization of data, including those instances that may initially appear illogical, that the final learner model can be constructed [9, 10]. Based on the above analysis, we can get the general pattern diagram in the overall learner model as shown in Fig. 1.

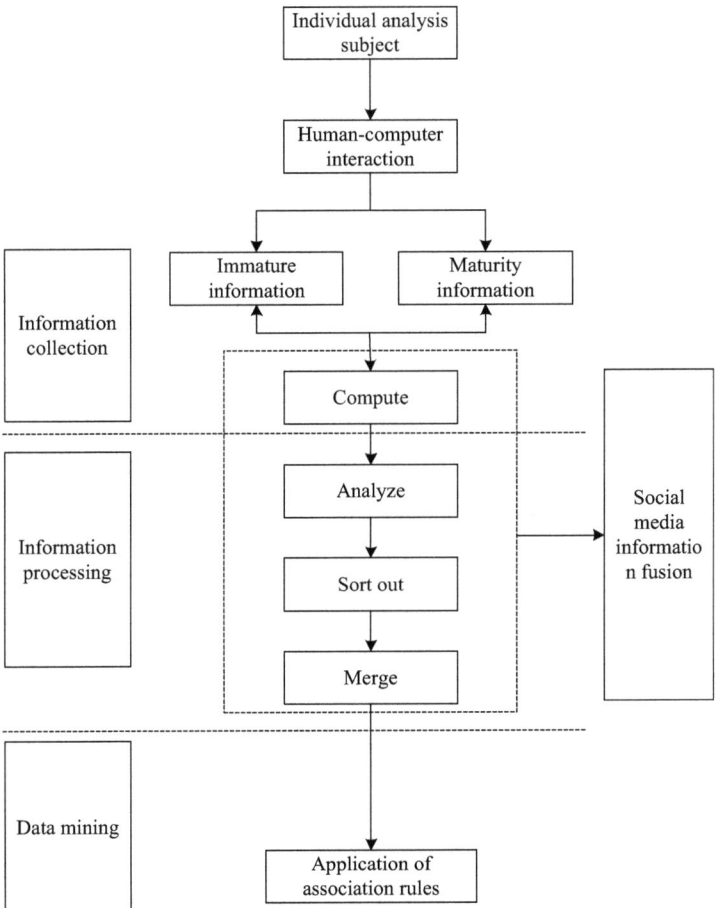

Fig. 1. Schematic diagram of the learner model

Based on the learner model, social media information fusion was utilized to capture learner information, which consisted of four main components.

(1) Basic personal information of the learner, which will be used for the management of the student in the system and is not directly related to the creation of the learner

model. For example, specific information such as name, gender, and ethnicity. This information is basic and necessary to differentiate and categorize learners, and is the basis for initial learner modeling and initial resource delivery. The personal information can be supplemented with information about the student's ethnicity, background culture, customs, and so on. This personal information is clearly directed and can also rely on direct feedback from the learner, i.e., through enrolment information.

(2) Learners' data related to their learning status, such as courses, school registration information, grades, classes, grade rankings, and so on. The basic information here can be collected through the entrance collection before user login or direct information feedback.

(3) Learner's learning style information. It mainly contains: learning attitude, learning motivation and so on. Learning preference information mainly contains interest preference and resource type preference (e.g., text, chart, video, etc.); learning strategy preference (teacher's lecture, teamwork, innovation, etc.); learning time preference (i.e., learning time period); system function preference, i.e., what is the demand for the system's function and interface, etc. The aforementioned information can be conveyed through questionnaires or directly by the learners themselves. Concurrently, this information serves as the foundational basis for the initial characterization of learning styles. Similarly, it can be obtained through learner-completed questionnaires or direct self-reporting, and it forms the essential basis for the preliminary construction of the learner model.

(4) The level of knowledge and cognitive ability level test that learners already have. Generally obtained through questionnaires and cognitive ability questionnaires, in the questionnaire, you can learn to grasp the knowledge is divided into five grades, respectively: do not know at all, know part of the knowledge, know most of the knowledge, familiarity and knowledge transfer, etc., the learner to choose the option that best meets their own cognitive level can be. On this basis, in order to better assess students' mastery of knowledge, generally through the conduct of multiple tests to synthesize the assessment. The specific calculation formula is as follows:

$$\bar{c} = \sum_{i=1}^{m} w_i c_i \qquad (1)$$

Among them, $\bar{c}$ denotes the weighted sum of the scores of multiple tests. c_i indicates the i results of the following tests. w_i indicates discipline test the i proportion of the test that is taken by the learner is the same as the proportion of the test that is taken by the test itself. After obtaining the above information, the acquired learner information is analyzed and processed, and a preliminary learner model is constructed based on a certain method of summarization. This usually involves clustering or classifying user behavioral data to identify the user's topics of interest or areas of preference [11–13].

2.2 Filtering Information Based on Learner Preferences

User preference-based information filtering consists of a user preference module, an information processing/filtering module, and a user requirement module. The user preference module records the user's preference information in the registration information, the user requirement module records, extracts and updates the user's inquiry behavior characteristics, and the information processing/filtering component completes the filtering of the original information. Figure 2 shows the flow of the information filtering process based on the learner's preference designed in this paper.

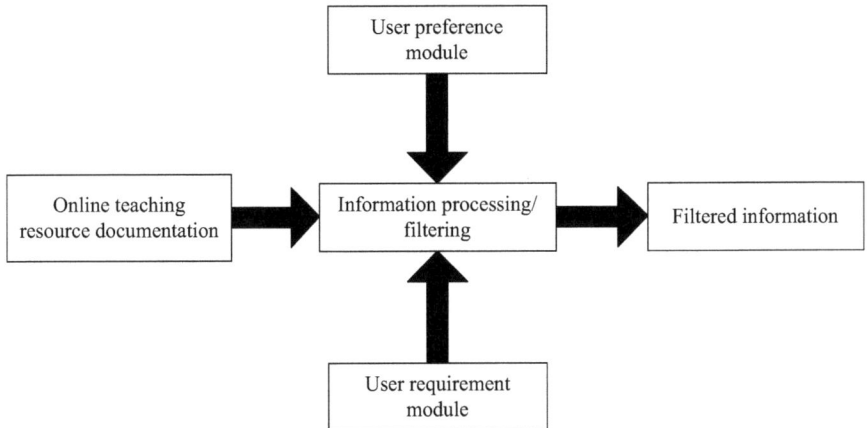

Fig. 2. Information filtering based on user preferences

The merit of this type of information filtering lies in the fact that the overall user preference information [14, 15] can be regarded as the superposition and combination of individual filtering behaviors. This facilitates problem simplification and future maintenance and upgrades. In information filtering, each user's online learning interest model can be represented by the following set of feature items.

$$S = S(T_1, W_1; T_2, W_2; ...; T_n, W_n) \tag{2}$$

Among them, S denotes an online teaching resource text. T_n denotes the feature term, i.e., the text of an online teaching resource S the basic linguistic unit in the text and capable of representing the content of the text. W_n denotes weight of T_n. During the information - filtering process, the user - interest model has an adaptive feedback evaluation mechanism. Specifically, in the information - filtering process, the user - interest model is equipped with an adaptive feedback evaluation mechanism. That is to say, the user - interest model will analyze user behaviors along with the operation of information resources and adjust its own interest model by modifying the weights of feature items. The formula for adjusting the weights of feature items is as follows.

$$W_i = W_i + B(T_i) \tag{3}$$

Among them, B represents the behavioral analysis function. Drawing upon the user's interaction patterns with information resources, a value spectrum spanning from -1 to 1 is established. Subsequently, acknowledging the unique blend of group mentality and knowledge foundations inherent in various college and university communities, the information resources are categorized into distinct channels, tailored to disciplines.

Within each channel, information is filtered in accordance with the individualized interest profiles of learners. In this way, we can avoid the large amount of information and lack of knowledge being pushed to learners. D_1 denotes a vector of interest models for a particular user, the D_2 denotes the feature vector of a certain online teaching resource document to be compared, then the formula for measuring the similarity between the user interest and the online teaching resource document can be expressed as follows.

$$Sim(D_1, D_2) = \cos \theta = \frac{\sum_{i=1}^{n} W_{1k} W_{2k}}{\sqrt{\sum_{i=1}^{n} W_{1k}^2 \sum_{i=1}^{n} W_{2k}^2}} \quad (4)$$

Among them, W_{1k}, W_{2k} denote weights of the individual feature terms for the set of feature terms D_1, D_2 of section k; the θ denotes the angle between the vectors. The formula shows that the greater the cosine of the angle between the vectors, the greater the similarity between them. A threshold value is set for the similarity, greater than this value it is considered that D_2 the information resources represented meet the characteristics of the learner's information needs, i.e., the information is filtered out and sent to the representation layer.

2.3 Personalized and Accurate Push of Online Teaching Resources

Subsequent to the aforementioned filtering of online educational resources tailored to learners' preferences, the subsequent endeavor involves crafting a tailored and precise online resource dissemination strategy. The objective of this refined resource allocation plan is to comprehend the learners' behavioral patterns, leveraging the student-centric model established by the platform, and subsequently pinpoint the educational materials that precisely align with their requirements. Initially, the salient attributes outlined in Table 1 below are extracted from the previously filtered online educational resource pool.

Based on user behavioral characteristics, personal information and social relationship characteristics, a user profile is constructed to comprehensively reflect user interests, needs and behaviors. On this basis, the basic operation principle of online teaching resources personalized and accurate pushing scheme is designed, as shown in Fig. 3.

In Fig. 3, leveraging the amassed data pertaining to students' learning behaviors throughout their academic pursuits, learning analytics technology is employed to gain a comprehensive understanding of their learning landscape and individual traits. This technology then dynamically furnishes students with tailored learning resources that cater to their unique characteristics. The data generated in the process of students' learning include quantifiable data such as the learners' test scores, the cumulative length of online learning, the frequency of knowledge learning, and the differences in performance between different courses. Quantifiable data such as the learner's test scores, cumulative online learning hours, the frequency of learning knowledge points, and the differences in student performance between courses. As well as the types of resources that students have browsed, the questions that learners have done, and the records of learners' wrong

Table 1. Feature extraction

Trait	Specific description
User behavior characteristics	Utilizing engagements like likes, comments, shares, and other behaviors displayed by users on social media platforms, their individual interests and preferences can be extracted
Characteristics of teaching resources	Extract text information such as title, description, label and keywords of teaching resources, as well as statistics such as click-through rate and completion rate of resources
Social relationship characteristics	Analyze users' attention, fans, friends and other relationships on social media, and extract the characteristics of social relationships

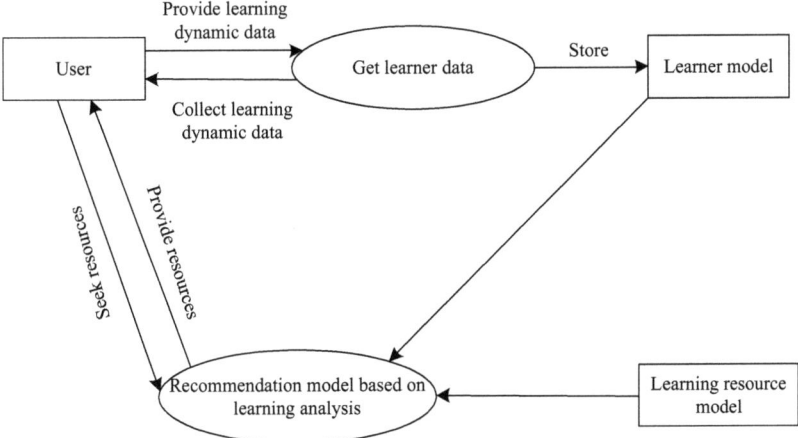

Fig. 3. The basic operation principle of personalized and accurate push of online teaching resources

questions. The resources pushed by this program include videos of teachers' explanations, exercises, cases, and extended knowledge. As shown in Fig. 4, the personalized and accurate push contains three levels of structure.

The hierarchical framework depicted in Fig. 4 facilitates the realization of a tailored and precise dissemination of online educational resources. Leveraging the user's profile and the unique attributes of the teaching materials, the pre-trained model generates a personalized compilation of recommended educational resources. These recommendations are then disseminated to users via various channels such as email, SMS, and app notifications. The recommendations are prominently showcased on the app or webpage interface, accompanied by concise resource details and intuitive operational controls. By incorporating user feedback and behavioral data as fresh training inputs, we continuously refine user profiles and pushing strategies, thereby optimizing the recommendation

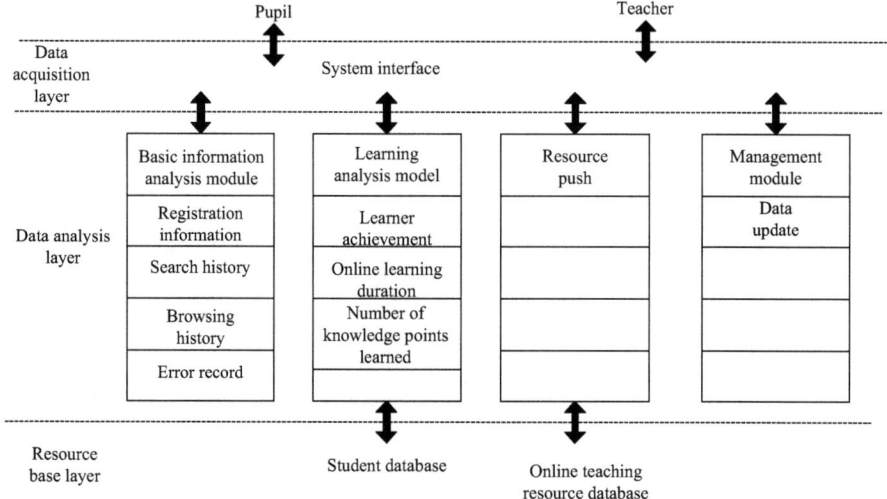

Fig. 4. Personalized and accurate push hierarchy

algorithms. This iterative process culminates in the creation of a dynamic, personalized, and accurate push model for online educational resources, as illustrated in Fig. 5.

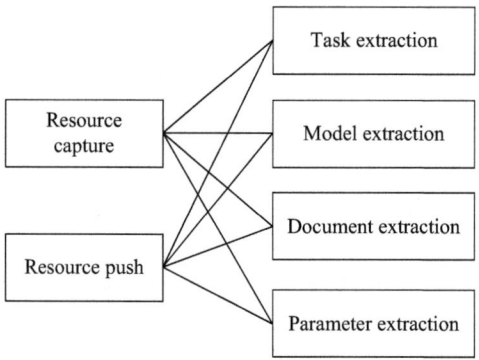

Fig. 5. Personalized and accurate push model of online teaching resources

The push model incorporates software integration techniques to promptly gather students' learning data, construct learning contexts, and subsequently transmit pertinent elements such as tasks, models, documentation, and parameters to the resource dissemination hub.

The aforementioned process enables the delivery of tailored and precise online educational resources, factoring in the multi-faceted attributes of the user, thereby enhancing the personalization and accuracy of teaching resource recommendation services offered to users.

3 Test Analysis

3.1 Test Preparation

Prior to embarking on the experimentation of the personalized and accurate dissemination method for online educational resources grounded in social media information integration, meticulous preparatory measures are indispensable. The primary objective of this experiment is to rigorously assess the efficacy and efficiency of the proposed personalized and precise dissemination approach that harnesses social media information fusion. We must gather a comprehensive array of user behavioral data, preference insights, and online educational resources spanning multiple social media platforms, ensuring the diversity and depth of the data to thoroughly validate the effectiveness of the dissemination method. To guarantee the seamless execution of the experiment, an apt experimental environment must be established and configured. The subsequent table outlines the specific details of the experimental environment's configuration.

Table 2. Configuration of the experimental environment

Disposition	Argument
Server model	Dell PowerEdge R740
CPU	Intel Xeon Gold 630, 2.10ghz, 18 cores, 36 threads
Internal memory	512GB DDR4 ECC 2933MHz
Store	1TB NVMe SSD (system disk) + 4TB SATA HDD (Data disk)
Graphics card	NVIDIA Tesla V100, 32GB GDDR3 video memory
Operating system	Ubuntu 20.04 LTS
Python version	Python 3.8
Deep learning framework	TensorFlow 2.5.0
Machine learning library	scikit-learn 0.24.2
Data processing tool	Pandas 1.2.4, NumPy 1.19.5

Based on the specifications detailed in Table 2, meticulously complete the setup of the experimental environment. Secondly, to streamline the acquisition of experimental data, we leverage Python to craft an efficient Scrapy scraping framework. This framework adeptly extracts pivotal resources from interfaces and compiles the necessary resource package for the experiment's execution. The resource retrieval architecture is illustrated in Fig. 6.

As depicted in Fig. 6, the execution of the Scrapy framework necessitates initially acquiring resource scheduling directives via the crawler for subsequent sorting and categorization. Subsequently, these directives are relayed to the downloader to facilitate the saving of downloaded data. Upon completion, the resource extraction process halts, resulting in the creation of multiple resource packages, each assigned a number proportional to the total quantity of its internal resources, with higher numbers indicating

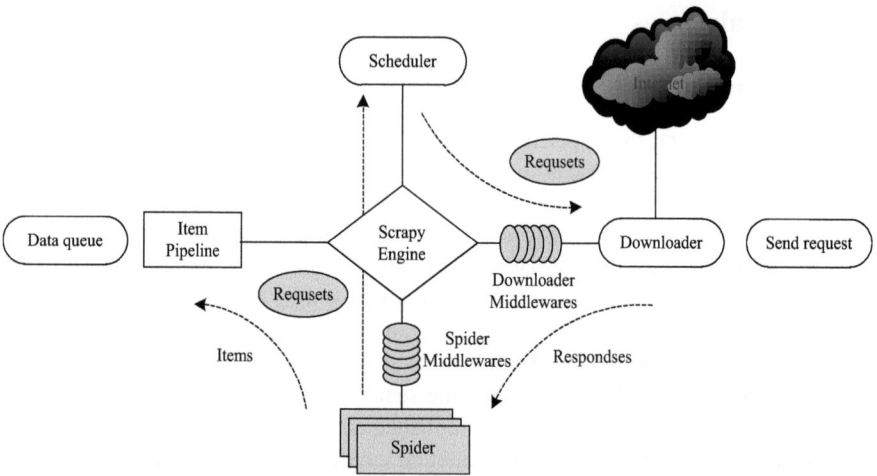

Fig. 6. Framework for capturing online teaching resources

a greater abundance. Leveraging this resource capturing architecture, approximately 1 million user behavior records were procured from a fusion of user behavior data and online educational resource data sourced across multiple social media platforms. This raw data underwent rigorous cleaning, filtering, and standardization procedures to eliminate noise and redundancy, thereby ensuring data quality and consistency. In adherence to experimental prerequisites, pertinent features were extracted from the preprocessed data, potentially encompassing user behavior frequency, preference tags, social connections, among others. These refined features served as input for training the chosen personalized and precise push algorithm model. Throughout the training phase, model parameters and hyperparameters were iteratively adjusted to optimize its performance.

3.2 Analysis of Results

For the purpose of this experiment, we opted to employ the accuracy rate of personalized recommendations for online educational resources as the primary evaluation metric. Its mathematical expression is outlined in the subsequent formula.

$$A = \frac{TP + TN}{TP + FP + FN + TN} \tag{5}$$

Among these, "*TP* (True Positive)" quantifies the number of genuine positives that have been accurately distinguished; "*TN* (True Negative)" denotes the count of actual negatives that have been correctly classified. Conversely, "*FP* (False Positive)" signifies the number of negatives erroneously labeled as positives, while *FN* (False Negative)" represents the quantity of positives incorrectly identified as negatives. An increase in the value of *A* indicates a stronger alignment between online teaching resources and individual learning profiles, translating to improved accuracy in personalized recommendations, and the opposite is also true.

For the purpose of comparison, the experimental group was comprised of the personalized push method for teaching resources, rooted in social media information fusion, as introduced in the preceding text. Meanwhile, two control groups were established: control group 1 featured the push method stemming from the online teaching modality presented in reference [3], and control group 2 encompassed the push approach based on optimized configurations, as outlined in reference [4].

Within the same experimental environment, as the quantity of online teaching resource data progressively increases, the three methods are utilized to push online teaching resources and gather user feedback on the recommended results. By applying the aforementioned evaluation index formula, the accuracy rate of resource pushing is calculated and compared, with the findings presented in Table 3.

Table 3. Comparison results of personalized push accuracy of online teaching resources

Data quantity of online teaching resources/GB	Experimental group A value	Control group 1 A value	Control group 2 A value
200	98.52	90.12	89.49
400	99.01	92.34	96.34
600	99.76	89.67	95.06
800	98.67	94.63	94.26
1000	98.99	95.16	94.58
1200	99.24	90.38	96.31
1400	98.76	88.26	94.28

The comparative outcomes presented in Table 3 reveal that upon implementation of the methodology introduced in this study, as the volume of online teaching resource data escalates, the personalized push accuracy rate notably surpasses that of the two contrasting methods, both exceeding 98% thresholds. This advanced strategy demonstrates a heightened capability in delivering tailored and meticulous teaching resource recommendations to users, tailored to their multifaceted attributes.

To enhance experimental efficiency and mitigate the time-consuming nature of the process, multi-process technology is employed for data parsing and subsequent distribution, as illustrated in Fig. 7, streamlining the workflow and expediting the experimental phase.

As depicted in Fig. 7, the experimental course data, subsequent to undergoing the aforementioned data parsing and allocation procedures, adheres to the prerequisites of this experiment. Subsequently, to further substantiate the viability of the methodology introduced in this paper, the resource extraction rate of the resource packages is adopted as the experimental metric, computed in accordance with the formula outlined below.

$$U = \frac{L_E}{F} \times 100\% \tag{6}$$

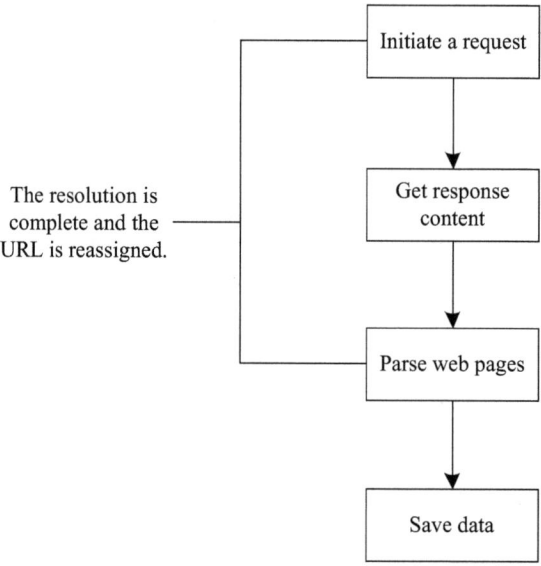

Fig. 7. Flowchart of the data parsing distribution process

Among them, L_E, which designates the quantity of targeted resources, is juxtaposed against F, signifying the total resource count within the package. An elevated extraction ratio of these resources underscores a more favorable extraction performance, whereas a diminished figure implies a less satisfactory outcome. To attenuate the sway of variability in experimental parameters on the final results, the course data were subjected to normalization, with the normalizing transformation function, denoted as x_0, formulated in the manner outlined below:

$$x_0 = \frac{x - x_{\min}}{x_{\max} - x_{\min}} \tag{7}$$

Among them, x measures the time interval between user accesses to the resource, while $x_{\min}$ and $x_{\max}$ represent the boundaries of creation times, marking the minimum and maximum points respectively. Post-normalization, it's possible that some data may not contribute meaningfully to the experiment, necessitating a filtering step to prune irrelevant characters for subsequent analyses. Consequently, the efficacies of resource extraction via the three methods are quantified, as visually presented in Fig. 8.

The comparison depicted in Fig. 8 reveals a noteworthy trend: the personalized and precise push method for online teaching resources, grounded in social media information fusion as introduced in this paper, exhibits a resource extraction rate that surpasses the rates observed in the two control groups by a considerable margin, achieving over 96% efficiency across varying resource package sizes. This outstanding performance underscores the superior push effectiveness of the proposed approach.

Upon executing the aforementioned experimental sequence, the validity and proficiency of the personalized and precise dissemination strategy for online teaching

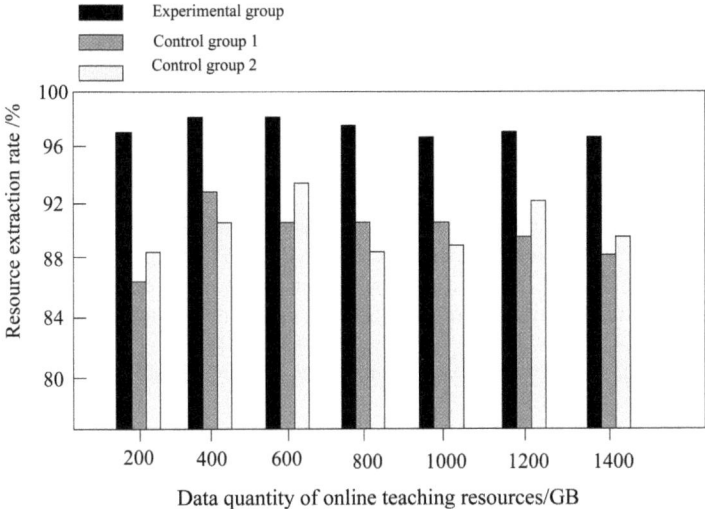

Fig. 8. Comparative results of the extraction rate of teaching resources

resources, rooted in social media information fusion, are firmly established, offering robust reinforcement for its implementation in practical settings.

4 Conclusion

Amidst the swift advancements in information technology and the pervasive influence of social media, online teaching resources have emerged as a cornerstone in contemporary education and self-directed learning. Yet, the challenge persists in sifting through the vast expanse of teaching resources to identify content tailored to individual requirements and learning traits, achieving personalized and targeted delivery. Drawing upon the amalgamation of social media information, this research delves into the methodology for personalized and precise dissemination of online teaching resources, achieving precise discernment of user needs and individualized resource recommendations. This endeavor presents a novel approach to enhancing learners' efficiency and satisfaction, albeit with room for further exploration and innovation to better cater to learners' personalized demands and optimize the utilization of online teaching resources. Future iterations of intelligent recommendation systems will delve deeper into learners' behavioral patterns, interests, preferences, and learning modalities, thereby crafting even more precise and tailored learning journeys and resource suggestions.

References

1. Peng, C., Zhou, X., Liu, S.: An introduction to artificial intelligence and machine learning for online education. Mobile Netw. Appl. **27**(3), 1147–1150 (2022)
2. Meng, F., Zhai, J.L.: Learning simulation of network database based on big data visualization of learning behavior. Comput. Simul. **38**(09), 216–220 (2021)

3. Bag, S., Sinha, A., Aich, P.: Determinants of using online mode of teaching: evidence from higher educational institutions. Int. Soc. Sci. J. **73**(248), 415–434 (2023)
4. Liu, C., Xia, J., Leung, M.F.: Research on optimization and allocation of English teaching resources. Math. Probl. Eng. Theory Methods Appl. **56**(24), 1–8 (2022)
5. Xia, T., Ahmad, M.T., Jan, N.: Method of ideological and political teaching resources in universities based on school-enterprise cooperation mode. Math. Probl. Eng. Theory Methods Appl. **23**(8), 1–9 (2022)
6. Li, H.J., Lu, J.Q., Wu, J.M.: Deep knowledge tracing method incorporating learning process characteristics. J. Zhejiang Univ. Technol. **50**(03), 245–252 (2022)
7. Yang, H., Peng, L., Liu, Q.T., et al.: An online anomaly learning behavior detection method based on negative selection algorithm. J. South-Central Univ. National. Nat. Sci. Ed. **42**(05), 664–671 (2023)
8. Li, Y.Y., Zhou, J.M.: Learners' self-regulation in online English learning. Foreign Lang. Teach. **314**(05), 45–54 (2020)
9. Liu, F., Tian, F., Li, X., et al.: A collaborative filtering recommendation method for online learning resources incorporating the learner model. CAAI Trans. Intell. Syst. **16**(06), 1117–1125 (2021)
10. He, S.H., Yin, S.Q., Jing, Z.Y., et al.: Educational resources recommendation method based on learner model. Comput. Digital Eng. **50**(04), 697–702 (2022)
11. Liu, H.J., Ma, H.F., Zhao, Q.Q., et al.: Target community detection with user interest preferences and influence. J. Comput. Res. Dev. **58**(01), 70–82 (2021)
12. Hu, X.R., Deng, J.W., Luo, R.Q., et al.: Multi-relationship recommendation model based on user interest-aware. Comput. Eng. Appl. **59**(11), 231–240 (2023)
13. Xu, T.Y., Liu, X.H., Zhao, W.D.: Knowledge graph and user interest based recommendation algorithm. Comput. Sci. **51**(02), 55–62 (2024)
14. Yan, M.M., Wang, H.T., He, J.F., et al.: Sequence recommendation algorithm based on fusion of hierarchical attention mechanism and user dynamic preference. J. Chin. Comput. Syst. **45**(03), 621–628 (2024)
15. Zhong, Y.Y., Ding, X.J., Yang, F.: Recommendation algorithm combined with user preference and item attribute extension. Comput. Syst. Appl. **30**(09), 192–199 (2021)

Evaluating the Effectiveness of the Application of an Interactive Teaching Model for English Courses in Undergraduate Colleges and Universities Based on Graph Neural Networks

Yalin Sun[✉] and Ping Huang

School of Economics and Business Foreign Languages, Wuhan Technology and Business University, Wuhan 430065, China
sunyalin2083@163.com

Abstract. Due to traditional classroom teaching often being teacher centered, students passively receive knowledge and lack initiative and creativity. And classroom teaching is difficult to fully consider individual differences and learning needs of students, resulting in uneven teaching effectiveness. To address this issue, an evaluation of the effectiveness of an interactive teaching model for undergraduate English courses based on graph neural networks is proposed. Firstly, based on the application of graph neural networks, the evaluation indicators for undergraduate English classroom teaching are pre-set through the design of graph neural network model structure and the establishment of an evaluation index system for interactive teaching mode. Secondly, the design is completed in three steps: inputting the interactive teaching data indicators of undergraduate English courses into a graph neural network, constructing an evaluation model for the application effect of the interactive teaching mode, calculating the membership degree of the teaching effect, and evaluating the application effect of the interactive teaching mode of undergraduate English courses.

Keywords: Evaluation of the effectiveness of teaching applications · Interactive teaching model · Undergraduate universities · English courses · Graph neural networks

1 Introduction

In today's globalized educational environment, the teaching mode of English courses in undergraduate universities is undergoing profound changes. Traditional teaching methods often focus on one-way instruction by teachers, while neglecting the active participation of students and the cultivation of their practical language skills. With the continuous advancement of educational technology, especially the integration of artificial intelligence (AI) technology, interactive teaching mode has gradually become an important way to improve teaching effectiveness and student learning experience. Artificial intelligence technology [1, 2], with its powerful data processing capabilities and personalized learning support, provides new possibilities for interactive teaching of English courses. Therefore, this study aims to explore the application effect of interactive teaching mode for undergraduate English courses supported by artificial intelligence, in order to provide scientific basis and improvement direction for teaching practice, promote the improvement of teaching quality and the comprehensive development of students' language ability.

By leveraging the graph neural network, educators gain a deeper insight into students' learning progress and requirements, enabling them to offer more precise guidance and support. The teaching approach, underpinned by interactive elements, fosters heightened student engagement in the learning process. Leveraging online platforms and intelligent learning tools facilitates real-time interaction between students, teachers, and peers, fostering knowledge sharing, inquiries, and collaborative problem-solving. This pedagogical approach not only sparks students' curiosity and passion for learning but also nurtures their cooperative ethos and capacity for innovation. Consequently, evaluating the interactive teaching model for undergraduate English courses using the graph neural network is of paramount importance. Through this evaluation, we can appraise the practical efficacy of the teaching model and identify any extant issues and deficiencies, thereby enhancing and refining the teaching approach. During the evaluation, it is imperative to consider several key indicators. Students' academic performance and advancements are pivotal metrics for assessing the effectiveness of the teaching model. By scrutinizing the learning achievement data of both the experimental and control groups, we can preliminarily determine whether the teaching methodology has contributed to enhancing students' learning outcomes. Furthermore, students' attitudes towards learning and their satisfaction are vital elements in evaluating the teaching model's efficacy. Through surveys and interviews, we can glean insights into students' reception and contentment with the teaching approach. Moreover, considering educators' experiences and sentiments is also indispensable when evaluating the teaching model. Teachers' proactive involvement and positive disposition are vital for advocating and implementing the teaching model.

Aiming at the demand for teaching effectiveness evaluation, a method for evaluating the application effectiveness of interactive teaching mode in undergraduate English courses based on graph neural network is proposed. Graph neural networks can effectively capture the complex interactions between students and teaching content, providing support for personalized feedback and dynamic evaluation. In addition, graph neural networks have adaptability and can handle non Euclidean data structures such as social networks and learning interaction networks, thereby facilitating in-depth analysis of interactive behavior.

2 Applications of Graph Neural Networks

Graph Neural Networks (GNNs) [3–5] is a deep learning method based on graph structure, which mainly consists of two parts: graph data structure in graph theory and neural network structure in deep learning. The "graph" here refers to the graph data structure in graph theory, which is used to represent the relationship between objects, where nodes represent the object and edges represent the relationship between objects. "Neural networks" are familiar deep learning structures such as multi-layer perceptrons (MLPS), convolutional neural networks (CNNS), and recurrent neural networks (RNNS).

The design of a graphical neural network model structure.

Graph Neural Networks (GNNs) offer a novel paradigm in deep learning, harnessing the power of graph structures. These structures, rooted in graph theory, represent relationships between objects using nodes and edges. GNNs combine this graph data structure with neural network architectures commonly found in deep learning, such as the multi-layer perceptron (MLP), convolutional neural network (CNN), and recurrent neural network (RNN). In contrast to traditional graph embedding methods, the graph neural network model not only maps the graph into a lower-dimensional space but also preserves its structural information. Through this neural network framework, it can extract intricate features, enhancing efficiency and accuracy in handling complex graph data structures. Consequently, GNNs find applications across various domains, including social network analysis, recommendation systems, bioinformatics, and natural language processing. As technology advances, GNNs are poised to extend their influence into even more fields, catalyzing new breakthroughs in data processing and machine learning. Thus, in this study, we opt for the graph neural network algorithm to assess the effectiveness of an interactive teaching mode in undergraduate English courses. The distinctive traits of graph neural networks include:

1. The input order of the nodes has no effect on the calculation.
2. While performing calculations, the nodes acquire information from neighboring nodes in close proximity, ensuring the graph's structural consistency remains unaltered.
3. There is help in reasoning about structural graphs between structural nodes in the graph.

With the goal of solving non-Euclidean data, the researchers designed the architecture of computational graph neural network by extending the graph data with the ideas of convolutional network, recurrent neural network and autoencoder. In the field of images, each pixel in an image is regarded as a node, and the computation with the neighboring nodes is determined by the filter of the neural network, while the graph neural network also gathers the information of the surrounding nodes with one node as the center.

The idea of the graph neural network model designed in this paper is mainly to diffuse and aggregate information in non-Euclidean space using spectral methods. The whole network graph is calculated and transformed as a whole. The graph neural network model can also be constructed by multiple graph convolution layers, each layer of which is still the feature matrix of the node: the $X \in R^{n*F}$, of which n is the number of nodes in the network graph, the F is the feature dimension of each node. The Laplace transform based graph convolution layer can be generally represented as:

$$X' = f(X, A) = f\left(aggregate(A, X)W^T + b\right) \tag{1}$$

In the formula: A is the self-ring adjacency matrix of this social network, the $W \in R^{n \times F}$ is the model parameters, the b denotes the node error.

The above equation represents a process of information dissemination, through the aggregation function $aggregate(\cdot)$ one time information aggregation, nodes can aggregate attribute features of neighboring nodes, after a nonlinear activation function $f(\cdot)$ transformation, the result is a new node identity matrix '$X' \in R^{n \times F}$'.

Aggregation function $aggregate(A, X)$ is a $n \times n$ matrix, which is used for information aggregation. This function is a static matrix closely related to the normalized Laplace transform, and the aggregation function is specifically expressed as follows:

$$aggregate(A, X) = D^{-0.5}AD^{-0.5}X \qquad (2)$$

In the formula: D is the self-ring adjacency matrix A of the degree matrix, which is a diagonal matrix, the N is the number of rows A, $i = 1, 2, ..., N, j = 1, 2, ..., N$, X is the user feature matrix.

In computing the features, in the adjacency matrix, the A left and right side-by-side $D^{-0.5}$. This method gives higher weights to low-level neighbors and lower weights to high-level neighbors, and is more reasonable when low-level neighbors can provide more information than high-level neighbors.

The graph neural network model also embodies three fundamental traits of deep learning: 1) hierarchical arrangement, where features are progressively extracted layer by layer, with deeper layers encapsulating more abstract and sophisticated information compared to shallower ones. 2) non-linear transformation capability, augmenting the model's expressive prowess. 3) end-to-end training methodology, obviating the necessity for pre-defined rules; solely requiring node labeling, facilitating autonomous learning of model parameters while seamlessly fusing structural and attribute feature information. Figure 1 depicts the architecture of the graph convolution model comprising two graph convolution layers.

In the two-layer GNN model shown in Fig. 1, the input is a complete graph that includes nodes (vertices) and edges (edges). Each node may have a series of attributes attached (such as eigenvectors), while edges define the connectivity between nodes. The goal of the model is to output the updated feature representation of each node or some global feature of the entire graph through layer by layer processing. In each layer, information aggregation is the core operation. It involves the exchange and integration of information between nodes and their neighboring nodes. This process usually includes the following steps:

1. Feature extraction: Firstly, the model extracts features for each node and its adjacent nodes. These features may include the attributes of the node itself, as well as information about other nodes connected through edges.
2. Information dissemination: By utilizing the connection relationships between nodes, the characteristics of nodes are transmitted to their neighbors through information dissemination mechanisms. This can be achieved through direct information exchange (such as message passing) or graph based propagation rules (such as attention mechanisms).
3. Information aggregation: In the process of information dissemination, the model aggregates information from neighboring nodes to form updated node features. Multiple aggregation methods can be used here, such as average aggregation, weighted

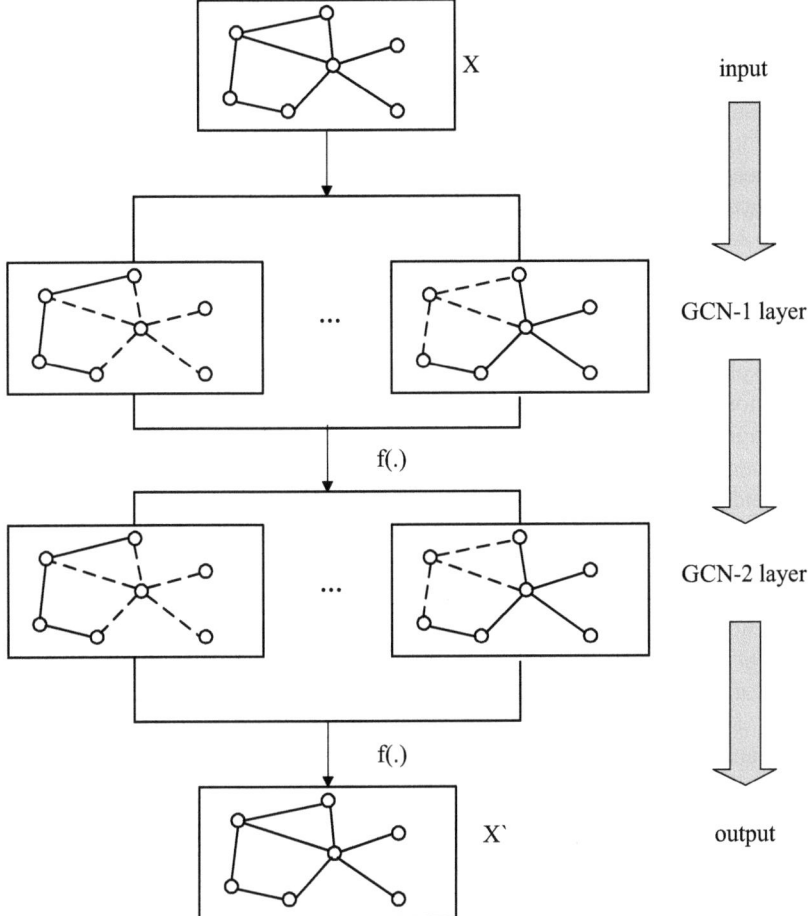

Fig. 1. Structure of the 2-layer graph neural network model

aggregation, pooling aggregation, or LSTM aggregation, depending on the task requirements and the structural characteristics of the graph.
4. Feature transformation: Finally, the aggregated node features are transformed using transformation functions (such as fully connected layers, convolutional layers, etc.) to generate new node feature representations. These transformation functions help to further extract and abstract advanced features in graph structures.

Convolution operation [6, 7] is the core part of graph neural network, which defines how to aggregate the neighbor information of nodes. Common convolution operations include average pooling, maximum pooling and so on. The purpose of this step is to extract useful information from the node's neighbors for subsequent node classification or graph-level tasks.

After the model is established, the graph neural network model is trained with the training data. In the training process, it is necessary to adjust the parameters of the model constantly to achieve the best training effect. After the training, the model also needs to be tested using test data to evaluate the accuracy and generalization ability of the model. The graph convolutional neural network aggregates the information of adjacent nodes and encapsulates it into the hidden layer through the graph convolutional neural network. According to whether the information extracted by the graph convolutional neural network needs to collect the information of the entire graph, the graph convolutional neural network is further divided into the graph convolutional neural network based on frequency domain and the graph convolutional neural network based on airspace [8].

2.1 Establishing the Evaluation Index System of Interactive Teaching Mode

Before developing the assessment program, it's crucial to establish clear objectives and standards for the evaluation. These objectives may include measuring the effectiveness of the instructional approach in improving students' English proficiency, assessing the level of student engagement in learning activities, and evaluating the interaction between teachers and students. Evaluation criteria involve putting these objectives into practice through metrics such as student performance on assessments, participation in class activities, and feedback from satisfaction surveys. At the undergraduate level, English courses increasingly employ interactive teaching methodologies, which prioritize bidirectional communication between students and teachers. These approaches aim to foster active engagement during class discussions and enhance students' language application and communication skills through collaborative tasks, role-playing exercises, simulated presentations, and similar activities. This shift towards more participatory learning environments encourages student curiosity and autonomy in the learning process, moving away from traditional teacher-centered instruction.

The establishment of an evaluation index system for interactive teaching modes forms the essential foundation for appraising the efficacy of these methods. This study constructs a comprehensive assessment index framework by examining and synthesizing literature pertaining to the evaluation of interactive teaching modes, considering a wide range of factors that influence their effectiveness. Designing such a comprehensive evaluation index system is immensely significant for accurately gauging the educational impact of English courses, and it also furnishes a crucial point of reference for decision-making regarding interactive teaching strategies.

Evaluation of English courses in undergraduate institutions involves converting abstract evaluation aspects into concrete criteria for judgment that emphasize tangible and substantive factors. To ensure the reliability of these criteria, it is essential to clarify the guiding principles for selecting them in a manner that enhances their practicality, scientific rigor, and overall validity.

(1) The principle of wholeness

There are many factors affecting the evaluation of English classroom teaching in secondary schools, and it is necessary to select indicators from various aspects and directions at the same time, and in the process of selecting indicators, the results of the evaluation of English classroom teaching should be taken into account as much as possible, so as to provide reference for the evaluation of classroom teaching in the future.

(2) The principle of authenticity

In selecting classroom teaching evaluation indicators, it is necessary to assess the overall performance of teachers in classroom teaching based on the actual situation, avoiding personal bias towards certain indicators and selecting appropriate and fair indicators.

(3) The principle of maneuverability

In the process of selecting indicators, the evaluation of classroom teaching should have a certain degree of feasibility, taking into account whether the data of certain indicators are easily accessible and whether the factors of the indicators are clear, so whether the indicators are reasonable or not, the evaluation data collected will not have any valuable role.

Through the undergraduate colleges and universities English course teaching evaluation related literature review, such as the literature "Teacher teaching quality comprehensive evaluation of exploration", "mutual promotion of English classroom teaching evaluation mode construction" and other analysis and research, and the college English experts (mainly new English teachers and students engaged in the teaching of English) research, they have a unique insight into the evaluation of English teaching and a wealth of experience in teaching, undergraduate colleges and universities They have unique views and rich teaching experience on English teaching evaluation, and the evaluation indexes of undergraduate colleges and universities are predetermined, and the specific indexes are shown in Table 1.

Table 1 shows a comprehensive pre-set system of evaluation indicators for evaluating the teaching effectiveness of undergraduate English classrooms. Starting from the overall goal of "Teaching Effectiveness in College English Classrooms", the system is divided into four main indicators: teaching plan, teaching equipment, teaching attitude and classroom performance. Under each major indicator, there are specific secondary indicators, such as the scientific and rational teaching plan, the use of multimedia teaching, the adequate preparation of teaching content, the management of classroom discipline, the clarity of English pronunciation, and the proportion of spoken English and Chinese. These indicators are more comprehensive, scientific, interactive, practical and responsive than ever before, and can reflect all aspects of teaching activities in a more detailed manner, thus providing an effective framework to improve the quality of teaching and student learning.

3 A Design for Evaluating the Effectiveness of Applying Interactive Teaching Models in English Courses in Undergraduate Colleges and Universities

The interactive teaching mode, assisted by graph neural network, makes the interaction between teachers and students more frequent and deeper. Through the online plaorm, virtual classroom and other forms, students can communicate and discuss with teachers at any time, and put forward their own questions and insights. This instant feedback mechanism not only helps teachers to adjust teaching strategies in time, but also helps students to better understand and master knowledge. The use of graph neural networks

Table 1. Predefined system of evaluation indicators

Overall goal	Primary indicators	Secondary indicators
Teaching Effectiveness in College English Classrooms	Teaching plan	Scientific and reasonable classroom teaching
		Moderately adjust the teaching plan to ensure timely completion
	Teaching devices	Using multimedia teaching to explain
		Helping students expand their horizons in English classroom teaching
		Teaching is meticulous, easy to understand, and provides multiple examples
	Teaching attitude	The content of English classroom teaching in universities should be fully prepared
		Strictly manage classroom discipline
		Timely feedback on questions raised by students
	Classroom performance Evaluation of Interactive	English speaking clarity
		The proportion of spoken Chinese and English
		The approximate proportion of new words in spoken language

also enables real-time tracking and assessment of students' learning progress. This approach involves analyzing the data generated during a student's learning process to gain insights into their performance and identify areas where they may require additional support or guidance. By leveraging this data-driven methodology for teaching evaluation, educators can provide more accurate and fair assessments that are grounded in objective scientific evidence, as opposed to subjective criteria.

3.1 Constructing a Model for Evaluating the Application Effect of Interactive Teaching Mode

In order to evaluate interactive teaching of English courses in a more scientific way, researchers have proposed many methods, such as mathematical statistics, genetic algorithm, artificial intelligence and other methods, but for undergraduate English classroom evaluation, some are too rough, although some can get accurate conclusions, but the calculation process is complicated, not suitable for the actual needs of work. Based on the advice of experts and practice, this paper uses the graph neural network method to evaluate the effect of interactive teaching of English courses in colleges and universities. In this paper, based on the evaluation index system established in the above chapter, a new evaluation model of interactive teaching mode application effect based on graph neural network will be constructed. Graph neural network is a relatively mature network model. According to neural network theory, the network can approximate the continuous function of any square integrable space by setting the number of hidden layer nodes appropriately, and it is a uniform approximator of square integrable space.

The convolutional neural network model is shown in Fig. 2.

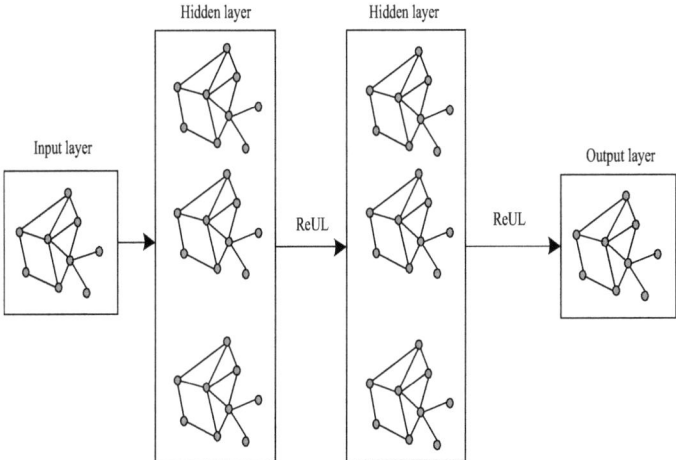

Fig. 2. Convolutional neural network model

The evaluation model of interactive teaching model based on graph neural network includes three main parts: first, the network structure is determined; Secondly, the training sample is constructed. Finally, the graph neural network is trained. After these three steps, the teaching effect evaluation model based on graph neural network can be established, and can be directly used to evaluate the teaching effect of other universities or courses of different years in the same university, rather than limited to the scope covered by the training sample. The specific construction method is described as follows:

The most important part of the evaluation model of the application effect of interactive teaching mode is the attention layer and the graph convolution layer, both of which aim to automatically extract the node features of the input nodes. For the second part of the attention layer, after the input of the cleaned data graph, the 2-head attention layer is used to aggregate and absorb the multi-attribute information of the data in the network, and output the attribute feature extraction results. The graph convolution layer uses the output result of the attention layer to make the source node completely aggregate the information of the neighbor node, and automatically extract the feature vector of the node. In this paper, the multi-head attention mechanism is adopted to absorb the weight of messages in the network. Since the self-attention mechanism is easy to ignore some important information, the multi-head attention mechanism [9, 10] is adopted to replace the self-attention mechanism, which can enable nodes to pay maximum attention to and absorb the messages in the network.

The inputs to the attention layer are the adjacency matrix and the node feature matrix of the graph structure data, and the output is the node feature vector matrix. The attention layer defines a self-attention matrix that $A_{GAT} = [a_{ij}]_{N \times N}$ to represent the attention weight from node i to node j, a_{ij} indicates the absorption weight of the edge (i, j). For a node pair (i, j), define a parameter matrix W to calculate the cross correlation coefficient between node i and node j e_{ij}, using a mapping function $\alpha(\cdot)$ to perform the calculations.

$$e_{ij} = \alpha(Wx_i, Wx_j) \tag{3}$$

Different from the traditional self attention calculation method, the traditional attention calculation method needs to calculate the cross correlation coefficient of all nodes. In GAT, only the coefficient of adjacent node pairs needs to be calculated $e_{ij}(i, j) \in E$ or $e_{ji}(i, j) \in E$, does not need to perform all calculations, and can be better used in large-scale graph structures. In order to make the cross correlation coefficient of each node pair easier to calculate and compare, a softmax function is used to calculate the row regularized attention coefficient, as shown in Formula (4):

$$a_{ij} = soft\max(e_{ij}) = \frac{\exp(e_{ij})}{\sum_{k \in N_i} \exp(e_{ik})} \tag{4}$$

A simple schematic of this is shown in Fig. 3.

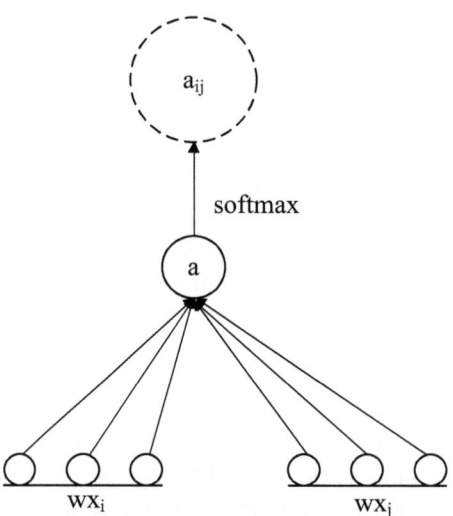

Fig. 3. Schematic representation of the attention function

To sum up, calculate the attention correlation coefficient a_{ij} of node pair (i, j), the formula is:

$$a_{ij} = \frac{\exp(soft\max([Wx_i][Wx_j]))}{\sum_{k \in N_i} \exp[Wx_i][Wx_j]} \quad (5)$$

Since then, the attention matrix of GAT layer can be calculated $A_{GAT} = [a_{ij}]_{N \times N}$. The information aggregation of GAT layer is similar to that of GCN, and the calculation expression is:

$$X' = \alpha \left(arrgregate \left(A_{GAT} X W^T \right) + b \right) \quad (6)$$

In this paper, 2-head attention mechanism is used to fully capture multiple messages W 2-head attention layer is used for feature aggregation, focusing on different features. Although the amount of calculation will be large, it avoids the loss of some information.

Based on the above analysis, the expression formula of the evaluation model of the application effect of interactive teaching mode can be derived as follows:

$$\partial = \alpha \left(arrgregate \left(A_{GAT-1} X W_1^T, A_{GAT-2} X W_2^T \right) + b \right) \quad (7)$$

Use the specified formula for extracting the feature information related to teaching effectiveness on the network. This process aims to bolster the focus and assimilation of various information by the nodes, generating a distinctive node feature vector. This resultant vector preserves pertinent data while attenuating extraneous details. Following this, utilize the updated feature vector matrix to enable automated feature extraction via a graph convolutional network, thereby enabling the exploration of more intricate representational attributes.

3.2 Calculating the Affiliation of Teaching Effectiveness

First of all, to determine the evaluation level, the set of comments in the classroom teaching evaluation is based on the evaluation indicators given in the table of secondary school English classroom teaching effect evaluation index system, resulting in the set of secondary school English classroom evaluation results, expressed as $V = \{V_1, V_2, .., V_m\}$, of which $V_j = (j = 1, 2, ..., m)$ is the level of rubrics, and different experts, students and teachers have different levels of satisfaction with the indicators. Through the questionnaire survey of experts, teachers and students, the number of evaluations of each indicator rubric is obtained, and thus the relationship matrix of experts, students and teachers can be obtained R elements r_{ij} values. Each indicator is rated and a comment $\{V_1, V_2, ...V_n\}$ is given. In response m questionnaires, yielding a vector of the number of rubric evaluations for each indicator as $N = \{N_{i1}, N_{i2}, ..., N_{ij}\} (i = 1, 2, 3)$, of which N_{ij} is the numerical value of the j-th evaluation language in the i-th and first-level evaluation index, shown in Table 2.

Constructing the affiliation, relationship matrix is obtained from the evaluation vectors in the evaluation of English classroom teaching in colleges and universities. By determining the affiliation matrix, in the case of experts, the i th evaluation indicators

Table 2. Scale of numerical values for rating

Index	C1	C2	...	Cn
V1	N11	N21	...	Nn1
V2	N12	N22	...	Nn2
V3	N31	N23	...	Nn3
...	...	...	...	...
Vn	N1n	N2n	...	Nnn

to primary evaluation indicators U_i evaluate, and obtain a rubric relative to the set of comments V the fuzzy vectors of the R. All rows in the expert matrix are the results of evaluations for all indicators, and the matrix includes the set of evaluation results V, for the set of evaluation indicators U evaluation is carried out and all the information obtained, as shown in Eq. (8).

$$r_{ijz} = \frac{N_{ijz}}{m} \qquad (8)$$

In the formula, N_{ij} respectively represent the third level of the jth second level indicator under the i-th first level indicator among experts, students and teachers z number of evaluations of each comment grade, m is the number of questionnaires of experts, students and teachers, r_{ijz} is C_{ij} indicator to V_z affiliation of fuzzy subsets.

Obtaining the evaluation results of experts, students and teachers, respectively, and establishing a fuzzy relationship matrix between the secondary indicators and the rubrics of the primary indicators R_i. There is p evaluation levels, q second-level evaluation indicators, resulting in a fuzzy matrix of relationships for the first-level indicators R_i based on the weighting of the individual indicators for experts, students and teachers C_i and the relationship matrix R_i, calculated i the judgmental values of the indicators, of which r_{ijz} is C_{ij} indicator to V_z the fuzzy subset affiliation, computed to obtain the row vector B_i, as shown in Eq. (9).

$$B_i = C_i \circ R_i = (c_{i1}, c_{i2}, ..., c_{iqi}) \qquad (9)$$

In summary, it can be concluded n of the first-tier indicators B_i combination establishes a fuzzy relationship matrix $R = \begin{bmatrix} B_1 \\ B_2 \\ ... \\ B_n \end{bmatrix}$ for the overall goal.

3.3 Evaluating the Effectiveness of the Application of Interactive Teaching Models in English Courses in Undergraduate Colleges and Universities

Amidst the swift pace of China's educational reform, English majors at various undergraduate institutions are placing greater emphasis on incorporating interactive teaching methods into their English instruction. This paradigm shift is fundamentally altering traditional teaching approaches and serving as a strong foundation for enhancing students' English language abilities. This teaching mode is created with constructivism as the core, which puts students in the leading position and gradually stimulates students' interest in learning and enhances students' participation through corresponding teaching methods, thus building more efficient English teaching activities and continuously strengthening students' English ability. Modern English teaching usually consists of three aspects of ability, namely: cognitive ability, communicative ability and language ability. Effective application of the interactive teaching mode will inevitably improve students' abilities in these three aspects. In this teaching mode, students take the leading role, all teaching activities are centered on students' actual needs, and teachers use appropriate teaching methods to gradually improve students' English proficiency. Teachers guide the students so that they can learn independently and autonomously, and then according to the problems encountered by the students in their independent learning, they take effective ways to explain to deepen the students' impression of the knowledge points. In interactive teaching activities, the emphasis is on communication between teachers and students as well as between students. Through the communication between teachers and students, students' English communication skills can be improved. Encouraging student communication and dialogue can deepen knowledge understanding and enhance teamwork skills, thus establishing a strong basis for future practical applications. Additionally, integrating contextual teaching is pivotal to the instructional framework. Implementing scenario-based learning environments offers students an optimal setting for active participation, enriching their learning experience and boosting overall academic performance.

The expert-level indicators' weights and the fuzzy relationship matrix pertaining to these indicators were identified in the aforementioned sections R. As a result, intermediate variables B for experts can be derived. The same holds true for determining intermediate variables B for students and teachers, as depicted in equation

$$B = C \circ R = (b_1, b_2, ..., b_p) \tag{10}$$

Within the formula, $b_k (k = 1, 2, ..., p)$ represents the overall evaluation value assigned to each assessment index. Specifically, it signifies the conclusive rating provided by experts, students, and teachers during the evaluation of interactive English classroom teaching in higher education institutions.

Normalization was performed to b_k obtain Eq b'_k, as shown in (11)

$$b'_k = \frac{b_k}{\sum_{k=1}^{p} b_k} \tag{11}$$

Based on the above formula, the judgment value can be calculated, namely. ε value, the evaluation level of interactive English classroom teaching in colleges and universities can be obtained. As shown in Eq. (12).

$$\varepsilon = b'_k \times v^T \tag{12}$$

As a result, the evaluation ratings ε for evaluating the effectiveness of the interactive teaching mode of English courses in undergraduate colleges and universities were obtained. If the value of ε is between which level of the evaluation level, then the overall comprehensive evaluation is at which evaluation level.

4 Example Analysis

4.1 Experimental Preparation

To validate the effectiveness of the assessment methodology employing graph neural networks for implementing interactive teaching methodologies in English courses at the undergraduate level, as proposed in this paper, an experimental assessment is recommended. Initially, a test environment will be established, with the parameters of this environment provided in Table 3 below, to facilitate the experimental evaluation.

Table 3. Configuration of the experimental environment

Configuration items	configuration parameter
operating system	Windows 10
Development tools	JetBrains PyCharm CE 2018.2.4
programming language	Python 3.7
Computer model	X550CC
processor	Intel(R) Core(TM) i7-3537U CPU 2.5GHZ
Memory	12GB RAM
Graphics card	NVIDUA GeForce GT 720M

4.2 Steps and Processes

The steps of this experiment are as follows.

Selection of research participants: The experimental group consists of 40 full-time undergraduate students from Class 1, 2018 in the Translation Major at University A, while the control group comprises 41 full-time undergraduate students from Class 2, 2018.

Research methodology selection: In line with the English course syllabus requirements, both sets of undergraduate classes undertook a combined total of 66 practical class hours. These student cohorts share similar academic standing, displaying no noticeable statistical variations in age, gender, enrollment scores, overall competence, etc. Moreover, they were exposed to identical textbooks, study materials, and instructional personnel. The experimental group embraced a blended teaching strategy merging physical and virtual elements, whereas the control group adhered solely to traditional lecture-based methods. Notably, both groups received comparable instruction in terms of hours, procedures, and subject matter.

Assessment and evaluation strategy outline: at the conclusion of each experimental session, students will undergo assessment based on their performance in the experiment, accuracy of results, quality of lab reports, and responses to critical thinking questions. Individual scores will be calculated at the close of the semester. Post-course, a comparative analysis of learning outcomes between the two student groups will be conducted. Additionally, questionnaires will be administered to measure teaching satisfaction and effectiveness. Total score data will be statistically analyzed using SPSS17.0, with group comparisons conducted via t-tests. Questionnaire responses will be presented as percentages for ease of interpretation.

4.3 Results and Conclusions

According to the above experimental preparation, the experimental group and the control group are now tested, and the comparative results of the evaluation of the teaching effect by the students in the two groups are shown in Table 4.

Table 4. Comparison of the two groups of students' evaluation of the effectiveness of teaching

Evaluation content	Experimental group (%)	Control group (%)
Improve operational skills	100	50.23
Improving self-directed learning ability	82.19	56.18
Knowledge extension application	81.42	74.06
Stimulate learning interest	84.62	60.58
Increase academic burden	97.13	36.94

According to the above test results, it can be seen that the method in this paper is obviously superior to the traditional method in the test. After adopting the interactive experimental teaching mode combining virtual and real, the experimental group organically combined virtual experiment and laboratory teaching. After the course, the experimental test scores were significantly better than the control group, and the experimental teaching satisfaction was significantly better than the control group, which improved students' hands-on ability and learning initiative. The understanding of learning content and the ability to analyze and solve practical problems were significantly better than the control group.

In order to further test the efficiency of the evaluation method of the application effect of interactive teaching mode, the evaluation time is taken as the index, and the method in this paper is compared with the traditional method. The evaluation time test results are shown in Table 5.

Table 5. Evaluation of time-consuming test results

Number of tests	Evaluation time/min	
	experimental group	control group
1	3.2	14.5
2	3.1	16.3
3	3.3	15.4
4	3.5	14.5
5	3.3	16.4
average value	3.28	15.42

According to the data analysis in Table 5, the experimental group showed a significant improvement in evaluation time compared to the control group. The average evaluation time for the experimental group is about 3.28 min, while the average evaluation time for the control group is as high as 15.42 min. This significant time saving indicates that the method used in this article greatly improves efficiency and reduces the required time in the evaluation process.

5 Conclusion

Following an extensive examination of the effectiveness of implementing a graph neural network-based interactive teaching mode for English courses in undergraduate colleges and universities, it becomes evident that this pioneering pedagogical approach delivers a highly individualized and astute learning experience to students. Capitalizing on its robust capacity for processing information and acquiring representational knowledge, the graph neural network notably enhances communication and engagement between educators and learners, thereby rendering English instruction more dynamic and effective.

By examining real teaching data, this paper evaluates the effectiveness of an instructional model that utilizes graph neural networks in improving students' English proficiency and cultivating their passion for learning. The findings indicate that the interactive teaching approach based on graph neural networks offers substantial advantages over traditional teaching methods in various areas. This is evident not only in the enhancement of students' English grades but also in the holistic development of their language skills, critical thinking, and creativity. It is important to recognize that no teaching approach is without its imperfections, and continuous refinement of algorithm models and pedagogical practices is essential to better address students' learning requirements.

Acknowledgment. 1. The Scientific Research Team Plan of Wuhan Technology and Business University (WPT2023041);

2. Teaching Reform Research Project of Wuhan Technology and Business University-"Design and Application of Blended Teaching for College English from the Perspective of Learning Experience" (2022Y01);

3. Research on the Teaching Model Construction and Practice of College English "Golden Course" based on OBE (2019ZA10), one of the key projects supported by the Educational Science Planning Program of Hubei Province with Special Funding, 2019;

4. One of the First Class Undergraduate Courses of Hubei Province–College English II of Wuhan Technology and Business University.

References

1. Junhao, S., Yonggui, L.: Research on learning design empowered by generative artificial intelligence. E-educ. Res. **45**(07), 73–80 (2024)
2. Peng, C., Zhou, X., Li, S.: An introduction to artificial intelligence and machine learning for online education. Mobile Networks Appl. **27**(3), 1147–1150 (2022)
3. Shengli, H., Chun, C.: Session-based recommendation analysis of fusing graph neural networks and sparse self-attention. J. Lanzhou Institute Technol. **30**(06), 13–18 (2023)
4. Wei, J., Yang, W., Jing, Y., et al.: Few shot image classification algorithm based on semantic feature propagation graph neural network. Laser & Infrared **53**(12), 1944–1952 (2023)
5. Yue, S., Yuhong, Z., Ting, X.: Overlapping community discovery method based on heterogeneous graph attention network. Comput. Eng. Design **44**(12), 3649–3655 (2023)
6. Xiong, W., Zifan, W.: Human activity recognition model based on multi-scale convolution and BiGRU network. J. Comput. Appl. **43**(S2), 72–76 (2023)
7. Chungeng, G., Junli, L., Kai, L.: Research on verification code recognition algorithm based on improved convolutional neural network. J. Beijing Inst. Fashion Technol.: Natural Sci. Ed. **43**(04), 67–72 (2023)
8. Chao, Y., Xuanzhu, S., Yilong, C.: Distributed and semi-supervised automatic annotation method based on graph convolutional neural network. Cyber Secur. Data Gov. **42**(S2), 231–235 (2023)
9. Daozhu, X., Cheng, J., Chao, M., et al.: Geographic named-entity recognition based on BERT-BiGRU-CRF and multi-head attention mechanism. Cyber Secur. Data Gov. **42**(S1), 169–173 (2023)
10. Krei, S., Yingying, W.: Cross-domain recommendation model based on neighbor interaction enhancement and multi-head attention mechanism. J. Hubei Minzu Univ.: Nat. Sci. Ed. **41**(04), 454–461 (2023)

Research on Software Test Data Optimization Using Adaptive Differential Evolution Algorithm

Zheheng Liang[1,2(✉)], Wuqiang Shen[1,2], and Chaosheng Yao[1,2]

[1] Joint Laboratory on Cyberspace Security of China Southern Power Grid, Guangzhou 510000, China
liangzhaoheng23213@163.com, shenwuqiang@gdxx.csg.cn

[2] Guangdong Power Grid, Guangzhou 510000, China

Abstract. In order to improve the coverage of the target path corresponding to the generated software test data after optimization, and make the data better adapt to the software test process, the adaptive differential evolution algorithm is introduced to design the optimization method for software test data. Using the PSO algorithm to simulate the biological evolution mechanism in nature, and with the help of computer programming, the generated software test data are preliminarily trained; Draw on the path correlation based regression test data evolution of adaptive differential evolution algorithm to generate the relevant path representation method, mark the program to be tested, and construct the test data fitness function based on this; A hybrid model MPSO is proposed to select the best individual data in software test data; The selection criteria of scaling individuals are introduced into the adaptive scaling factor to cluster the optimal data individuals, so as to realize the design of optimization methods. The comparison experiment results show that the designed method has a good effect in practical application. This method can improve the target path coverage corresponding to the generated software test data on the basis of controlling the time length and evolution times required for software test data optimization.

Keywords: Adaptive differential evolution algorithm · Data selection · Individual clustering · Fitness function · Software test data · Optimization

1 Introduction

In order to ensure the quality of software, not only need advanced technical means, perfect R & D process, but also need continuous testing, no matter how advanced the current technology, how perfect the R & D process, can not 100% guarantee the software developed zero defects. Software testing can help scientific research and technology developers find out the errors or potential defects in the process of software development, and reduce the damage caused by software errors and potential defects as much as possible. In order to ensure the high reliability and integrity of software products, it is necessary to design enough test cases. However, the scale of software products is increasing day by day, the structure is becoming more and more complex, and the update cycle is becoming shorter and shorter. It is a huge challenge for software testing to carry out comprehensive testing to find the problems and potential defects in the increasingly tight software development cycle. Therefore, it is necessary to use effective test cases within a limited test time to find out the errors and potential defects of the product as far as possible to ensure the accuracy and reliability of the software product, and the basis of this test is a high degree of target path coverage.

In recent years, the use of artificial intelligence algorithms such as ant colony algorithm, particle swarm optimization algorithm and genetic algorithm to automatically generate test cases has attracted more and more researchers' attention, and has achieved good research results. For example, Li Lu et al. [1] studied the defect optimization method of high-performance traffic analysis software based on particle swarm optimization algorithm, which introduced particle swarm optimization algorithm to analyze the current defect status of high-performance traffic analysis software, reset defect parameters, and build the defect optimization model of high-performance traffic analysis software. However, this method has the problem of long running time. Yang Bo et al. [2] studied a software fault location assisted test case generation method using improved genetic algorithm. Based on genetic algorithm, this method assisted the generation of test cases in the process of software fault location through the ranking of suspected faults in software fault location, but the marking path coverage of this method was low. The research shows that the adaptive differential evolution algorithm has the advantages of fast yield with fewer parameters and fast yield in high dimensions [3, 4]. Therefore, this paper takes the program code of unit test in software testing as the research object, introduces adaptive differential evolution algorithm to calculate all the tested paths, and uses the fitness function of the tested paths to improve the process of output optimal solution by optimizing the mutation operator, and improves the adaptive differential evolution algorithm by improving the mutation operator. Test cases covering each path are generated automatically to optimize software test data. This paper makes use of the convenient and efficient characteristics of artificial intelligence algorithm, which not only reduces the test cost, but also obtains a high efficiency of automatic case generation, moreover, the target path coverage of the generated software test data is more than 90%, which is more than 10% higher than that of the comparison method, which is of great significance to the improvement of product quality.

2 Software Test Data Generation

A lot of work has proved that PSO algorithm can solve the problem of software test data generation better than GA algorithm. Moreover, the PSO algorithm is simple, with fewer parameters to configure and easy to implement [5]. Therefore, this chapter introduces PSO algorithm to design the generation of software test data.

In this process, with the help of computer programming, the problem to be solved is expressed into strings (or chromosomes), that is, binary codes or digital strings, by using the PSO algorithm to simulate the biological evolution mechanism in nature, so as to form a group of strings, and they are placed in the problem solving environment. According to the principle of survival of the fittest, the strings that adapt to the environment are selected for replication, and the two gene operations are crossover and mutation, Create a new generation of clusters that are more adaptable to the environment. After such continuous changes from generation to generation, finally converge to a string that is most suitable for the environment, and obtain the optimal solution of the problem [6]. That is, the training of test software requirement data is preliminarily realized.

On this basis, it should be clear that the essence of PSO algorithm is an optimization algorithm based on swarm intelligence, and its idea comes from artificial life and evolutionary computing theory. The algorithm simulates the behavior of birds flying to find food, and makes the group achieve the optimal goal [7] through the collective cooperation between birds. When using the PSO algorithm to generate software test data, it is necessary to input the software test data into the test environment first, and express the test environment as D, on D In the target retrieval space of dimension, each software test data is regarded as a particle, and the particle is represented as m, then m Particles form a population, where the i Particles in the d. The position of the dimension is x_{id}, x_{id} The corresponding flight speed in space is v_{id}, the optimal location searched by this particle is p_{id}, the current maximum value of the entire particle swarm p_{gd} In this way, the particle position in space is updated, and this process is taken as the data update process, as shown in the following calculation formula.

$$v_{id}^{t+1} = wv_{id}^{t} + c_1 r_1 \left(p_{id} - x_{id}^{t}\right) + c_2 r_2 \left(p_{gd} - x_{id}^{t}\right) \tag{1}$$

In formula (1): v_{id}^{t+1} Means on $t+1$ The corresponding flight speed of particles in space at time; w Is the inertial factor, w The larger the value of, the more suitable for large-scale exploration of the solution space, w The smaller the value of is, the more suitable it is to explore the solution space in a small range; v_{id}^{t} Means on t The corresponding flight speed of particles in space at time; x_{id}^{t} Means on t At the moment i Particles in the d Dimension position; c_1, c_2 Represents normal number, called acceleration factor; r_1, r_2 Represents a random number with a value between 0 and 1 [8]. On this basis x_{id}^{t+1} the calculation formula is as follows:

$$x_{id}^{t+1} = x_{id}^{t} + v_{id}^{t+1} \tag{2}$$

On the basis of the above contents, it should be clear that, d The value range of is $1 \leq d \leq D$, for the d Position of dimension x Constrain the range of change and make it clear x The value range of is $[x_{d\,min}, x_{d\,max}]$, for the d Dimensional velocity v Constrain the range of change and make it clear v The value range of is $[v_{d\,min}, v_{d\,max}]$.

In the iteration, if the position and speed exceed the boundary range, the boundary value shall be taken. The particle position in the space can be updated according to Fig. 1. In this way, the spatial position of the software test data can be updated to generate more complete and global adaptive test data [9].

The termination condition of the PSO algorithm takes the maximum number of iterations or the predetermined minimum fitness threshold that the optimal position searched by the particle swarm optimization meets according to the specific problem.

Because p_{gd} It is the optimal position of the whole particle swarm. Therefore, the software test data can be generated according to the above steps, or it can be used as the retrieval process of global particles. The global PSO algorithm has a fast convergence speed, but sometimes it falls into a local optimum;The local PSO algorithm converges slowly, but it is relatively difficult to fall into the local optimal value.Therefore,

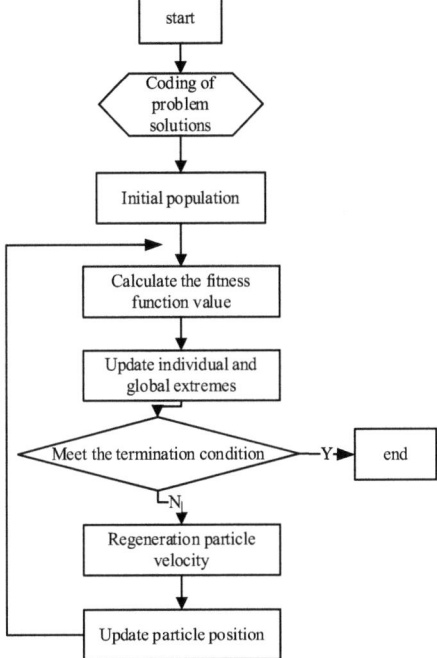

Fig. 1. Software test data generation and update process

corresponding measures can be taken for data processing according to the specific requirements of software test data generation in practical applications.

3 Construct Test Data Fitness Function based on Adaptive Differential Evolution Algorithm

On the basis of the above contents, in order to achieve the extraction of the optimal data in the generated software test data and master the fitness of different data in the generated data and the software test process, the adaptive differential evolution algorithm is introduced to construct and design the fitness function of the test data [10].

In this process, it should be clear that the establishment of the mathematical model of the multi-path coverage test data generation problem is closely related to the path representation method, and the tested program generally has multiple paths. In structured programming, the program usually consists of three structures: sequence, selection, and cycle.Among them, selection and cycle structure determine the trend of different branches of the program; For cyclic structure, by introducing Z Path coverage can decompose it into multiple selection structures. The existing path representation methods include: using the statement number sequence of the program to represent the path, using the branch node sequence to represent the path, and using Huffman code to represent the target path.Using the path correlation based regression test data evolution generation correlation path representation method in adaptive differential evolution algorithm for reference, the program to be tested is marked.

Assume that the program to be tested is Q, insert piles into each branch node in the program, and set the value of "true" to be 1 when crossing the corresponding branch node. If it is false, it is -1, and if it does not pass through the branch node, it is 0. Therefore, the code sequence of 1, -1, and 0 can be used to represent the path of program execution. Run the test program after inserting piles with a set of input data, and the branch node sequence is $D_1, D_2, \cdots, D_n$ The corresponding execution path is expressed as $q(T_i)$, where n Is the path $q(T_i)$ Length of. In addition, the program to be tested Q The target path of is $q_1, q_2 \cdots, q_N$, N Is the number of target paths.

Hypothetical procedure Q The input field of is α, input data is $T_1, \cdots, T_N$, then $T_N \in \alpha$, the path will be executed $q(T_i)$ Compare the coding sequence of and the target path from left to right, find the first different coding position, record the number of the same coding before, and record it as $|q(T_i) \cap q_j|$, take the ratio between it and the target path code, that is, the total length of the target path, to obtain the path similarity, and express it as $f_i(T)$, then $f_i(T)$ It can be calculated by the following formula.

$$f_i(T) = \frac{|q(T_i) \cap q_j|}{\max\{q(T_i), q_j\}} \tag{3}$$

In formula (3): $f_i(T)$ Indicates the path similarity. After the above calculation is completed, the problem of solving test data generation can be transformed into solving $f_1(T), f_2(T), f_N(T)$ The fitness function of the maximum value problem is as follows.

$$H = \max(f_1(T), f_2(T), \cdots, f_N(T)) \tag{4}$$

In formula (4): H Represents the fitness function of test data. According to the above method, the research on the construction of test data fitness function based on adaptive differential evolution algorithm is completed.

4 Selection of Optimal Individual Data in Software Test Data

After completing the above design, in order to select the best individual data in the software test data and seek the best test data, it should be clear that in the GPSO, particles evolve in the direction of the global optimal particles, and more obviously converge to the current optimal solution. Each time the particle position is updated, the characteristics of all particles in the population are integrated, and the information transmission speed is fast, but the population diversity is easy to lose, and the probability of falling into the local extreme value is large. In the local model LPSO, each particle only shares information with its neighbor nodes, which slows down the speed of information transmission, but the diversity of the population is guaranteed and it is not easy to fall into local extremum.

Therefore, based on the analysis of the impact of different fixed forms of neighbor patterns on the algorithm performance, it is found that the higher the average connectivity of social interactions between particle individuals, the faster the information transmission speed in the population, and the premature convergence phenomenon is more prone to occur. In view of this, a hybrid model MPSO is proposed, that is, by observing the diversity index of particle swarm, each generation of particles selects GPSO or LPSO for evolutionary optimization, so as to realize the selection of the best individual data in software test data. In this process, describe the diversity of particles, that is, analyze the diversity of software test data in space. This process is shown in the following calculation formula.

$$I = \frac{1}{|k| \times |L|} \sum_{i>1}^{N} \sqrt{\sum_{d>1}^{D} \left(x_{id}^t - x_d^t\right)^2} \tag{5}$$

In formula (5): I Represent the diversity of software test data in the space; k Represents the size of particles in the search space; L Indicates the maximum diagonal length in the search space. After calculation, output I The calculation results of I It represents the dispersion level of each particle in the community, I The larger the value, the more dispersed the population; I The smaller the value, the more concentrated the population, and the lower the diversity of the population.

Therefore, by observing the group I The feedback information automatically adjusts the topology of particle swarm to maintain the diversity of the community, maintain a certain population density, and avoid the algorithm falling into the local optimal value. On the basis of the above I The population with higher values will be globally and locally optimized, and the process is shown in the following calculation formula.

$$W(I) = \frac{I_i - l_{\min}}{e^{l_{\max} - l_{\min}}} \tag{6}$$

In formula (6): $W(I)$ Represents the best individual data in software test data; $l_{\min}$ Represents the minimum branch nesting depth; $l_{\max}$ Represents the maximum nesting depth; e Indicates the depth of the current branch. Complete the selection of the best individual data in the software test data according to the above method.

5 Optimal Individual Clustering and Optimization of Software Test Data Set

On the basis of the above contents, extract the optimal individual data from the software test data selected in the above way, introduce the selection criteria of scaling individuals into the adaptive scaling factor, and cluster the optimal data individuals. The process is shown in the following calculation formula:

$$r = \sum_{i>1}^{n} Mr_i / \sum_{i>1}^{n} r_n \tag{7}$$

In formula (7): r Represents the optimal individual clustering; M Represents the cluster center. After completing the above calculation, the adaptive differential evolution algorithm is applied to automatically generate the model of test data, as shown in Fig. 2.

The construction of the test environment is the basis for the generation of test data. In this stage, the source program is statically analyzed, and the coded test program flowchart

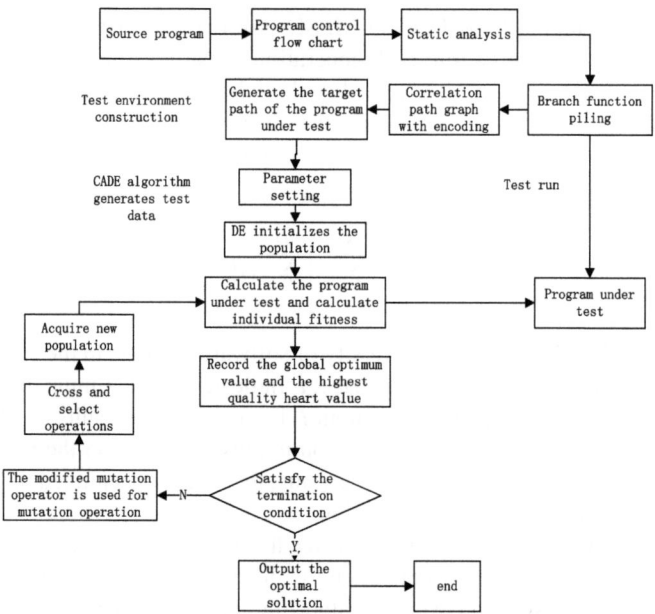

Fig. 2. Optimal solution of output software test data

generated by branch function instrumentation is used to generate the target path. This method is more concise than the traditional data flowchart.

However, as far as the current research is concerned, the existing methods assume that the mass of each particle in the population is the same, and do not take into account the differences between the particles. Therefore, in the optimization process, the process of outputting the optimal solution is improved. On the basis of the existing algorithms, a mutation operator based on the center of mass is proposed. The improved mutation operator can be calculated by the following formula.

$$A(t) = x_i(t) + F(r + c_1 r_1 + c_2 r_2) \tag{8}$$

In formula (8): $A(t)$ Represents the improved mutation operator; F Represents the mass of individual particles. After the selection of the improved mutation operator is completed, in order to break the stagnation of evolution, jump out of the local best, let the operator fluctuate in size, and by limiting its value range, make the difference dynamic scaling, and search for optimization in a large range, theoretically it can improve the ability of global optimization. In the early stage of evolution, the operator tries to maintain the diversity of the population, and takes the small value first; in the later stage, the drill takes the large value to speed up the convergence speed. In this paper, the corresponding generation operator of normal distribution random number is used as follows:

$$\beta = normrnd(U \times F) \cdot A(t) \tag{9}$$

In formula (9): β Represents the generating operator process; *normrnd* Represents the deviation compensation value; U Represents a normally distributed random number.

By compensating the deviation of the operator generation process, the software test data can be better optimized within the optimal range, which is helpful to jump out of the region where the local extreme value is located quickly.

The cross factor plays a fine tuning role, and the appropriate value can maintain the diversity of the population. If the crossover probability decreases, the candidate will contain more target individuals. If the crossover probability increases, the weight of variant individuals in the newly generated candidate will increase.

At the early stage of the iteration, the diversity of the population is relatively high, so cross operation with a small probability can maintain the diversity of the population. As the number of iterations increases, the diversity of the population is gradually losing and approaching the extreme point. Therefore, local search should be increased to accelerate the convergence of the algorithm, so as to approach the extreme point at a faster speed.

In this process, it should be clear that generating test data is the core part of the whole model based on the adaptive differential evolution algorithm. In this stage, the population is initialized according to the target path and test data range constructed by the test environment, combined with the setting of parameters required by the algorithm, and the population is updated by running the improved mutation operator to guide the population to evolve towards the optimal value; And use the test data to run the program to be tested after pile insertion to calculate the fitness value. When the fitness value meets the needs of software testing, complete the data optimization design.

According to the above method, complete the optimal individual clustering and software test data set optimization, and realize the design and research of software test data optimization method based on adaptive differential evolution algorithm.

6 Comparison Experiment

The research on optimizing software test data using adaptive differential evolution algorithm has been completed from four aspects above. In order to test the application effect of software test data optimization methods, the following will take the tested software program provided by a teaching and research institute as an example to design a comparison experiment and carry out the research as shown below.

In order to ensure that the designed method can play its expected role in the test, the environment in which the software program under test runs is described before the experiment, and the relevant contents are shown in Table 1.

According to the above contents, after completing the design of the technical parameters of the comparative experimental environment, the method designed in this paper is used to optimize the data of the tested software. In the optimization process, it assists modern intelligent algorithms such as GA to generate software test data. On this basis, the adaptive differential evolution algorithm is introduced to construct the fitness function of the test data. In order to ensure that the constructed fitness function can be used as the basis for evaluating the software test data, the parameters of the algorithm in the computer can be set according to Table 2 below during the iteration of the adaptive differential evolution algorithm.

After setting the experimental parameters, select the best individual data in the software test data, and optimize the software test data set by clustering the best individual data.

Table 1. Technical Parameters of Comparative Experimental Environment

S/N	Project	Parameter
1	Experimental programming language	Java Language
2	Experimental program running environment	Eclipse environment
3	Microcomputer environment	Windows operating system Intel (R) Core (TM) i3 CPU 2.0GHz
4	Computer running memory	2GB RAM

Table 2. Training parameter settings of adaptive differential evolution algorithm

S/N	Project	parameter
1	Population iterations (times)	1000
2	Operator crossover probability (%)	0.6
3	Operator compilation probability (%)	0.01
4	Number of independent runs for each case of different programs (times)	50
5	Data cycle branch (s)	6
6	Select nesting quantity (pcs)	3
7	search space	3
8	Learning factor	1.5
9	Inertia weight	0.9
10	Diversity threshold in algorithm	0.1

On the basis of the above content, software test data optimization method based on particle swarm optimization (reference [1] method) and software test data optimization method based on genetic algorithm (reference [2] method) are introduced, and the introduced method is regarded as traditional method 1 and traditional method 2. The method in this paper and the two traditional methods are used to optimize software test data.

In the optimization process, three software test data populations of different scales are selected, and three methods are used to evolve the software test data respectively. The average length of time required by the three methods to optimize the software test data is compared, which is the key basis for testing the application effect of the method in this paper. The statistical experimental results are shown in Table 3.

From the experimental results shown in Table 3 above, it can be seen that the average time required for optimization of software test data using this method is less than 1s, while the average time required for optimization of software test data using traditional methods is significantly higher than the average time required for optimization of data using this method.

Table 3. Average time required for three different methods to optimize software test data

S/N	Population size	Average duration (s)		
		Methods in this paper	Traditional method 1	Traditional method 2
1	50	0.125	0.569	0.639
2	100	0.156	0.693	0.678
3	150	0.269	0.854	1.069
4	200	0.489	0.910	1.598
5	300	0.526	1.025	1.756
6	400	0.963	1.156	1.963
7	500	0.956	1.569	3.051

After completing the above experiment, set the input range of software test data to [0,64], [0512], [1024], use three methods to optimize the software test data, and compare the average evolution times of the three methods on the input data during the optimization process. The results are shown in the following figure:

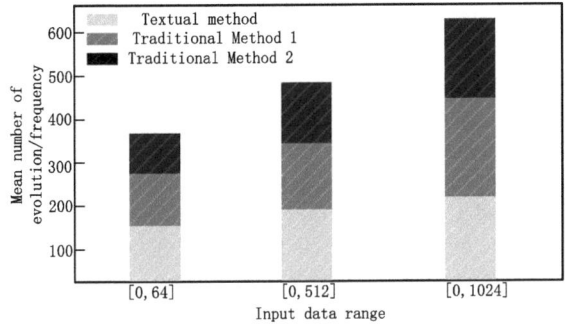

Fig. 3. Average evolution times of three methods for input data in different ranges during optimization

It can be seen from Fig. 3 above that the three methods optimize the training of input data in different ranges. With the increase of input data range, the average evolution times of the three methods show an increasing trend. Use this method to optimize [1024] software test data, and the average number of evolutions in the optimization process is less than 300;The traditional method 1 is used to optimize software test data, and the average evolution times are less than 500;The traditional method 2 is used to optimize the software test data, and the average evolution times are less than 700.Based on the above results, it can be seen that among the three methods, only using this method to optimize software test data can ensure that the average evolution times are at a relatively low level, while using traditional method 1 and traditional method 2 to optimize software test data, the corresponding method has more average evolution times, that is, the data optimization process is more complex.

After the above design is completed, several software test paths are set according to the software test requirements. It is known that different software test paths require different software test data sizes and categories. After comparing the three methods to optimize the software test data, the generated test data can cover the extent of the target test path, that is, compare the availability of the optimized corresponding data in the software testing process. According to the above method, carry out a comparison experiment and make statistics of the experimental results, as shown in Fig. 4.

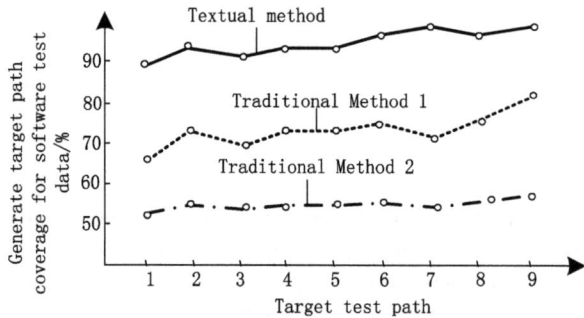

Fig. 4. Target path coverage corresponding to software test data generated by the three optimized methods (%)

From the experimental results shown in Fig. 4 above, it can be seen that the method in this paper is used to optimize the software test data, and the target path coverage corresponding to the generated software test data after optimization is > 90%, while the target path coverage corresponding to the generated software test data after optimization in the traditional method 1 is between 60% and 80%;The target path coverage corresponding to the software test data generated after optimization of traditional method 2 is between 50% and 60%.

Based on the above experimental results, the following experimental conclusions are obtained: compared with the traditional methods, the software test data optimization method designed in this paper using adaptive differential evolution algorithm has a good effect in practical application. This method can improve the target path coverage corresponding to the generated software test data on the basis of controlling the time required for software test data optimization and the number of evolutions. In this way, it provides further support and guidance for the standardized implementation of software program testing and other related scientific research work.

7 Conclusion

In order to find potential software defects as much as possible and improve the coverage of target paths corresponding to software test data generation, this paper conducts testing research on the optimization method of software test data through software test data generation, construction of test data fitness function based on adaptive differential evolution algorithm, selection of optimal individual data in software test data, optimal individual clustering and optimization of software test data set. After completing the design of this method, taking the software program under test provided by a teaching and research unit as an example, a comparative experiment was designed to prove that the method designed in this paper has a good effect in practical application. This method can improve the coverage of the target path corresponding to the generated software test data on the basis of controlling the length of time required for software test data optimization and the number of improvements.

References

1. Li, L.: Defect optimization method of high-performance traffic analysis software based on particle swarm optimization. J. Anhui Tech. College Water Resour. Hydroelectric Power **21**(4), 56–59 (2021)
2. Bo, Y., He Yuze, X., et al.: IGA: Software fault location assisted test case generation method using improved genetic algorithm. J. Beijing Univ. Aeronaut. Astronaut. **48**(3), 1–13 (2022)
3. Xiaoyan, Z., Qianqian, L., Yawei, Q.: An adaptive differential evolution algorithm with population size reduction strategy for unconstrained optimization problem. Appl. Soft Comput. J., 138 (2023)
4. Dapeng, N., Xudong, L., Yudong, T.: Operation optimization of circulating cooling water system based on adaptive differential evolution algorithm. Int. J. Comput. Intell. Syst. **16**(1) (2023)
5. Wei, L., Yafeng, S., Ying, H., et al.: An adaptive differential evolution algorithm using fitness distance correlation and neighbourhood-based mutation strategy. Connection Sci. **34**(1) (2022)
6. Kumar, G.S., Kumar, M.A.: A self-adaptive differential evolution using a new adaption based operator for software cost estimation. J. Instit. Eng. (India): Ser. B **104**(1) (2022)
7. Jiali, C., Xiaojie, C., Tao, Z., et al.: An automatic generation of software test data based on improved Markov model. Web Intell. **20**(4) (2022)
8. Ricardo, F.V., João, C.N., Hugo, S.C.P.V., de Souza Paulo Sérgio, L., de Souza Simone, R.S.: Bio-inspired optimization to support the test data generation of concurrent software. Concurr. Comput. Pract. Exper. **35**(2) (2022)
9. Kattuva, M.A.K., Narayanan, N., Mangottiri, V., et al.: Parameter evaluation of a nonlinear Muskingum model using a constrained self-adaptive differential evolution algorithm. Water Pract. Technol. **17**(11) (2022)
10. Mehdi, E., Hossein, D.A.: Automation of software test data generation using genetic algorithm and reinforcement learning. Expert Syst. Appl., 183 (2021)

A Hybrid Swarm Intelligence Algorithm Based Approach for Information Integration of English Language Database Calls

Xue Che[1,2(✉)] and Yuefei Wang[1,2]

[1] School of General Education, Changchun College of Electronic Technology, Changchun 130000, China
Colleen0104@163.com
[2] Liaodong University, Dandong 118000, China

Abstract. The information in English databases is scattered and formatted differently, making it difficult to integrate and utilize. This paper proposes an information integration method for English language database calls based on hybrid swarm intelligence algorithm. Build a middleware based information retrieval system model to retrieve information in the English language database, obtain information retrieval results of different member systems, combine Bat Algorithm(BA) and Cuckoo Search(CS) to form a hybrid group intelligent algorithm, find the optimal solution of the weight of each member system of the information retrieval system, and realize the weight distribution of member systems. The linear combination method is used to integrate the retrieval information under each system according to the weight of different member systems, and the information call results in the English language database are obtained. The experiment shows that the hybrid swarm intelligence algorithm integrates English database call information, improves performance, and balances timeliness.

Keywords: Hybrid Swarm Intelligence Algorithm · English Language Database · Database Calling Information · Calling Information Integration · Information Integration Methods

1 Introduction

Under the background of global informatization, English language database, as an important carrier for storing and disseminating knowledge, has become more and more prominent in various fields, such as natural language processing, machine translation, intelligent question and answer, etc. The demand for effective calling and integration of English language database information in these fields is growing, therefore, it is especially urgent to study the method of calling information integration in English language database. Traditional methods of information integration for English language database calling, such as keyword matching, Boolean logic operation, vector space modeling, etc., are mainly based on text content for information retrieval. Although these methods can show a certain degree of efficiency and accuracy when dealing with small-scale data, their limitations gradually appear with the explosive growth of data volume. For example, when dealing with large-scale data, traditional information integration methods are difficult to meet real-time requirements due to their high computational complexity; At the same time, these methods tend to focus only on the text content itself, ignoring the in-depth understanding and utilization of semantic information, resulting in a serious impact on the accuracy and comprehensiveness of information retrieval.

The rapid advancement of AI and big data technology has led to significant breakthroughs in the field. Especially, technologies based on convolutional neural network, and variational self encoder, provide new ideas and methods for the integration of English language database call information. These technologies can better capture the semantic information of text and improve the accuracy of information retrieval; At the same time, they can also automatically extract text features to achieve effective modeling of complex semantic relations. However, although these new technologies have great potential in theory, they still face many challenges in practical application. For example, Ahmad et al. [2] determined the information needs of software professionals in the process of continuous integration and delivery through the overall study of multiple cases, and discussed the challenges of developing visualization tools and expert suggestions. However, there may be other specific information needs and challenges in the integration of English language database calls, which need further research and exploration. Li et al. [3] Propose a digital library archive information integration scheme based on multidimensional data mining, and use support vector mechanism modeling to achieve integration. Faced with diverse data sources and varying quality, additional processing strategies are required. Kushwaha A et al. [4] Propose an efficient convolutional network architecture for identifying human activities and multi-level fusion of information. Integrating complex data into English databases requires comprehensive processing of multimedia such as text, images, and audio. Gebremichael et al. [5] proposed a method to integrate facility lifecycle information by introducing the main decomposition structure. Experiments have proved that this is an effective method, and can minimize the time and energy required by practitioners in information transformation and exchange. At the same time, this method also helps to simplify the integration process of the secondary subdivision structure in a complex range, and helps the industry effectively use advanced technology. In the information integration process of English language database calls, different fields and standards are involved, and the main decomposition structure cannot unify and standardize the information format and content of various data sources.

Propose a hybrid swarm intelligence algorithm to integrate English database call information. By combining BA and CS algorithms to optimize the weight allocation of member systems, efficient integration and utilization of information are achieved. Its superiority and innovation lie in significantly improving the performance of integration results and effectively solving the problems of information dispersion and inconsistent formats.

2 Searching for Information in the English Language Database

The first step in English database integration is to obtain data through an information retrieval system. The information retrieval system model of the English language database is based on the existing network layout, the use of English language data that have been completed or under construction, built on a distributed, heterogeneous network database system model, the system model should have the following characteristics:

(1) Transparently realize distributed and heterogeneous database access. Each database system may run in Windows, Linux or Unix environments. Clients should be able to access heterogeneous databases distributed in various regions with a unified interface.
(2) The client application software should be as simple as possible.
(3) It can carry out effective load balancing and security protection for the client's database access request.
(4) The system software should have good scalability, reusability and portability.

Therefore, the information retrieval system model designed in this paper adopts a four tier architecture based on middleware. The system model includes two layers of middleware, data service middleware and communication middleware. Data service middleware, which is composed of various service objects, provides services for customers to request database, acts as the intermediary between customers and database interaction, and realizes the transparency of database access to customers; The communication middleware provides services for the communication between customers and servers. Customers can transparently access the objects and database servers in the data service middleware through it. The communication middleware communicates between different platforms to achieve reliable, efficient and real-time data transmission in the distributed system. Therefore, the information retrieval system model is divided into four relatively independent logical units: representation part, functional logic part, data service part and communication service part.

Presentation layer (client): mainly deals with the interaction between the system and users. This system model will provide users with a unified Web search interface, and the user search results will also be presented in the Web mode. By using the Web mode, the client does not need to install any applications, which provides convenience for users in the network environment and forms a thin client working mode.

Communication service layer (communication agent): use the communication agent to provide a transparent path search service for the customer to locate the service object, and mask the object location, object status and other information for the user, making it as convenient for the user to call the local process or access the local database. The user only needs to make a request, without paying attention to how to find the object to be accessed.

Functional service layer (application server): responsible for the implementation of specific business, it is the medium for interaction between the presentation layer and the data service layer.

The components of this layer execute specific transaction logic, receive data retrieval request packets from the presentation layer, and request data and other resources to the data service layer through SQL, API, etc., and receive the return results from the data service laye.

Data service layer (database server): complete data storage, data integrity constraints, process data requests and access from the application service layer, return data result packets to the application service layer, and communicate between the data service layer and the application service layer in database API mode.

The core of the above four layer model based on middleware is how to realize the automatic positioning of distributed objects by customers, so that customers and application servers can interact transparently. Therefore, this system model uses CORBA of OMG to realize this core function. As a communication standard of distributed systems, CORBA provides technical support for the construction of the four layer model. Then, according to the powerful functions provided by CORBA, this paper constructs the four layer model of information retrieval system based on CORBA middleware as shown in Fig. 1.

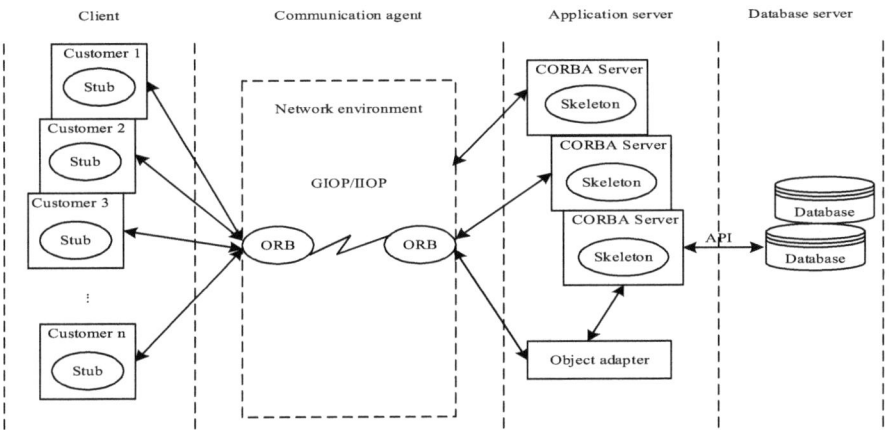

Fig. 1. Model of a middleware-based information retrieval system

As shown in Fig. 1, the central component of CORBA is the Object Request Broker (ORB), which facilitates the seamless transmission of requests and receipt of responses among objects. Based on ORB mechanism, we can make full use of distributed and interoperable objects to construct interoperable system models. The ORB core provides a method for transparent communication between client and object implementations. It can mask object locations, implementation methods, states, and communication mechanisms, and pass method calls and information back and forth between the client and the server. Its responsibilities include registering service objects and generating and interpreting object references, the activation and freezing of service processes implemented by objects, the activation and freezing of object implementations, and the distribution of customer requests.

IDL STUB (stub) is responsible for encoding the customer's request, sending the request to the server, explaining the received processing results, and returning the results or error information to the customer. IDL SKELETON (framework) decodes the customer's request, calls the method of the requested object, executes the method, and encodes the execution result or error information and returns it to the customer. In this model, the application service middleware is composed of service objects that comply with CORBA specifications, and provides mediation services for client data requests.

On the server side, define a unique name identifier on the interface for each object providing services, and use this identifier as the index for the client to access the server object. Find the matching object through the navigation of the transparent communication agent. When building a client, only the server IP and service interface are known, without the need to know implementation details or machine names.

The specific process of realizing information retrieval in English language database based on the four layer model of information retrieval system based on CORBA middleware is as Table 1:

Table 1. The process of information retrieval in English language databases

Step Number	Algorithm Step Description
1	Identify Requirement and Locate Object
2	Construct Request and Load IDL STUB
3	Send Request to ORB Core
4	ORB Core Forwards to Object Adapter
5	Object Adapter Invokes IDL SKELETON
6	Service Execution and Database Interaction
7	Result Processing and Return to Client

Since the information retrieval system model based on CORBA middleware can be integrated in many clients, there are different methods and steps in the member system of the retrieval system model carried by each client, such as parsing rules, stemming, phrase processing, relevant feedback and query expansion; At the same time, different use processes of internal components or services of information retrieval systems running on different clients will also produce different retrieval results. Therefore, under the operation of the four tier model of information retrieval system based on CORBA middleware, information retrieval results in English language databases of different member systems can be obtained. Therefore, the integration of English language database call information studied in this paper is a technology that can combine and reorder the retrieval results returned by different member systems to generate a new retrieval result.

3 Integrating Retrieval Results based on Hybrid Swarm Intelligence Algorithms

The information retrieval system model based on CORBA middleware can be used to retrieve the required information in English language databases from many clients. However, due to the large amount of information in English language databases and the existence of semantic diversity and ambiguity, the retrieval results may not be accurate enough. The bat algorithm is good at adjusting the position of the bat at a specific frequency to achieve global search and local search, while the cuckoo algorithm focuses on cooperation and competition between individuals, which is conducive to accelerating convergence to the optimal solution. Hybrid bat and cuckoo algorithms, complementary search strategies, optimized information integration.

3.1 Hybrid Swarm Intelligence Algorithm for Assigning Weights to Member Systems

Information integration aims to integrate retrieval results from different systems and improve retrieval efficiency. There are two types of integration methods: same processing and different processing. Same processing as CombSUM, uniformly process the scores of each system, sum up to obtain the global score, and sort to obtain the document order.

When using the CombMNZ method, the document scores of each system's retrieval results are accumulated and multiplied by the number of systems to obtain the global score, which is then sorted to obtain the comprehensive result. However, the reliability of this method is affected by differences in system contributions. This article innovatively adopts a differential processing optimization integration strategy. The differential processing method implements differentiated weighting and optimizes data integration based on the differences in the characteristics of each retrieval system. In the data integration methods with different processing methods, weight allocation is the most important factor affecting its integration performance, which is mainly weighted according to the specific characteristics of member systems and their contributions in the fusion process. Because the characteristics of the selected member systems are different, many different weight allocation strategies are generated. This paper mainly studies a weight allocation strategy based on intelligent optimization algorithm.

Intelligent optimization algorithm is a kind of optimization algorithm using "intelligence", which is mainly inspired by human intelligence, biological community sociality or natural phenomenon laws. These algorithms have a solid theoretical foundation that goes beyond simple experience and theoretically ensures reaching the optimal or approximate optimal solution within a finite time. Although intelligent optimization algorithm has many advantages, it also has some challenges and limitations. For example, some intelligent optimization algorithms may require a lot of computing resources and time to find the optimal solution, especially when dealing with large-scale or high-dimensional problems. This article combines bat and cuckoo algorithms, innovates hybrid swarm intelligence algorithms, and optimizes the weight allocation of member systems.

The BA algorithm was proposed by Professor Yang Xinshe in 2010 to simulate bat echolocation behavior and achieve heuristic search [6]. In the BA algorithm, each virtual bat flies in the solution space and has random flight speed, frequency, loudness and pulse emissivity. The parameters are iteratively adjusted with the algorithm to find the optimal solution, and the formula is:

$$g_i^{t+1} = g_{\min} + (g_{\max} - g_{\min}) \times \delta \tag{1}$$

$$v_i^{t+1} = v_i^t + (x_i^t - x^*) \times g_i^{t+1} \tag{2}$$

$$x_i^{t+1} = x_i^t + v_i^{t+1} \tag{3}$$

g_i^{t+1} is the $t+1$-generation search frequency of bat i, $g_{\min}$ and $g_{\max}$ are the upper and lower limits of frequency, and $\delta \in [0,1]$. v_i^{t+1}. v_i^t is the speed of i in $t+1$ and t generations; x_i^{t+1}. x_i^t represents i at positions $t+1$ and t; x^* is the optimal solution position. Together, these formulations of the bat algorithm simulate the behavior of bats when searching for prey, allowing the algorithm to search and optimize efficiently in the solution space.

CS algorithm is an optimization algorithm to simulate the behavior of cuckoo's brood, which combines the search mode of Levy Flights to conduct efficient global optimization [7]. In the CS algorithm, cuckoo simulates parasitism and Levy flight to update the position and find the optimal solution. The optimal position during iteration is given by the following equation:

$$X_i^{t+1} = X_i^t x_i^{t+1} + \beta \oplus L(\lambda) \tag{4}$$

In the formula, X_i^{t+1} and X_i^t are the position vectors of i during $t+1$ and t iterations; β is the step factor (usually taken as 0.01), $\oplus$ is the dot product; $L(\lambda)$ is a random path. CS integrates global and local search: Levy flight helps to escape from local optima, and parasitic behavior enables local search.

This article combines the advantages of BA and CS algorithms to create a hybrid swarm intelligence algorithm for finding the optimal solution of member system weights [8], as shown in Table 2.

According to the above steps, it can be seen that through continuous iteration and optimization, the hybrid swarm intelligence algorithm can iteratively converge towards the optimal solution for the weights of the individual systems, enabling the rational allocation of their weights. This in turn establishes a strong basis for the subsequent differentiated processing within the information integration method.

3.2 Linear Combination Method for Integrating Member System Search Results

In the information retrieval of English language databases, the invocation of information integration technology is also known as meta-search. In the information integration of this article, the member system retrieves user queries and integrates the results to generate new results [9], which serve as the final English database call results. This article adopts the differential processing integration method, based on the mixed swarm intelligence algorithm to obtain member weights, and uses linear combination method to integrate information. This technology is concise and efficient.

Its basic principles are according to the weight of each information source, the information provided by them is linearly weighted, so as to obtain a comprehensive, integrated the advantages of each information source results [10]. The specific English language database call information integration process is shown in Fig. 2.

As shown in Fig. 2, in the context of this paper, the linear combination method is used to integrate the retrieval results of different member systems into a unified and comprehensive calling result. This method not only fully considers the characteristics of each member system, but also ensures the comprehensiveness of the integration result

Table 2. Hybrid Swarm Intelligence Algorithm

Step Number	Algorithm Step Description
Step 1	Initialization: Set the parameters for the BACS algorithm, including bat flight velocity, search pulse frequency, number of nests, foreign egg discovery probability, etc. Define the set of member system weight values as the search solution space
Step 2	Fitness Calculation: Calculate the fitness value of each nest in the CS algorithm based on the fitness function, and determine the current optimal nest position and optimal value
Step 3	Nest Position Update (CS Part): Update the nest positions using the Levy Flight mechanism and apply boundary constraints
Step 4	Evaluation and Update (CS Part): Calculate the fitness values of the new nest positions obtained in Step 3, compare them with the previous generation, and update the nest positions and optimal values if necessary
Step 5	Foreign Egg Mechanism (CS Part): Compare the foreign egg discovery probability with a uniformly distributed random number. If the random number is greater, randomly change the position of some nests to obtain a new set of nest positions
Step 6	Conversion to BA Algorithm Initial Points: Use the new nest positions obtained in Step 5 as the initial points for the BA algorithm, and further update the nest positions using the BA algorithm rules (Eqs. (1) to (3))
Step 7	Evaluate BA Algorithm Results: Calculate and evaluate the fitness values of the nest positions updated by the BA algorithm in Step 6, and determine the current optimal nest position and optimal value
Step 8	Iteration Check: Check if the iteration termination conditions are met (e.g., maximum number of iterations reached, fitness value convergence). If so, exit the program and output the optimal solution and optimal value; otherwise, return to Step 3 for the next iteration

by weighted summation [11]. After setting the weights, the linear combination method multiplies the weights by the scores of each system document and then sums them up to obtain the global score. The formula is as follows:

$$G(d) = X_i^{t+1} \sum_{i=1}^{n} \omega_i s_i(d) \tag{5}$$

In the formula, the $G(d)$ denotes the result of the information retrieval system's return of the information integration call in the English language database, where d is document information. ω_i indicates the weight values of the i-th member systems. The $s_i(d)$ represents the score of the document information d from the i-th member system to the English language database; n indicates the number of member systems in the information retrieval system. Through the weighted summation shown in Formula (5), a comprehensive information result is obtained in this paper. This result is the integration of the retrieval results of each member system. It comprehensively considers the weight of each member system and the information they retrieve [12, 13]. To sum

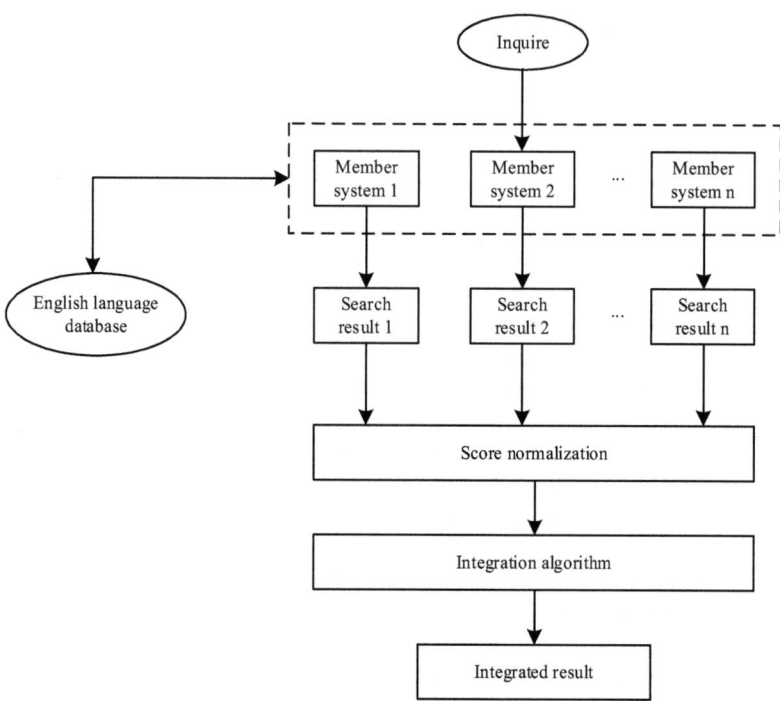

Fig. 2. The process of integrating information from English language database calls

up, this paper designs a differential processing information integration method based on linear combination method. In this method, BA-CS hybrid group [14] intelligent algorithm is used to obtain the best weight value of each member system, and then the retrieval information under each system is integrated according to the weight of different member systems, so as to obtain the comprehensive call results of information in the English language database.

4 Experimental Analysis

4.1 Experimental Setup

This chapter experimentally verifies the feasibility of applying the hybrid swarm intelligence algorithm to database call information integration. The methods involved in the experiments are: CombSUM, CombMNZ, database call information integration method based on multiple linear regression, database call information integration method based on differential evolution algorithm Database call information integration method based on particle swarm optimization, database call information integration method based on genetic algorithm, database call information integration method based on adaptive simplified particle swarm optimization Database call information integration method based on hybrid swarm intelligence algorithm (CombSUM, CombMNZ, MR Fusion, DE Fusion, PSO Fusion, GA Fusion, AESSPO Fusion and BA-CS will be replaced in the following experiment)to evaluate the effectiveness of the method proposed in this article, the TREC conference dataset was used, which is an authoritative database for retrieving and evaluating conference outcomes. The dataset used in this paper is TREC 2004 Robust Task. The main reason for selecting it is that the list of retrieval results documents of TREC 2004 Robust Task contains 249 queries, which is far more than other TREC conferences. This is more conducive to testing the reliability of intelligent optimization algorithms in data integration than using other TREC datasets. In the TREC 2004 Robust Task, a total of 14 academic organizations in the field of information retrieval participated in the competition and submitted the final call results, with a total of 110 results. It is divided into 14 groups according to the submitted organizations. The search results of each member system include 249 queries, and the query number range is 301 ~ 450, 450 ~ 700 (query number 672 has no list of documents related to the query, so it is not listed as experimental data).

The parameter settings of the hybrid swarm intelligence algorithm in this article are detailed in Table 3.

Table 3. Parameters of the hybrid swarm intelligence algorithm

Algorithm	Parameters	Numeric setting
The bat algorithm	Number of individual bats	50
	Number of bat iterations	100
	The self-positioning update factor	0.5
	Frequency adjustment factor	0.5
	Minimum frequency	0
	Maximum frequency	2
The cuckoo algorithm	Cuckoo population size	30
	The number of Cuckoo iterations that	50
	Migration rates	0.2
	Individual foraging ranges	[-100, 100]
	Migratory step ratio	0.01

4.2 Experimental Methods

In the experiment, the first step is to standardize the scores of all the system's result list documents by using the 1/(rank + 60) reciprocal ranking method. After normalization, the query in each group of experimental data is divided into two groups by number, odd array and even group, using the binary cross validation method. First, use several weight optimization methods described in this article to train the weights of the first 100 documents in the search results in the even number group, and then use the linear combination method to integrate the obtained system weights on the document results in the odd number group, and then, conversely, train the odd array and test the even array, and finally obtain the search results under all queries. The evaluation index MAP formula is:

$$MAP = \frac{1}{|K|} \sum_{i=1}^{|K|} \frac{1}{N} \sum_{j=1}^{N_i} \Pr e(F_{ij}) \qquad (6)$$

In the formula, the MAP denotes the average precision mean of the integration results of the database call information; the K indicates the total number of database information calls. $\Pr e(F_{ij})$ represents the accuracy of the i-th database information retrieval results of the j-th member system. N is the number of systems. The MAP value MAP ranges from 0 to 1, and the closer it is to 1, the better the performance of database call information integration.

The experiment was divided into 3 parts:

(1) To investigate the impact of iteration times on information integration, this article conducted 6 experiments with cycle times ranging from 50 to 300. In each group of experiments, from 14 groups of 100 member systems in TREC 2004 Robust Task, each group randomly selected one member system, and a total of 14 member systems were randomly integrated for 50 times. Finally, the results were averaged.
(2) The impact of the number of member systems participating in the integration on the results of information integration. Test the impact of the number of retrieval systems on the performance of integration methods, and conduct integration experiments on different number of member systems. Select different groups from 14 groups of 110 member systems in TREC 2004 Robust Task for integration experiments (from 4 to 14 groups). For the same group of experiments, randomly select a member system from each group to conduct 50 random integration experiments, and then average the performance indicators of the results.
(3) Comparison and analysis of the time spent in integrating algorithms. In this experiment, the time consumed by eight integration methods in the experiment was compared and analyzed. The computer configuration used in the experiment was: CPU: Intel Core i7–6700, RAM: 32GB.

4.3 Experimental Results

4.3.1 The Effect of the Number of Iterations on the Results of Information Integration

According to the experimental setup, select data for testing, and the results are shown in Table 4.

50, 75, 100, 150, 200, and 300 represent the number of iterations of the linear combination method based on the weight allocation strategy of the intelligent optimization algorithm, while the CombSUM, CombMNZ, and MR Fusion algorithms do not need to perform cyclic iteration operations. Table 4 shows that the performance of DE Fusion and other three methods is superior to the best member and traditional ensemble methods. As the number of iterations increases, the performance of the intelligent optimization ensemble method improves, and the BA-CS algorithm has the highest MAP value.

Table 4. Integration results of eight information integration methods with different number of iterations

	50	75	100	150	200	300
DE-Fusion	0.6724	0.6729	0.6725	0.6727	0.6731	0.6734
PSO-Fusion	0.6481	0.6454	0.6436	0.6437	0.6389	0.6457
GA-Fusion	0.6465	0.6493	0.6457	0.6477	0.6451	0.6463
AESPSO-Fusion	0.6732	0.6773	0.6741	0.6749	0.6753	0.6768
BA-CS	0.7129	0.7273	0.7156	0.7198	0.7172	0.7138
CombSUM	0.6133					
CombMNZ	0.6147					
MR-Fusion	0.6386					
Best comp	0.6573					

4.3.2 The Effect of the Number of Member Systems Involved in Integration on the Results of Information Integration

Experimental control of the number of member systems and analysis of their impact on the performance of ensemble methods, including CombSUM, etc. The results are shown in Fig. 3.

As shown in Fig. 3, among all information integration methods, the BA-CS based linear combination method has the best performance, followed by DE Fusion, AESSPO Fusion and MR Fusion, PSO Fusion and GA Fusion, CombSUM and CombMNZ. Among the eight integration methods, except CombSUM and CombMNZ, the performance of the other six integration methods is better than the average value of the optimal member system. Moreover, with the gradual increase of the number of member systems participating in the integration, it is conducive to the improvement of BA-CS integration performance.

4.3.3 Comparison and Analysis of Time Spent on Information Integration Methods

The time consumption for the training and integration phases of each of the eight database information integration methods is shown in Table 5.

Table 5 shows that the intelligent optimization algorithm requires training data weights, and the BA-CS algorithm has the shortest time consumption.

To sum up, this chapter mainly examines and analyzes the feasibility and efficiency of the intelligent optimization algorithm applied to database information integration through three sets of experiments, and compares and analyzes the intelligent optimization method with the traditional information integration method. Experiments have shown that intelligent optimization algorithms improve database retrieval performance, with the BA-CS algorithm being particularly effective. Considering both time and accuracy, it is the optimal method.

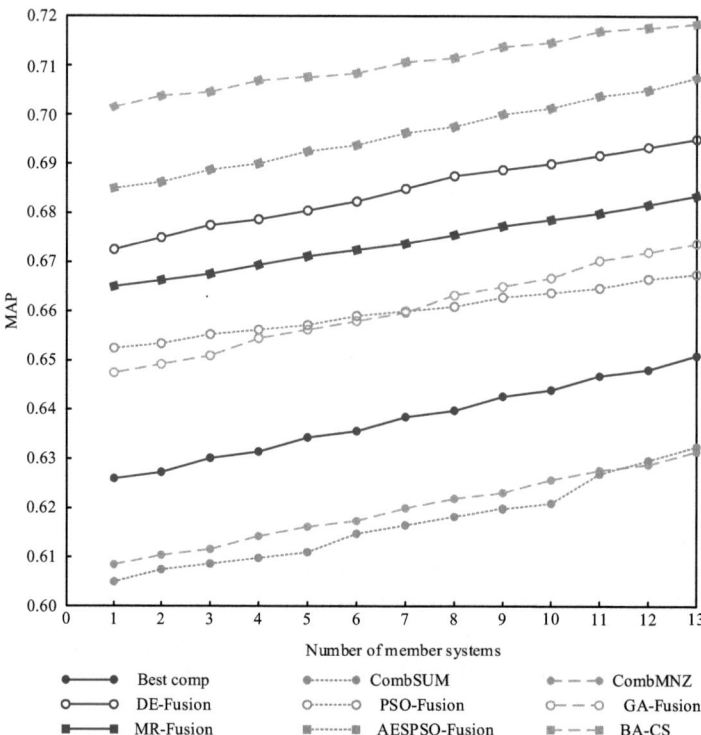

Fig. 3. Curve of integration results of database call information for different number of member systems

Table 5. Time taken to integrate the methods

Information integration method	Training consumes time/s	Integration cost time/s
DE-Fusion	263.81	1.821
PSO-Fusion	403.21	1.932
GA-Fusion	396.82	1.916
AESPSO-Fusion	22.64	1.724
BA-CS	210.55	1.503
CombSUM	/	1.528
CombMNZ	/	1.569
MR-Fusion	256.30	1.799

5 Conclusion

Research on integrating English database information with hybrid swarm intelligence algorithms to improve call efficiency and accuracy. Combining bat and cuckoo algorithms, build a middleware model to achieve fast retrieval and integration. Experimental results have shown that this method significantly optimizes the efficiency and accuracy of information transmission. However, this study still has some shortcomings. For example, the efficiency and stability of the algorithm in processing large-scale complex data need to be further improved. In addition, the applicability and universality of this method need to be further verified for English language databases in different fields and types. In the future, we will continue to optimize hybrid swarm intelligence algorithms and explore more efficient information retrieval integration strategies. At the same time, this paper will expand the application field of this method and provide more comprehensive and accurate information calling and integration solutions for English language databases in different fields and types.

References

1. Peng, C., Zhou, X., Liu, S.: An introduction to artificial intelligence and machine learning for online education. Mobile Networks Appl. **27**(3), 1147–1150 (2022)
2. Ahmad, A., Leifler, O., Sandahl, K.: Data visualisation in continuous integration and delivery: information needs, challenges, and recommendations. IET Software **16**(3), 331–349 (2022)
3. Li, Z.: Digital library archives information integration based on multidimensional data mining. Int. J. Reason.-Based Intell. Syst. **14**(4), 169–175 (2022)
4. Kushwaha, A., Srivastava, P., Khare, A.: Human activity recognition based on integration of multilayer information of convolutional neural network architecture. Concurr. Comput. Pract. Exper. **35**(5), e7571.1--e7571.14 (2023)
5. Gebremichael, D.D., Lee, H., Lee, Y., et al.: Unified breakdown structure for information integration to enhance interoperability in the built environment. J. Comput. Civ. Eng. **36**(6), 1–15 (2022)
6. Wang, F., Cao, Z., Li, Z., et al.: Path generation for a given performance evaluation value interval by modifying bat algorithm with heuristic. Int. J. Software Eng. Knowl. Eng. **33**(05), 787–814 (2023)
7. Chowdhury, A., Roy, R., Mandal, K.: Techno-socio-economic improvements due to optimal location of EV charging station at IEEE 22 bus using cuckoo search algorithm. In: 2022 22nd National Power Systems Conference (NPSC), pp. 77–81 (2022)
8. Sahu, B., Das, P.K., Kumar, R.: A modified cuckoo search algorithm implemented with SCA and PSO for multi-robot cooperation and path planning. Cogn. Syst. Res. **79**, 24–42 (2023)
9. Liang, M., Zhou, K.: A hierarchical deep learning framework for combined rolling bearing fault localization and identification with data fusion. J. Vib. Control **29**(13–14), 3165–3174 (2023)
10. Munnaf, M.A., Haesaert, G., Mouazen, A.M.: Site-specific seeding for maize production using management zone maps delineated with multi-sensors data fusion scheme. Soil Tillage Res. **220**, 105377–105381 (2022)
11. Al-Qassas, R.S., Qasaimeh, M.: Data fusion and the impact of group mobility on load distribution on MRHOF and OF0. Cybern. Inf. Technol. **22**, 77–94 (2022)
12. Huang, S., Fu, W., Zhang, Z., et al.: Global-local fusion based on adversarial sample generation for image-text matching. Inf. Fusion **103**, 102084 (2024)

13. Jia, Y., Mei, F., Sun, P.: Software reliability verification simulation based on dynamic integration of prior information. Comput. Simul. **40**(10), 377–380, 430 (2023)
14. John, J.C.G., Deepika, M., Elavarasan, B.: Hybrid interior ideals and hybrid bi-ideals in ternary semigroups. J. Intell. Fuzzy Syst. Appl. Eng. Technol. **45**(6), 10865–10872 (2023)

A Study on Fuzzy Comprehensive Evaluation of Blended Teaching Quality Based on Multi-source Information Fusion

Lei Ma[1]([✉]), Hongxue Yang[1], Jingyu Li[1], and Jiatong Wei[2]

[1] Beijing Polytechnic University, Beijing 100176, China
malei12669@163.com
[2] College of Automotive Engineering, Jilin Communications Polytechnic, Changchun 130012, China

Abstract. Fuzzy comprehensive evaluation of teaching quality is an important part of blended teaching quality management, which has positive significance and important role in improving the quality of blended teaching, for this reason, we put forward the research on fuzzy comprehensive evaluation of blended teaching quality based on the fusion of multi-source information. By considering learning attitude, process, and effect as key first-level evaluation metrics, we establish a framework for a fuzzy comprehensive evaluation of blended teaching quality. This framework is further expanded by selecting second and third-level indicators. To determine the importance of these indicators, we apply the hierarchical analysis method. Additionally, we incorporate multi-source information fusion technology to conduct a thorough analysis of the indicators, ultimately leading to the calculation of scores and grades for blended teaching quality. This approach offers a robust and comprehensive method for evaluating blended teaching quality using fuzzy comprehensive evaluation techniques combined with multi-source information fusion. The experiment proves that the Spearman's correlation coefficient and consistency coefficient of the evaluation results of the design method are above 0.9, which can realize the fuzzy comprehensive and accurate evaluation of the blended teaching quality.

Keywords: Multi-source information fusion · Blended teaching · Fuzzy comprehensive evaluation · Learning attitude · Learning process · Learning effect

1 Introduction

Since 2014, China has passed the document "Opinions on Comprehensively Deepening Classroom Reform and Implementing the Fundamental Task of Moral Education", which clarifies the direction of comprehensive reform of teaching, textbooks, and evaluation systems, emphasizes the importance of diversified and developmental evaluation, and aims to promote the common progress of teachers, students, and educational practice. In 2019, the Opinion on Deepening Education and Improving the Quality of Compulsory

Education further pointed out that strengthening the quality of basic education, especially the leading role of classroom teaching, requires improving the evaluation system, including quality evaluation, teacher professional skills, and student comprehensive evaluation, and advocates the introduction of process oriented and formative evaluation methods. This series of policies highlights the increasing emphasis on teaching evaluation in China's basic education, highlighting its core position in ensuring teaching quality. With the updating of educational concepts and the development of information technology, traditional teaching models are facing challenges, and the demand for personalized and customized learning is becoming increasingly urgent. In this context, the blended online and offline teaching model has emerged and been successfully implemented abroad, effectively promoting students' independent exploration and deep learning. The release of China's 2020 Key Points of Education Informatization and Network Security Project provides policy and technical support for the promotion of this model, aiming to create online learning conditions for primary and secondary school students across the country, improve the information literacy of teachers and students, and pave the way for the exploration of mixed teaching in primary mathematics and other disciplines through the construction of digital education. However, although blended learning has achieved significant results internationally, its practical application in primary school mathematics education in China is still in its early stages, with many unresolved problems and weak links. Therefore, this study focuses on the fuzzy comprehensive evaluation of blended teaching quality based on multi-source information fusion. The main research ideas of this method are as follows:

(1) Firstly, establish a quality evaluation index system for blended learning. This system takes learning attitude, learning process, and learning effectiveness as the core primary indicators. By refining the secondary and tertiary indicators, it comprehensively covers the key links of blended learning, provides a clear and systematic framework for subsequent evaluation, and ensures the comprehensiveness and pertinence of the evaluation.
(2) Then, the Analytic Hierarchy Process is used to scientifically determine the weight of indicators. Through expert scoring and calculation, the importance of different indicators in teaching quality evaluation is reflected, ensuring the objectivity and accuracy of the evaluation results, and providing effective guidance for the evaluation work.
(3) Finally, combining multi-source information fusion technology, comprehensively analyzing learning data from different channels, comprehensively and deeply analyzing students' learning status and teaching effectiveness, overcoming the limitations of a single information source, and improving the comprehensiveness and accuracy of evaluation. And present the evaluation results intuitively through quantitative scoring and grading, thereby achieving a fuzzy comprehensive evaluation of blended teaching quality based on multi-source information fusion.

Through the research of the above methods, the aim is to provide theoretical support and practical guidance for the effective application of blended learning in primary school mathematics education in China, seize the opportunities brought by educational informatization, respond to challenges, and promote the comprehensive improvement of educational quality.

2 Establishing a Blended Teaching Quality Evaluation Index System

Teaching quality evaluation must establish a scientific, reasonable and practical evaluation index system. Teaching in the institution for many years, personally experienced the students to evaluate their own teaching, supervisory teachers, peer teachers on their own teaching quality evaluation, but also peer teachers have been evaluated, has been familiar with the school teachers teaching quality evaluation standards, evaluation indicators and evaluation system. Drawing upon the educational objectives of schooling and referencing the existing teacher teaching quality evaluation system within the institution, the principles guiding the analysis and determination of teaching quality evaluation indicators include the integration of theory and practice, the balance of moral and ability considerations, and the emphasis on practical application. The quality of blended teaching is primarily evident in three aspects: learning attitudes, the learning process, and the learning outcomes. Consequently, these three elements serve as the primary level of evaluation indicators [1].

Learning attitude includes four levels of evaluation. It generally includes attendance, mental state of classroom learning, invalid learning state and submission of homework. Among them, attendance mainly refers to students' lateness, early leave and leave; The learning state in the classroom refers to the mental state and emotional state of students in the classroom and computer learning; Invalid learning state refers to the situation of playing with irrelevant objects, wandering, inattention or frequent chat in class, and browsing irrelevant web pages during online learning. In addition, chatting on WeChat or QQ, watching videos, listening to music and other things can also be regarded as invalid learning state; The submission of homework specifically refers to whether students can hand in homework to the teacher within the specified time, and also needs to pay attention to the quality and authenticity of homework.

The learning process mainly refers to the use of learning resources, students' independent learning and the interaction and expression between students and teachers in the teaching process. Learning effect can be evaluated through skills and knowledge, comprehensive quality ability development, knowledge and skills are mainly composed of three dimensions of evaluation indicators [2]. The quality of daily work completion refers to the quality of completing the week's work, whether it is detailed and accurate, whether it has its own understanding, whether it is original, whether there is plagiarism; the learning gains and questions are the procedures for answering the questions in learning, and the daily quiz is a kind of stage score. The three key competencies of collaborative leadership, teamwork, and independent research and learning of the study group members are matched with the learning objectives of each stage, reflecting their comprehensive competence [3]. Combining the above analysis to select secondary and tertiary indicators, the following table shows the blended teaching quality evaluation index table.

Integrating the evaluation indexes selected above, the fuzzy comprehensive evaluation index system of blended teaching quality is established, as shown in the Fig. 1.

Table 1. Indicators for evaluating the quality of blended teaching and learning

Level 1 indicators	Level 2 indicators	Level 3 indicators
Learning attitude	Learning status	Student attendance rate
		The mental outlook displayed in the classroom
		The frequency and duration of useless learning
	Learning intervention	Intervention in classroom learning
		Online learning situation
		After class group learning, mutual learning, and participation situation
		Learning micro course situation
learning	Use of learning resources	Personal access to online resources
		The use of textbooks
		Sketching of Notes
		Login status on the learning platform
	autonomous learning	Autonomous learning duration
		Online learning focus
		In class learning focus
	Interaction and expression	Activity level in the discussion group
		Communication with teachers and classmates after class
		Report and feedback from the class representative group
learning effect	Skills and knowledge	Regular evaluation
		Knowledge acquisition and questioning
	Comprehensive quality and ability	Leadership collaboration ability
		Self study ability
		teamwork

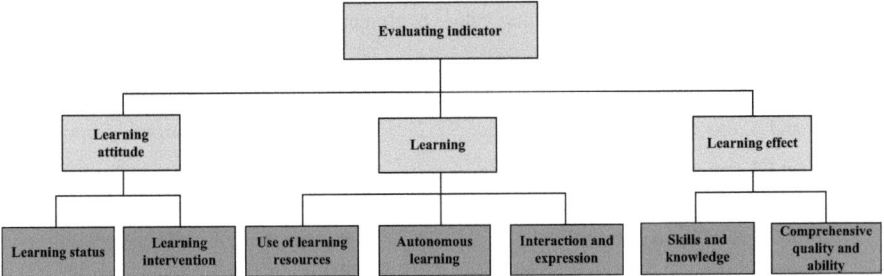

Fig. 1. Diagram of the fuzzy comprehensive evaluation index system of blended teaching quality

3 The Weights of the Blended Teaching Quality Evaluation Indicators are Determined

After the process evaluation indicators of students in blended learning are determined, the weights of each indicator are assigned, i.e., the weights of each indicator and the proportion of each indicator in each indicator need to be calculated. The weights indicate the importance of each indicator in the whole, and the corresponding weight coefficients are given [4]. The weights are determined by the position and magnitude of the role of each element of the constituent system in the overall system, and the weights refer to the impact of changes in the index on the indicator when other indicators remain constant. To enhance the scientific rigor of the research, the hierarchical analysis method was selected for assigning weights to various indicators. The process for determining the weights of specific indicators involves the following steps.

First of all, it is necessary to construct a judgment matrix, and constructing a judgment matrix is a key step in the hierarchical analysis method. Will affect the quality of blended teaching and learning, the n Indicators in the same layer are compared two by two in terms of their degree of influence on the factors in the previous layer of indicators, forming a judgment matrix. The judgment matrix compares the factors in the same tier with the previous tier and assigns certain judgments to the relative importance of the factors in each tier. The basis or source of these assignments is usually given independently by experts familiar with the evaluation of the quality of classroom instruction [5]. Using a 1–9 scale, we make relative comparisons between two factors. Now, let's assess the impact of factor m on a specific factor in relation to factor X, where X is represented as:

$$X = \{x_1, \cdots, x_m\} \quad (1)$$

In the formula, the x_m indicates the m-th evaluation indicators. In order to provide more credible data, the factors were compared two by two and a pairwise comparison matrix was created, i.e., two factors were selected at a time A_i and A_j, a_{ij} denotes the ratio of the magnitude of the effect on the factors A_i and A_j, with all comparisons expressed in a matrix as follows:

$$A = (a_{ij})_{m \times m} \quad (2)$$

In the formula, the A denotes the comparison matrix, i.e., the A is the judgment matrix between factors and indicators, scales 1–9 and their reciprocals were used, and the meaning of scales 1–9 is presented in the Table 1.

Table 2. Criteria for assigning values to elements in the judgment matrix

a_{ij}	Definition	a_{ij}	Definition
1	A_i and A_j are equally important	2	Between equal and slightly important
3	A_i and A_j are slightly important	4	Between slightly and significantly important
5	A_i and A_j are clearly important	6	Between obvious and very obvious importance
7	A_i and A_j are very clearly important	8	Between very obvious and absolutely important
9	A_i is absolutely more important than A_j		

Construct m order judgment matrix is shown in the following Table 3.

Table 3. m order judgment matrices.

C	C_1	C_2	C_3	...	C_m
C_1	1	a_{12}	a_{13}	...	a_{1m}
C_2	$1/a_{12}$	1	a_{23}	...	a_{2m}
C_3	$1/a_{13}$	$1/a_{23}$	1	...	a_{3m}
...	...	...	...	1	...
C_m	$1/a_{1m}$	$1/a_{2m}$	$1/a_{3m}$	...	1

Among them, C indicates an evaluation indicator. The weights of the judgment results a is given according to the quantitative rules of the judgment results. Taking "learning status" in learning attitude as an example, the judgment matrix is constructed by experts familiar with teaching according to the assignment criteria in Table 2.

$$C_{11} = \begin{pmatrix} 1 & 5 & 3 & 7 \\ \frac{1}{5} & 1 & \frac{1}{3} & 2 \\ \frac{1}{3} & 3 & 1 & 4 \\ \frac{1}{7} & \frac{1}{2} & \frac{1}{4} & 2 \end{pmatrix} \qquad (3)$$

Utilizing the previously mentioned 1–9 scale theory, the values within the judgment matrix reflect the relative importance of indicators pertaining to "learning status" within the context of learning attitude [6]. Based on this matrix, a hierarchical single ranking and a hierarchical total ranking were conducted to ascertain both the evaluation factors and their corresponding weights. The purpose of the hierarchical single ranking is to rank the importance of the factors at this level, and the weights of the hierarchical single ranking can be obtained by solving the eigenvalues, namely:

$$AW = \gamma_{\max} W \varpi \quad (4)$$

In the formula, the W denotes the eigenvector of the judgment matrix for the largest eigenroot; the $\gamma_{\max}$ denotes the largest characteristic root; the ϖ indicates the weight value of the hierarchical single ordering of the corresponding element [7]. The 1 ~ 9 ratio criterion is used to reduce the bias that occurs when judgmental thinking is inconsistent; The ratio CR of the index CI for testing the consistency of the judgment matrix and the correction coefficient RI is introduced to test the consistency of the judgment thinking. Should the value of CR fall below 0.1, the judgment matrix demonstrates acceptable consistency. However, if it exceeds this threshold, adjustments to the initial values of the judgment matrix become necessary. As a first step, the maximum eigenvalue of the judgment matrix $\gamma_{\max}$ must be calculated.

Whether the W obtained from the above calculation can be used as the basis for the ranking of the lower elements to the upper elements, we need to test and judge the consistency between the a_{ij} values in the matrix, when the matrices are identical, there is $\gamma_{\max} = m$, and when the judgment matrix is in error with respect to consistency, there is $\gamma_{\max} \geq m$. In this way, the consistency of the judgment matrix is tested to ensure the accuracy of the weights of the indicators.

4 A Comprehensive Fuzzy Assessment of Teaching Quality Utilizing the Integration of Diverse Information Sources

As can be seen from Table 1, the blended teaching quality evaluation index system is an evaluation system with secondary and tertiary indexes, and each level contains multiple indexes, with different sources of different indexes, so the fuzzy comprehensive evaluation of blended teaching quality adopts the fusion of source information technology, fusing the information of the learning attitude, the learning process, and the learning effect, to realize the comprehensive evaluation of teaching quality [8]. The fusion of multi-source information, first of all, the decomposition of each level of indicators of the second level indicators, third level indicators to make a comprehensive judgment, and then with the weight of the first level indicators of the fuzzy fusion, you can get the blended teaching quality fuzzy comprehensive evaluation results.

Step 1: Establish the evaluation factor set, comment set, value set and weight set. The evaluation factor sets obtained according to Table 1 are as follows:

$$\begin{cases} C = \{C_{11}, C_{12}, C_{13}\} \\ C_{11} = \{C_{111}, C_{112}\} \\ C_{12} = \{C_{121}, C_{122}, C_{123}\} \\ C_{13} = \{C_{131}, C_{132}\} \end{cases} \quad (5)$$

In the formula, the C denotes the set of fuzzy evaluation indicators of blended teaching quality; the C_{11} indicates a learning attitude; and C_{12} represents the learning process. C_{13} indicates a learning effect. C_{111}, C_{112} denote learning states and learning interventions, respectively; the C_{121}, C_{122}, C_{123} indicate the use of learning resources, self-directed learning, and interaction and expression, respectively. C_{131}, C_{132} denote the skills and knowledge and comprehensive quality ability [9], respectively. Establishing the rubric set and value set: Given the disparities in qualifications and experience among teachers of various titles, it is imperative to avert mechanical evaluations using identical standards across different levels. Drawing upon the insights of teaching experts, distinct grade scores are assigned to teachers of varying titles to calculate the overall evaluation scores, as outlined in the subsequent table (Table 4).

Table 4. Scorecard

Serial Number	Grade	Score
1	Very good	80–100
2	Better	60–80
3	same as	40–60
4	Relatively poor	20–40
5	very poor	0–20

Combined scoring construction rubric set:

$$V = \{V_1, V_2, V_3, V_4, V_5\} \quad (6)$$

In the given formula, V represents a compilation of feedback. V_1 signifies exceptional quality, while V_2 denotes superior quality. Additionally, V_3 also stands for average quality, and V_4 indicates below-average quality. Lastly, V_5 signifies inadequate quality. To conduct a fuzzy comprehensive evaluation of blended teaching quality, the collected data from each index is integrated. The evaluation formula is outlined below:

$$F = V = E = \sum_{g=1} \varpi_g x_g \quad (7)$$

In the formula, F represents the outcome of the fuzzy comprehensive evaluation for blended teaching quality, E signifies the score pertaining to blended teaching quality, and g denotes the quantity of evaluation indicators. x_g represents the g-th index value [10]. Through the above calculation, we get the fuzzy comprehensive evaluation results of hybrid teaching quality, so as to realize the fuzzy comprehensive evaluation of hybrid teaching quality based on the fusion of multi-source information.

In summary, a three-dimensional evaluation system covering learning attitude, learning process, and learning effectiveness has been constructed in the fuzzy comprehensive evaluation of blended learning quality. Multiple indicators are subdivided under each dimension to comprehensively reflect teaching quality. By using multi-source information fusion technology, integrating evaluation indicators from different levels and aspects, and conducting fuzzy comprehensive evaluation based on weights, a fuzzy result that comprehensively reflects the quality of teaching is obtained. This process combines scoring and commenting, and by setting different levels of score ranges, it is possible to quantify the different levels of teaching quality and obtain a specific score. Meanwhile, the set of comments corresponding to the score table provides a more specific and vivid description of teaching quality, helping readers better understand the meaning behind the evaluation results. The evaluation results not only provide feedback to teachers and students, but also guide continuous improvement, promote the optimization and enhancement of blended learning mode, promote the comprehensive development of students and teacher teaching innovation.

5 Experimental Analysis

5.1 The Experimental Process

Taking the blended learning course of a certain university as an example, and combining with the characteristics of the course, a detailed and meticulous experimental setup was adopted to ensure the accuracy and effectiveness of the evaluation system. In the sample selection stage, a group of experts including teaching management personnel, senior professors, curriculum designers, as well as representative groups of teachers and students were selected. Through random sampling and comprehensive coverage, the diversity and representativeness of the sample were ensured. For the experimental conditions, the stability of the blended learning environment and the adequacy of teaching resources were ensured, providing students with an online learning platform and an offline teaching environment. Subsequently, a survey questionnaire with a well-designed structure and clear content was distributed, and opinions and suggestions from experts, instructors, and some students were solicited through channels such as email and online survey platforms. In order to improve the reliability of the data, strict screening and organization were carried out on the collected questionnaire data, invalid or incomplete questionnaires were removed, and statistical software was used for in-depth data analysis. In addition, to verify the stability and consistency of the evaluation results, it is considered to repeat experiments in different time periods or batches, and invite multiple experts for independent evaluation to comprehensively evaluate the quality of blended learning. After collecting feedback, weight coefficients were assigned based on the importance of the factors in the evaluation, covering key factors related to computer course experiments. After the development of the evaluation index system, extensive student evaluations were conducted and received recognition from the majority of students. This process ensures the comprehensiveness and objectivity of the evaluation index system. According to the above process to determine the weight of the indicators and judgment matrix, the following table is the hybrid teaching quality evaluation index weight table (Table 5).

Table 5. Weighting table of indicators for evaluating the quality of blended teaching and learning

Primary indicators	Secondary indicators	Third level indicators
Learning attitude 0.064	Learning status 0.014	0.562
		0.261
		0.154

(*continued*)

Table 5. (*continued*)

Primary indicators	Secondary indicators	Third level indicators
	Learning intervention 0.036	0.014
		0.126
		0.142
		0.264
learning 0.034	Use of learning resources 0.047	0.236
		0.485
		0.164
		0.526
	autonomous learning 0.025	0.362
		0.245
		0.246
	Interaction and expression 0.017	0.426
		0.612
		0.584
learning effect 0.016	Skills and knowledge 0.025	0.845
		0.362
	Comprehensive quality and ability 0.035	0.846
		0.565
		0.597

The indicators were evaluated according to their weights, and the resulting evaluation set are:

$$C_{11} = (56.15, 42.61)$$

$$C_{12} = (67.15, 75.15, 69.17)$$

$$C_{13} = (89.15, 96.14)$$

Teaching quality scores and grades were obtained through a fuzzy synthesized evaluation of the quality of blended teaching:

$$F = (87.16, V_1)$$

The evaluation shows that the results of the quality assessment of this blended teaching are exceptionally good.

5.2 Experimental Results and Discussion

In order to verify the performance of the proposed method, Spearman correlation coefficient and consistency coefficient are chosen as evaluation indexes. The Spearman correlation coefficient serves as a crucial metric for assessing the performance of evaluation systems, models, algorithms, and the like. The higher the value of Spearman correlation coefficient, the higher the reliability and credibility of the evaluation results, which is calculated as follows:

$$\varepsilon = \frac{\sum (s - s^*)(s + s^*)}{\sum (s - s^*)^2} \qquad (8)$$

In the formula, the ε denotes the Spearman's correlation coefficient of the blended teaching quality evaluation results; the s denotes the value of the blended teaching quality assessment; the s^* indicates the actual value of blended teaching quality. The Spearman correlation coefficient value ranges from 0–1. The consistency coefficient is an important indicator of the precision of the evaluation method or model, indicating the degree of consistency between the evaluation results and the actual situation. The value of the consistency coefficient is in the range of 0–1, and the larger the value is, the more the analytical results are consistent with the actual situation. The experiment selects two existing methods as the evaluation control of the design method, and the comparison charts of Spearman's correlation coefficient and consistency coefficient of the evaluation of the three methods are given in Figs. 2 and 3.

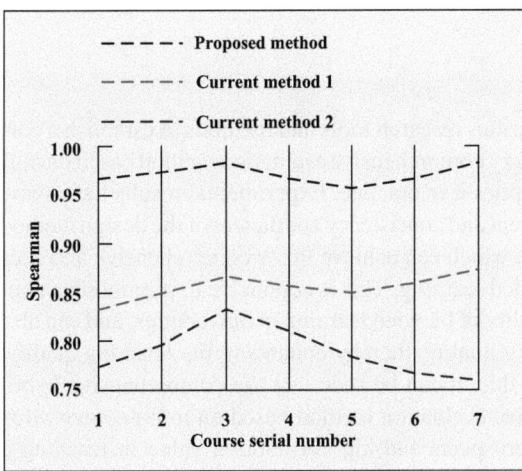

Fig. 2. Comparison of Spearman's correlation coefficients for fuzzy comprehensive evaluation of teaching quality

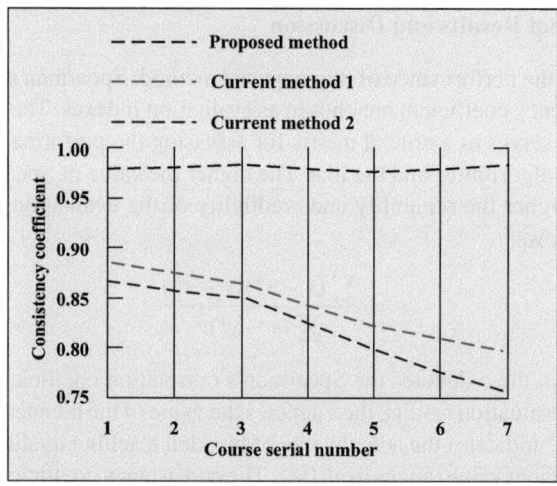

Fig. 3. Comparison of consistency coefficients of fuzzy comprehensive evaluation of the quality of teaching and learning

Comparing Figs. 2 and 3, the Spearman correlation coefficient and consistency coefficient of the evaluation results of this paper's method are much higher than those of the two existing methods, which proves that this paper's method has a greater advantage in terms of the evaluation accuracy, and it is more suitable for the fuzzy comprehensive evaluation of blended teaching quality.

6 Conclusion

This article uses various research tools and methods to establish a comprehensive mixed teaching quality fuzzy comprehensive evaluation method based on multi-source information fusion, and applies it in practice. Experimental results have shown that the Pearson correlation coefficient and consistency coefficient of the design method evaluation results are both above 0.9, which can achieve fuzzy comprehensive and accurate evaluation of the quality of blended learning. This is conducive to promoting a comprehensive understanding of the quality of blended learning in universities, and can also promote teachers to improve teaching quality, thereby enhancing the teaching quality and reputation of universities. From this, it can be seen that the comprehensive hybrid teaching quality fuzzy comprehensive evaluation method based on multi-source information fusion has good application prospects and high promotion value in teaching quality evaluation. Although this method has achieved certain research results, there are also shortcomings. For example, the current evaluation index system may still have certain limitations and cannot fully cover all aspects of blended learning. Although Analytic Hierarchy Process (AHP) can scientifically allocate weights, the determination of weights still depends on the subjective judgment of experts, which may have certain subjective biases. In the future, the evaluation index system will be further refined and improved based on the characteristics of different disciplines and teaching environments, ensuring the comprehensiveness and pertinence of the evaluation. With the development of educational philosophy and technology, new evaluation dimensions can be introduced, such as student innovation ability, team collaboration ability, etc., to more comprehensively reflect the quality of teaching.

Acknowledgment. 2024 Beijing Municipal Digital Education Research Project, Project Name: Research on Teaching Quality Evaluation System in Higher Vocational Colleges under the Multimodal Data Environment, project code:BDEC2024QN062.

References

1. Liu, S., He, T., Li, J., Li, Y., Kumar, A.: An effective learning evaluation method based on text data with real-time attribution - a case study for mathematical class with students of junior middle school in China. ACM Trans. Asian Low-Resour. Lang. Inf. Process. **22**(3), 63 (2023)
2. Xinggang, L., Zhiqing, W.: Enhanced CoCoSo method for picture 2-tuple linguistic MAGDM and applications to classroom teaching quality evaluation of college physical education. Int. J. Knowl.-Based Intell. Eng. Syst. **27**(3), 303–318 (2023)
3. Fan, X.: Method for classroom teaching quality evaluation in college English based on the probabilistic uncertain linguistic multiple-attribute group decision-making. Int. J. Knowl.-Based Intell. Eng. Syst. **27**(2), 245–257 (2023)
4. Yingying, C.: Optimizing decision trees for English teaching quality evaluation (ETQE) using artificial bee colony (ABC) optimization. Heliyon **9**(8), e19274–e19274 (2023)
5. Junying, C., Haoyu, Z., Lijuan, W.: Statistical analysis of advanced mathematics teaching quality evaluation based on quantile regression model. Appl. Math. Nonlinear Sci. **8**(2), 2549–2556 (2023)
6. Boyi, Z., Jun, Z., Shantian, W.: A textual and visual features-jointly driven hybrid intelligent system for digital physical education teaching quality evaluation. Math. Biosci. Eng. MBE **20**(8), 13581–13601 (2023)
7. Xujian, W., Minli, Y., Fenggan, Z.: Teaching quality evaluation-based differential evolution and its application on synthesis of linear sparse arrays. Soft. Comput. **27**(20), 14735–14758 (2023)
8. Junbo, Z., Cheng, Z.: Teaching quality monitoring and evaluation of physical education teaching in ordinary college based on edge computing optimization model. J. Supercomput. **79**(15), 16559–16579 (2023)
9. Aaron, S.: Feature data dimensionality reduction algorithm for online English teaching quality evaluation based on self-organizing feature map neural network. Comput. Inf. Mech. Syst. **6**(3), 10–18 (2023)
10. Tao, S.: Method for 2-tuple linguistic neutrosophic number MAGDM and applications to physical education teaching quality evaluation in colleges and universities. J. Intell. Fuzzy Syst. **44**(3), 4233–4244 (2023)

A Study on the Job Information Recommendation Method Based on Social Network Information Fusion

Shan Gao[1(✉)] and Xiangjun Shi[2]

[1] Changchun University of Technology, Changchun 130122, China
13944152234@163.com
[2] College of Humanities and Information, Changchun University of Technology, Changchun 130122, China

Abstract. Given the constantly changing job market, the recommendation of job information is inevitably hindered by factors such as the speed of recruitment information updates, leading to a lack of confidence in the suggestions provided. To address this challenge, social network data has been introduced to personalize job recommendations, solving the problem of insufficient updating speed of job information. By extracting user preferences from social networks, we have understood the interests and behaviors of job seekers, thereby improving job recommendations. The experimental results show that classifying work data and dynamically adjusting recommendations enhances confidence in work information and enhances the effectiveness of employment services.

Keywords: Social Network Information · Information Fusion · Job Posting Information · Job Recommendation Method

1 Introduction

As information technology continues to soar, social networks have solidified their position as a crucial aspect of modern life [1]. These networks extend beyond basic communication, serving as vast databases filled with diverse information. Embedded within this data are rich elements such as users' personal profiles, professional histories, interests, and social engagements, all of which offer profound understanding into individuals' employment preferences and requirements.

The traditional employment information service has gradually shown its insufficiency, especially in the face of the huge and diversified job market, the problem of matching information between job seekers and recruiters has become more and more obvious. Traditional employment information services, whether offline job fairs or online recruitment platforms, are essentially a passive, one-way information transfer. Job seekers need to sift through a large amount of job information on their own, which is not only time-consuming and labor-intensive, but also difficult to ensure the accuracy and timeliness of the information. Recruiters also face a similar dilemma, they need to select the talents that meet the requirements of the job in a large number of job seekers, and this process is also full of challenges. More importantly, the traditional employment information service can hardly achieve real personalized recommendation and cannot meet the increasingly diversified needs of job seekers and recruiters.

In this paper, the "job information recommendation method based on social network information fusion" is proposed to solve the above problems. First of all, through the deep mining of user information in social networks, we can more comprehensively and accurately understand the employment needs and preferences of users. Secondly, this study utilizes advanced information fusion technology to effectively integrate multi-source and heterogeneous information in social networks to construct a comprehensive and accurate employment information service model. This model not only realizes the precise identification of user needs, but also can be dynamically adjusted according to the changes of the market and user needs, so as to ensure the timeliness and accuracy of the recommendation results. Finally, through the algorithms of clustering analysis and association rule mining, the method in this paper can classify and recommend job information in a more refined way [2]. This can not only greatly improve the matching efficiency between job seekers and recruiters, but also realize the real personalized recommendation to meet the differentiated needs of different users.

To sum up, the "employment information recommendation method based on social network information fusion" researched in this paper aims to break the limitations of traditional employment information service and realize more accurate and efficient employment recommendation service by deeply mining and utilizing the user information in social network. The innovative technological roadmap of this method is as follows:

(1) Social network user preference mining: By analyzing user behavior, interests, and needs in social networks, potential preference information of users is mined to provide data support for personalized recommendations.
(2) Information Fusion and Intelligent Recommendation Model Construction: Integrating multiple sources of data, including users' social activities, occupational backgrounds, educational experiences, etc., to construct an intelligent recommendation model that accurately captures users' employment needs and expectations.
(3) Job data classification and management: Refine the classification of massive job data, organize it according to key characteristics such as industry, job type, and workplace, and form an orderly and easily retrievable information collection.
(4) Personalized job recommendation algorithm: Develop a personalized recommendation algorithm that can dynamically adjust recommendation strategies based on users' specific needs and market changes, achieving accurate matching of job information.
(5) Real time adjustment and optimization: Establish a feedback mechanism to adjust and optimize recommendation models in real-time based on user feedback and market dynamics, ensuring the timeliness and accuracy of recommendation results.
(6) Performance evaluation and iterative improvement: Verify the performance of the recommendation system through experiments, including indicators such as recommendation accuracy and user satisfaction, and iteratively improve the system's recommendation quality based on the evaluation results.

This innovative technological route integrates social network information to achieve personalized and accurate recommendations of employment positions, improving the efficiency and quality of employment services, and providing more intelligent and convenient solutions for job seekers and recruiters.

2 Mining Social Network User Preference Information

In order to realize more efficient and accurate job information recommendation, this study is dedicated to integrating multi-dimensional user information in social networks. A joint distribution model of time and location of user nodes in social networks is constructed based on the social relationships of users to comprehensively capture the traces and spatial-temporal characteristics of users. Mixed and global kernel functions are constructed to jointly detect the location, time and social information in the social network, so as to draw a richer behavioral portrait for the users. Utilizing the user's historical activity records to obtain the fuzzy decision function of social network information, constructing a sequence of activity records $\Phi : x \in R^n \to F$ to represent the personalized feature distribution mapping of social network users, further characterizes the personalized feature distribution mapping of users, combines the social activity information, establishes the intelligent swarm algorithm innovatively for the fusion recommendation design of social network, projects the recommendation information into the high-dimensional phase space for feature reorganization and information fusion processing, improves the automatic matching ability of the user information and the network information, and also enhances the accuracy of the identity recognition [3]. The specific steps are as follows:

Assuming that the initial sample set of recommended spatio-temporal information and the user's social information of the social network is $\{(x_1, y_1), (x_2, y_2), \cdots, (x_n, y_n)\}$, where the characteristic quantity of personalized evolution is $x_i \in R^n$, represents a fuzzy constraint independent variables of fusion recommendation, $y_i \in R^n$ is for the user's historical access information, n is the number of test samples. Combined with the association rule mining algorithm, the control objective function for fuzzy perception constraints and fusion recommendation of social networks is obtained as:

$$\begin{array}{l} \text{min } imize \ \frac{1}{2}\|\omega\|^2 + C \sum_{i=1}^{n} \left(\xi_i + \xi_i^*\right) \\ \text{subject to } \begin{cases} y_i - \left(\omega'\Phi(x_i) + b\right) \leqslant \varepsilon - \xi_i \\ \left(\omega'\Phi(x_i) + b\right) - y_i \leqslant \varepsilon - \xi_i^* \\ \xi_i, \xi_i^* \geqslant 0, i = 1, 2, \cdots, n; C > 0 \end{cases} \end{array} \quad (1)$$

In Eq. (1), ξ_i and ξ_i^* represent the similarity information between users, combining correlation detection methods to conduct the independent variable analysis of recommendation models, using time and location-aware methods to control the process of fusion recommendation in social networks, and carrying out adaptive learning of the recommendation process based on semantic and regular feature analysis methods [4]. Taking C as the cost factor of social network integration recommendation, and the user's privacy protection protocol is established by matrix decomposition, and the individual difference function of social network integration recommendation is obtained as:

$$f(x) = \sum_{i=1}^{n} (a_i - a_i^*) K(x_i, x_j) + b \qquad (2)$$

In Eq. (2), a_i and a_i^* represent the social network personalization attribute values and feature matching coefficients, $K(x_i, x_j)$ represents the amount of similarity features between users, b represents the recommended elastic random coefficient. Thus, the joint distribution model of social network user information is constructed, and the joint feature component of personalized preference recommendation x_1, x_2, x_3, x_4 are obtained and expressed as:

$$\begin{cases} x_1 = p_1 - m \\ x_2 = p_2 - m \\ x_3 = p_3 - m \\ x_4 = m \end{cases} \qquad (3)$$

In Eq. (3), m represents the associated attributes of geographic location, based on the above analysis, the joint distribution transfer model of social network users information is constructed as shown in Fig. 1.

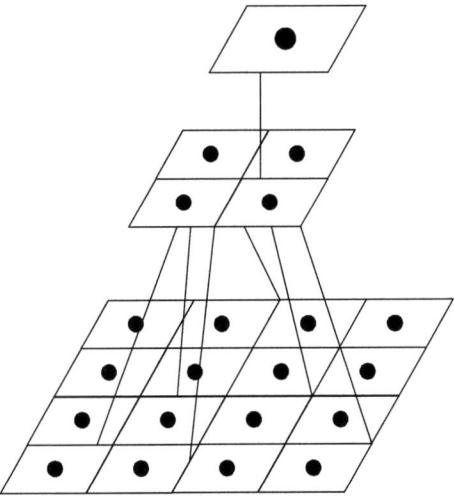

Fig. 1 A Joint Distribution Transfer Model of Social Network Users Information

On the basis of constructing the joint distribution model of social network user information, this paper adopts the cross-sectional adaptive learning method for social network user preference information mining and associated feature detection, and constructs the intelligent swarm evolution algorithm for the process control of user preference information mining and adaptive optimization, and obtains the expression of the optimization control function as follows:

$$y_i = \mu y_i(1 - y_i) \quad (4)$$

In Eq. (4), $y_i \in [0, 1]$ represents the weighted random number, μ represents the recommended control parameter for social network integration.

The intelligent particle swarm optimization method is used to analyze the optimal recommendation value, and the particle swarm search cross terms for the social network topology results in the d dimensional space are obtained as $S = \{P_1, P_2, \cdots, P_m\}$, notation columns $P_i^d(t)(i = 1, 2, \cdots, m)$ represents user-related subgraphs. In the process of intelligent swarm individual i optimization, the optimal feature quantity of information distribution center for mining preference feature information in the current d dimensional solution space is obtained, taking $V_i^d(t)(i = 1, 2, \cdots, S)$ to represent the user preference associated eigenvalues of the social network, looping over the individual i, to obtain the global extremum of the intelligent recommendation $G_{best}^d(t)$ and individual extremum $P_{best}^d(t)$, the weights are randomly walked to obtain the particle swarm evolutionary optimization of social network fusion recommendation [5]. The process expression is as follows:

$$\begin{cases} V_i^d(t+1) = W \cdot V_i^d(t) + C_1 \cdot R_1 \cdot \left(P_{best}^d(t) - P_i^d(t)\right) \\ +C_2 \cdot R_2 \cdot \left(G_{best}^d(t) - P_i^d(t)\right) \\ P_i^d(t+1) = P_i^d(t) + V_i^d(t+1) \end{cases} \quad (5)$$

In Eq. (5), $V_i^d(t)$, $V_i^d(t+1)$, $P_i^d(t)$, and $P_i^d(t+1)$ denote the particle i's optimal position for the current and subsequent moments, respectively, C_1 and C_2 represent the learning factor of user information mining in heterogeneous information networks, which generally takes values between 15 and 20, R_1 and R_2 represent the fuzzy matching value of particle swarm optimization, taking a random number between [0, 1], W is inertia weight, in the interval of the optimal analytic parameters $[W_{min}, W_{max}]$, the recommended fuzzy perception formula for social network integration is obtained as follows:

$$\begin{cases} W(t+1) = 4.0W(t)(1 - W(t)) \\ W(t) = W_{min} + (W_{max} - W_{min})W(t) \end{cases} \quad (6)$$

In Eq. (6), $[W_{min}, W_{max}]$ represent the value range of inertia factor, and according to the above swarm intelligence algorithm, the cross-sectional adaptive learning method is used for social network user preference information mining, and the associated feature detection method is used to realize the attribute extraction of social network users.

3 Constructing Intelligent Recommendation Models of Social Network Information

On the basis of the above constructed social network user information joint distribution transfer model, social network user preference information mining and recommendation algorithm optimization design, this paper proposes a fuzzy spatio-temporal information perception based on the fuzzy social network cross-layer fusion of intelligent recommendation algorithm. The spatial structure of the collected personal preference data of social network is reorganized, and the 4-tuple is expressed as follows: $\{S_1, S_2, \cdots, S_L\}$, the confidence level of personalized recommendation is taken as the control constraint parameter and the fuzzy perception method is used to analyze the location information and spatio-temporal information, the feature extraction conduction control model of fusion recommendation for social network operation is obtained as follows:

$$a_{desira}^i = a_1 \frac{Density_i}{\sum_i Density_i} + a_2 \frac{AP_i}{\sum_i AP_{init}} \tag{7}$$

Of which:

$$\begin{cases} a_1 + a_2 = 1, a_1, a_2 \in [0, 1] \\ a_2 = \frac{\max_i(AP_i) - \min_i(AP_i)}{AP_{init}} \end{cases} \tag{8}$$

Extracting the feature quantity of personalized preference-related attributes of social network, analyzing the interaction information within the group, searching for the optimal control in the particle swarm fuzzy clustering learning process, analyzing the characteristics of the user's interest over time by combining with the optimization learning process of the social relationship, and introducing the search radius of the fusion recommendation R_1 and R_2, the updated formula of the information recommendation process for a user to access information at a given point in time is as follows:

$$R_i(t+1) = 4.0 R_i(t)(1 - R_i(t)) \tag{9}$$

In Eq. (9), $R_i(t) \in (0, 1)$, $i = 1, 2$. Considering the influence of social relations, in the process of best user matching, the recommendation information is fused with topological information vectors by the joint prediction method of time and place, and the fuzzy clustering method is used to make it jump out of the local optimum, and the information fusion and fuzzy clustering process are carried out according to the characteristics of user identity matching of social network [6]. The process of this implementation is described as follows:

$$\begin{cases} V_i^d(t+1) = 4.0 V_i^d(t)\left(1 - V_i^d(t)\right) \\ V_i^d(t) = V_{\min} + (V_{\max} - V_{\min}) V_i^d(t) \end{cases} \tag{10}$$

In Eq. (10), $[V_{\min}, V_{\max}]$ represents the feature distribution interval of fuzzy recommendation of social network, under the control of association rule constraints and considering the relative error of user ratings, the clustering center of the associated feature of user recommendation of social network is obtained to be expressed as:

$$\varepsilon_t(i,j) = \frac{a_t(i) a_{ij} b_j(o_{t+1}) \beta_{t+1}(j)}{\sum_{i=1}^N \sum_{j=1}^N a_t(i) a_{ij} b_j(o_{t+1}) \beta_{t+1}(j)} \tag{11}$$

In Eq. (11), $b_j(o_{t+1})$ represents the user spatio-temporal similarity feature set, $t+1(j)$ represents the number of user nodes, a_{ij} represents the fuzzy measurement information recommended to users by social network, and the user spatio-temporal similarity variance δ^2 is calculated, and determine whether $\delta^2 < H$ is valid or not, the regression analysis model for intelligent recommendation of social networks is obtained to be expressed as follows:

$$\min F = R^2 + A \sum_i \xi_i$$
$$s.t. \|\varphi(x_i) - o\|^2 \leqslant R^2 + \xi_i \text{ and } \xi_i \geqslant 0, i = 1, 2, \cdots, n$$
$$\max \sum_i a_i K(x_i, x_i) - \sum_i \sum_j a_i a_j K(x_i, x_i) \qquad (12)$$
$$s.t. \sum_i a_i = 1 \text{ and } 0 \leqslant a_i \leqslant A, i = 1, 2, \cdots, n$$

Combined with the association rule mining method, the output of association feature detection is obtained to satisfy $\sum_i a_i = 1$, $K(x_i, x_i) = 1$, combining username attributes to extract cross-layer attribute feature sets of social network users to improve the confidence level of recommendation.

4 Clustering Job Information

Based on the above analysis of user's social network information, the job information is processed. In order to mine information data beneficial to users from massive employment information, random forest algorithm is used to realize mining and feature extraction. When using random forest algorithm for information mining and feature extraction, the Bootstrap method is usually used to train the dataset and generate multiple decision trees, so as to achieve the operation purpose through integrated learning [7]. However, the employment information is extremely unbalanced, so the desired results cannot be obtained by adopting the above methods. On the basis of the above methods, we adopt a hierarchical approach to sample many job information categories. When mining job information, we conduct territorial integration of job information collected from various job recruitment websites and employment announcements issued by enterprises. In a random forest iteration, bootstrap sampling is carried out for the classification with little information, and the same amount of information is randomly extracted from the main categories for replacement. The decision tree is directly generated from the employment information data until all decision trees are completely split. In this process, pruning is not required. Employ the refined Classification and Regression Tree (CART) algorithm to scout for the most discriminative features at each decision tree node. Through iterative execution of this process for n iterations, the ultimate outcomes are amalgamated using a combined voting approach. Figure 2 depicts the process of mining job information and extracting features using the random forest algorithm.

The process of job information mining and feature extraction can be expressed as follows:

$$R_f = \arg\max \sum_{m=1}^{M} rjH(X, \Theta_m) \qquad (13)$$

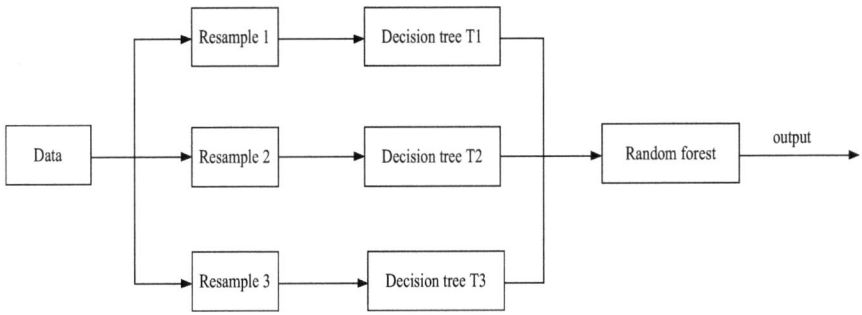

Fig. 2 Job Information Mining and Feature Extraction Process

In Eq. (13), R_f represents the features of job information mined and extracted by random forest algorithm, H represents the traversal function, r represents the decision tree prediction results, X represents the job information characteristics, j represents the decision tree voting results, Θ_m represents the number of features to complete the mining and feature extraction of job information according to the above way, and to provide conditions for the subsequent personalized recommendation.

After successfully mining job information and extracting its features, the redundant data is significantly diminished. Nevertheless, to further enhance the precision of recommendations, it's crucial to categorize different types of job information through clustering. To achieve this, we introduce the K-means clustering method to select k representative samples from the job information. These samples, coupled with the extracted job information features, are then used to train the clustering models. The training samples can be formulated in the following manner:

$$Y = \{y_1, y_2, \cdots, y_n\} \quad (14)$$

In Eq. (14), Y represents the clustered training samples, while n denotes the total number of job postings in the dataset, where each sample comprises features in m dimensions. The selection of the number of clusters, k, is based on the aforementioned operations, and the value of k is directly linked to the final clustering effectiveness. Through iterative training and comparison of clustering effects, the number of clusters k is adjusted until convergence of the final clustering is achieved. The resulting value at this point is then adopted as the number of clusters, denoted as k [8]. For the calculation of the distance from the feature projections of all job information to the center of the clusters, the distance from each job information sample to the center of mass is calculated by using the distance calculation method, and the expression is:

$$c = \arg\min \|s_i - \mu_j\|^2 \quad (15)$$

In Eq. (15), c stands for the distance measured from a given sample of job information to the centroid of a cluster. Meanwhile, $\|s_i - \mu_j\|$ represents the distance between a specific job posting sample, identified as s_i, and a particular set of centroids designated as μ_j. The centroid is computed using the following formula:

$$u_j = \frac{\iint_D \mu F d\sigma}{\iint_D F d\sigma} \qquad (16)$$

In Eq. (16), F represents a closed region. Since the job information itself has multi-dimensional characteristics, it is also necessary to determine the spatial distance from the job information sample to the center of mass by combining the way of multi-dimensional spatial calculation. Through the above steps, we get n attribute clusters corresponding to the training samples of job information, at this time, each cluster already contains a number of training samples. Repeat the operation until all the job information is classified into its corresponding clusters to complete the whole clustering process.

5 Associated with Job Information to Achieve Personalized Job Recommendation

After the completion of the clustering processing of the employment information, it is also necessary to correlate the employment information with the employment needs of each applicant and introduce the Apriori algorithm for correlation analysis, the employment demand information of each job seeker is called a data item. It's assumed that all job information sets in all clusters are employment position information data sets, and each employment record contains some data item components, and after each employment record contains a mark that combines the minimum support and minimum confidence, and confirm the Apriori format of job information as $X \Rightarrow Y$ after completing the above assumptions, the degree of support for job information was calculated using the formula:

$$s(X \Rightarrow Y) = \frac{d}{e} \qquad (17)$$

In Eq. (17), $s(X \Rightarrow Y)$ represents the degree of support for job information, d represents the number of concurrent occurrences of employment records X and Y, e represents the total number of records of job information and then calculate the confidence level of job information, the formula is:

$$c(X \Rightarrow Y) = \frac{s(X \Rightarrow Y)}{s(X)} \qquad (18)$$

Equation (18) designates $c(X \Rightarrow Y)$ as the confidence metric for job information. Meanwhile, $s(X)$ represents the frequency of a specific piece of information, labeled X, within employment records, normalized against the total count of job information based on our correlation analysis. Utilizing the Apriori algorithm, we establish connections between the job information and the employment preferences of relevant job seekers.

According to the above Apriori association results of job information, the current job seekers and job information are classified respectively, and the job information associated with the same type of job seekers is automatically recommended to the job seekers. At the same time, the enterprise automatically recommends each job information to the job seekers who meet their needs, which can not only recommend job information to job seekers, it can also recommend information to recruitment enterprises [9]. The job information recommendation module for job seekers is mainly aimed at users who are currently looking for jobs. According to the above clustering results and association results, job seekers are classified, and the job information associated with the same type of job seekers is automatically recommended to each job seeker. The personalized intelligent recommendation process for job seekers and recruitment enterprises is shown in Fig. 3.

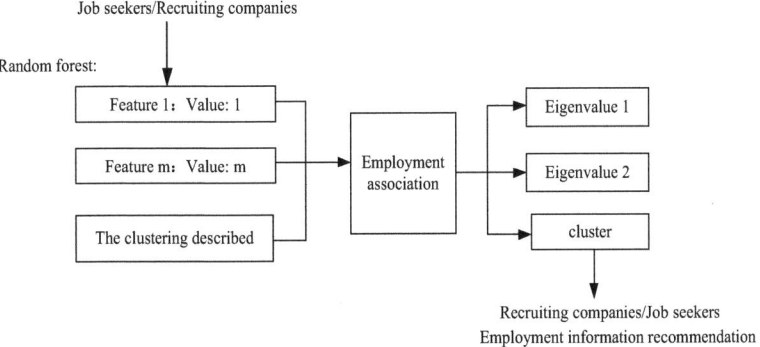

Fig. 3 Personalized Intelligent Recommendation Process for Job Seekers and Recruiters

The process of personalized intelligent recommendation of job information for job seekers and for recruiting companies is very similar, and only the identity of the two ends of the process is adjusted. The personalized recommendation of personal information for recruiting companies is to categorize the existing job seekers' job information through clustering results and relevant rules for different companies, and recommend them to different positions [10]. Through the above operations, the two-way job information recommendation between job seekers and recruiting enterprises is realized, so as to enhance the effectiveness of job information recommendation and further improve the probability of job seekers being employed. Thus, the design of job information recommendation method based on social network information fusion is completed [11, 12].

6 Experiment

6.1 Experiment Preparation

To evaluate the effectiveness of this method in job information recommendation, simulation experiments are conducted. MATLAB 7 is chosen as the design platform, and the Foursquare dataset, rich in user recommendation and IP location data, is employed as the social network data source. Furthermore, a web crawler approach is implemented to gather information on the temporal distribution of network convergence. The sampling process of the information crawler is visually presented in Fig. 4.

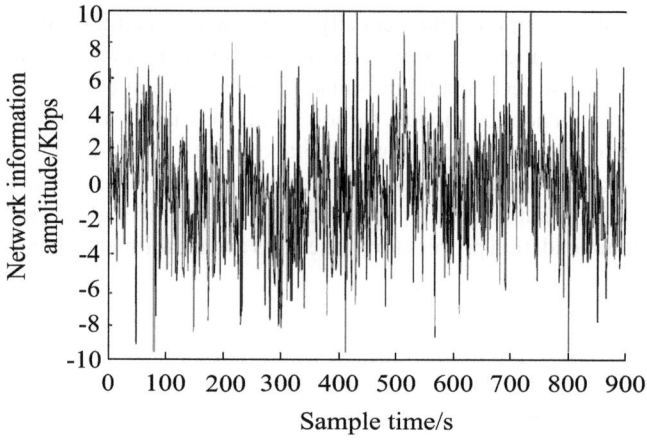

Fig. 4 Information Crawler Sampling

Utilizing the actual dataset from a prominent recruitment website, this paper focuses on job information, encompassing user details, job specifications, and the interactions between users and job postings. Following data preprocessing, the experimental dataset comprises 160,000 users, 40,000 job postings, and 640,000 delivery records. For the experiment, 80% of the data is randomly chosen as the training set, while the remaining 20% serves as the test set. Additionally, data samples from users who have delivered at least 3 jobs are filtered out. Tables 1, 2 and 3 provide a comprehensive overview of the experimental data format and quantity statistics.

Table 1. Statistical Results of the Number of Data Collected for Each Experiment

serial number	The amount of relevant data	Experimental datasets
(1)	Number of users	200000
(2)	Number of positions	50000
(3)	Number of delivery records	800000
(4)	The number of social relationships of the user	1500000

Table 2. Data Format for Social Relations in the Experimental Dataset

serial number	Data attributes	Experimental datasets
(1)	User ID (user_id)	123456
(2)	Friend ID (friend_id)	789012

Table 3. Format of User, Post and Interaction Data in the Experimental Dataset

serial number	Data attributes	Experimental datasets
(1)	user_id	123456
(2)	job_id	ABCDEFG
(3)	job_title	software engineer
(4)	location	Beijing
(5)	application_status	Delivered
(6)	user_rating	4.5

Table 4. Results of Three Methods Comparing the Recommended Confidence Levels

user_id	Recommended job_id	The confidence level is recommended for the methodology in this paper	Content-based recommendation method confidence	Confidence in the recommended method for collaborative filtering
User1	Job1	0.95	0.80	0.75
User2	Job2	0.88	0.75	0.70
User3	Job3	0.92	0.78	0.72
User4	Job4	0.85	0.70	0.65
User5	Job5	0.90	0.82	0.77
User6	Job6	0.87	0.76	0.70
User7	Job7	0.93	0.85	0.80
User8	Job8	0.86	0.72	0.68
User9	Job9	0.91	0.81	0.75
User10	Job10	0.94	0.83	0.78

Drawing from the extensive dataset provided by the recruitment website, we observe a rich trove of user information, job specifications, and interaction data between users and job postings. Utilizing this authentic data, the present study introduces a recommendation method that seamlessly integrates social network information. This approach integrates users' social network profiles with their job posting details, creating a holistic dataset that encompasses both social and job-related insights. Subsequently, we extract key features pertinent to the recommendation algorithm and divide the integrated dataset into distinct training and test sets. The training set is employed to train the recommendation model, while the test set serves to objectively evaluate its performance. By harmoniously combining multiple information sources, we arrive at a unified decision for the final recommendation outcome, thereby enhancing the precision of job recommendations. The pseudocode for running this algorithm is as follows:

```
#Initialization
Input: User social network data, employment position data
Output: Personalized job recommendation list
#Data preprocessing
function PreprocessData(user_data, job_data):
cleaned_user_data = CleanUserData(user_data)
cleaned_job_data = CleanJobData(job_data)
return cleaned_user_data, cleaned_job_data
#User preference mining
function ExtractUserPreferences(cleaned_user_data):
user_preferences = AnalyzeSocialActivity(cleaned_user_data)
return user_preferences
#Job data classification
function CategorizeJobData(cleaned_job_data):
categorized_job_data = SortByIndustry(cleaned_job_data)
categorized_job_data = SortByJobType(categorized_job_data)
categorized_job_data = SortByLocation(categorized_job_data)
return categorized_job_data
#Building an intelligent recommendation model
function BuildRecommendationModel(user_preferences, categorized_job_data):
model = InitializeModel()
model = TrainModel(model, user_preferences, categorized_job_data)
return model
#Personalized job recommendations
function PersonalizedJobRecommendation(model, user_id):
recommended_jobs = model.Predict(user_id)
return recommended_jobs
#Main process
function Main():
user_data, job_data = LoadData()
cleaned_user_data, cleaned_job_data = PreprocessData(user_data, job_data)
user_preferences = ExtractUserPreferences(cleaned_user_data)
categorized_job_data = CategorizeJobData(cleaned_job_data)
recommendation_model = BuildRecommendationModel(user_preferences, categorized_job_data)
user_id = GetCurrentUserId()
recommended_jobs = PersonalizedJobRecommendation(recommendation_model, user_id)
DisplayRecommendations(recommended_jobs)
#Execute the main process
Main()
```

6.2 Experimental Results and Analysis

6.2 Experimental Results and Analysis

To demonstrate the superior performance of the method outlined in this paper, a comparative analysis was conducted against both content-based and collaborative filtering recommendation techniques. The outcomes of this comparison, particularly with regards to recommendation confidence, are presented in the table that follows.

Upon examination of the provided data, it becomes apparent that the methodology introduced in this paper surpasses other techniques in terms of recommendation confidence for the dataset encompassing 10 distinct user groups seeking job recommendations. Specifically, our method registers an average recommendation confidence of 0.904, significantly higher than the 0.777 average confidence achieved by content-based recommendation methods and the 0.737 average confidence of collaborative filtering techniques. This superiority is primarily due to our method's ability to effectively capitalize on user and job characteristics, as well as the latent relationships between them. Conversely, content-based and collaborative filtering methods may be constrained by their narrower recommendation perspectives. Furthermore, even in cases where our method's recommendation confidence is not particularly high, such as for User2, User4, User6, and User8, it still manages to match or exceed the performance of alternative methods. This underscores the robustness of our approach in diverse scenarios. In essence, our method offers a comprehensive approach that takes into account various factors, including users' past behavior, preferences, and job attributes. As a result, it is able to deliver more precise job recommendations to users, thereby enhancing the overall accuracy of the recommendation process.

7 Conclusion

This paper presents a method for personalized job recommendations leveraging social network information fusion. The approach involves mining user preferences from social networks, developing intelligent recommendation models, and conducting cluster and correlation analyses on job data. By capitalizing on the rich, multi-source, and heterogeneous information in social networks and effectively integrating it through information fusion techniques, we strive to enhance the accuracy of our recommendations. However, it's crucial to acknowledge the study's limitations. Noise and redundant information during data acquisition and processing can compromise recommendation accuracy. Additionally, our model's performance may vary in specific scenarios, necessitating further refinement and enhancement. Looking forward, we aim to deepen our exploration of job recommeation fusion. We plan to refine our data preprocessing methods, optimize our models, and explore more personalized recommendation algorithms and techniques to cater to the diverse needs of users. Concurrently, we'll keep abreast of emerging technologies such as artificial intelligence andation through social network informnd big data, aiming to foster continuous innovation and development in employment information services.

References

1. Liang, J., Yao, J., Shi, L., et al.: A fusion recommendation model based on mutual information and attention learning in heterogeneous social networks. Future Gener. Comput. Syst. **148**(19), 128–138 (2023)

2. Chen, C., Dou, X., Zhang, J., et al.: Research and implementation of a knowledge graph-based job recommendation system. Comput. Inform. Mech. Syst. **6**(1), 20–24 (2023)
3. Wu, Q., Wei, L.: Research on biometric identity recognition method based on human communication. J. Elect. Measur. Instrum. **36**(05), 113–119 (2022)
4. Ke Weiwen, W., Sheng, K.R.: Analysis method of residents' travel characteristics based on OD flow semantic and spatiotemporal semantic clustering. J. Earth Inform. Sci. **25**(11), 2150–2163 (2023)
5. Jiaxin, Y., Huixiang, X., Ming, Y., et al.: Social network group recommendation method integrating influence dissemination. J. Intell. **41**(4), 11 (2022)
6. Xu, X., Ting, X., Chen, X.: A consensus model for emergency decision-making of large groups based on the fusion of collective intelligence knowledge in the social network environment. Chinese Manage. Sci. **32**(02), 285–297 (2024)
7. Dai, L., Dai, X., Cui, Y., et al.: Social network anomaly data mining algorithm based on deep ensemble learning. J. Jilin Univ. (Eng. Edn.) **52**(11), 2712–2717 (2022)
8. Xiaoying, Y., Mei, W., Rong, T., et al.: Social network clustering algorithm based on graph clustering and ant colony algorithm. Comput. Appl. Res. **37**(6), 2469–2475 (2020)
9. Ravita, M., Sheetal, R.: Enhanced DSSM (deep semantic structure modelling) technique for job recommendation.J. King Saud Univ. Comput. Inf. Sci. **34**(9), 7790–7802 (2022)
10. Ping, O.: The construction of college students' job recommendation model based on improved k-means-CF. Int. J. Comput. Syst. Eng. **7**(2), 190–198 (2023)
11. Ali, A.S., Minyar, S.H., Ahmed, E.H., et al.: Learning-based matched representation system for job recommendation. Comput. **11**(11), 161–167 (2022)
12. Dimos, N., Panagiota, M., Anastasios, K., et al.: Ontology-based personalized job recommendation framework for migrants and refugees. Big Data Cogn. Comput. **6**(4), 120–120 (2022)

A Study of Algorithms for Deep Integration of Information on Teaching Resources of Aerobics Course in the Context of Curriculum Thinking and Politics

Xiuyan Hong[1(✉)] and Yuanyuan Zhang[2]

[1] Xi'an Eurasia University, Xi'an 710065, China
jkhdfkdw656@aliyun.com
[2] Sanya Aviation and Tourism College, Sanya 572000, China

Abstract. To facilitate the seamless integration and optimal utilization of teaching resource information pertaining to the aerobics course, this paper introduces an innovative deep fusion algorithm for the compilation of such resources, aligned with the ideological and political framework of the course. Considering the characteristic path length of resource distribution topology structure, the data structure model of aerobics teaching resource information distribution is established, and the information characteristics of aerobics teaching resource are extracted. On this basis, the minimum editing distance of dynamic programming and Doc2vec model are used to fuse the information of teaching resources of aerobics courses with weights. Design the aerobics course teaching resource data table, store and manage the aerobics course teaching resource information after fusion processing. The test results show that the AUC-ROC value is 0.95 after the application of the algorithm, which is the best performance in the depth fusion of aerobics teaching resources information.

Keywords: Curriculum ideological and political · Aerobics course · Teaching resources · Deep information fusion · Doc2vec model

1 Introduction

In contemporary China's education system, as a brand new education concept, curriculum ideology is gradually infiltrated into the teaching of all kinds of courses. As an important part of physical education, aerobics program not only has the important function of shaping students' healthy body and enhancing their overall quality, but also fostering the cultivation of socialist core values and promoting Chinese excellent traditional culture [1]. Aerobics program is characterized by rich content, diverse forms and strong practicality, and its teaching resources cover multiple levels of sports skills, theoretical knowledge and values. With the rapid development of information technology, modernization of education has become an irreversible trend. The traditional teaching model is gradually revealing its limitations, such as scattered teaching resources, low efficiency in information transmission, and difficulty in meeting students' personalized needs. Bodybuilding operation is a comprehensive course that integrates fitness, entertainment, and education. The deep integration of teaching resources is an important part of the modernization process of education, which helps to improve teaching quality and meet the diverse learning needs of students. As a popular physical education course, the effective integration and utilization of teaching resources hold immense significance in enhancing teaching quality and fostering the comprehensive development of students.

The fusion of global and local features improves the matching of multimodal information [2, 3]. However, the current traditional algorithms for deep integration of teaching resource information have obvious deficiencies in data integration. Since the design of the algorithms is often based on fixed data structures and formats, it is difficult for the algorithms to integrate effectively when dealing with diversified teaching resource information [4]. At the same time, the processing ability of unstructured data is also weak, such as video, audio and other multimedia teaching resources, which cannot effectively extract their intrinsic features and information. Traditional algorithms also have obvious deficiencies in semantic understanding. Teaching resource information often contains a large number of text, pictures, videos and other multimedia content, which contains rich semantic information [5]. However, traditional algorithms can only extract and process information at a shallow level, and it is difficult to understand the inner meaning and association of teaching resource information, which leads to unsatisfactory fusion effect. Teaching resource information is a dynamic process, and new teaching resource information will emerge continuously. However, the traditional deep fusion algorithms of teaching resource information often lack the ability of dynamic updating, and are unable to incorporate new teaching resource information into the fusion process in a timely manner, as a consequence, both the timeliness and accuracy of the fusion results are compromised. With the continued advancement of education informatization, the type and quantity of teaching resource information are increasing. The traditional algorithms for deep fusion of teaching resource information often determine the type and quantity of data to be processed at the beginning of the design, which lacks scalability [6]. This makes it difficult for the algorithms to process and integrate effectively when facing new types and quantities of teaching resource information.

Deep fusion algorithm as a key technology to realize the efficient integration of teaching resources information, its research and the application holds great significance in enhancing the teaching quality of aerobics courses and fostering the comprehensive development of students. Through the optimization and innovation of the algorithm, the precise analysis, intelligent matching and efficient use of teaching resources can be realized, thus providing powerful support for the teaching of aerobic gymnastics course. Therefore, this paper carries out the research on the deep integration algorithm of teaching resources information of aerobics course under the background of curriculum ideology and politics. Analyze the teaching needs and characteristics of aerobics courses under the background of ideological and political education, and clarify the necessity and urgency of deep integration of teaching resource information. Design and implement an efficient and intelligent deep fusion algorithm that requires powerful data processing capabilities, accurate analysis models, and flexible matching mechanisms.

Applying algorithms to practical teaching, verifying their effectiveness through practice, and continuously optimizing and improving them to better serve the development of ideological and political education in aerobics courses.

2 Design of Integration Method for Teaching Resources of Aerobics Course

2.1 Establishing a Data Structure Model for the Distribution of Teaching Resource Information in Aerobics Courses

Analyze the distribution structure of aerobics teaching information and establish a grid structure model for the distribution of aerobics teaching resource information, as shown in Fig. 1.

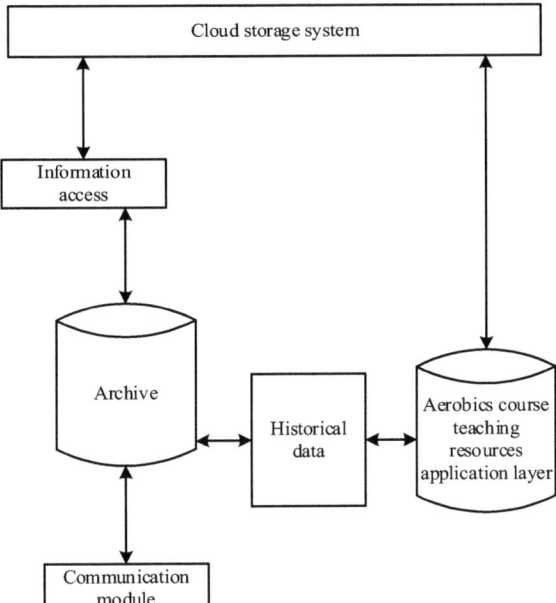

Fig. 1. Grid structure model for the distribution of teaching resource information in aerobics courses

Under the cloud porolio service platform for aerobics course teaching resource information collection, the collected aerobics course teaching resource information is configured with attributes using the association rule mining method [7, 8], and the initial clustering center of the resource distribution is S_0, the density of nodes for the distribution of information about teaching resources in aerobics courses is δ, the information distribution and storage nodes of aerobics curriculum teaching resources are modeled as energy equilibrium, and multiple temporal intervals are constructed to obtain the state space of resource integration as follows:

$$S_n = \frac{S_0}{\delta} \quad (1)$$

The dynamic properties of resource integration Y is the dependent variable, X, X_1, X_2, X_{m-1} is the $m-1$ independent variables with binding features for Y, and the resource distribution set is:

$$S = \{s_1, s_2, ..., s_m\}, m \in N \quad (2)$$

Using relevant statistical analysis methods to schedule resources online in multiple information streams, the information flow of course education resources under cloud computing is obtained as follows:

$$f_k = \{n_1, n_2, ..., n_q\}, q \in N \quad (3)$$

In the formula (3), the q denotes the node location of a collection of multiple aerobics course teaching resource information flows; n_q is a data sequence representing the flow of information about the teaching resources of an aerobics program; Using the optimal linear balance method to optimize resource allocation, combined with dynamic regression analysis, m transmission link model for resource information scheduling is obtained, with each output parameter being:

$$y = F(x) = (f_1(x), f_2(x), ..., f_m(x))^T \quad (4)$$

Among them, $y = F(x)$ denotes the balanced scale factor.

2.2 Teaching Resource Information Feature Extraction

In order to optimize and integrate course teaching resources, the collected resource information will be subjected to attribute configuration and balanced allocation using association rule mining methods, assuming that the teaching resource information of aerobics course, the n_i kind attribute of the species is u_j, then the prioritization attribute of the set of resource distributions can be expressed as follows:

$$S(n_i) = \{s_k | sr_{kj} = 1, k = 1, 2, ..., m\} \quad (5)$$

On this basis, the semantic concept set distributed scheduling method is used to obtain the resource transfer scheduling data set as $X = \{x_1, x_2, ..., x_n\}$, n is the number of slot allocation nodes X. Combined with the adaptive equilibrium configuration method, the control cluster center distribution function of resource information is as follows:

$$X_s(u) = p_c(t)e^{jt} \quad (6)$$

Among them, $p_c(t)$ indicates that the Aerobics Curriculum Teaching Resource Information Training Center belongs to the probability density of the elements of the class b_i; the e^{jt} denotes the independent correlation variables of teaching resource information of aerobics courses in cloud storage environment. The task sequence $\{x_n\}_{n=1}^N$ of scheduling the teaching resource information of aerobics course, the formation of new mappings under time-division multiple-access protocols:

$$x_n = [x(0), x(1), \ldots, x(N-1)]^T \tag{7}$$

with T_d as a unit, dividing the time axis into separate time slot allocation windows, through which the time slots are allocated. The time slot i allocation autocorrelation function for scheduling the teaching resource information of an aerobics course satisfies:

$$\overline{D} = \sum_{l_i=1}^{M-1} |D_{l_i}| / \sum_{j=1}^{M-1} |L_j| \tag{8}$$

Among them, D_{l_i} denotes the closeness metric of the cluster nodes for scheduling information transfer of teaching resources for aerobics courses; the L_j represents the width of time window of time division multiple access protocol.

2.3 Multi-information Weighted Integration of Teaching and Learning Resources

After the feature extraction of teaching resource information of aerobics course, next, the weighted fusion of teaching resource information is carried out. Standardize knowledge representation through data analysis, construct a dictionary, and introduce dynamic programming and Doc2vec models to calculate the similarity between structured and unstructured data [9, 10]. After integrating two types of similarity, entity alignment of teaching resources is achieved through weight adjustment and threshold setting, and a set of aligned entities is output.

For structured attributes and unstructured text of entities, the dynamic programming minimum editing distance and Doc2vec model depth mining feature vectors containing text structural information and semantic information are respectively used to solve their similarity. The similarity calculation formulas are as follows:

$$SimZ(E_a, E_b) = \frac{\sum_{i=1}^{t} w_i \cdot SimCP(P_{ai}, P_{bi})}{t} \tag{9}$$

$$SimF(E_a, E_b) = \frac{\sum_{i=1}^{t} Sim(F_{ai}, F_{bi})}{t} \tag{10}$$

Among them, $SimZ(E_a, E_b)$ denotes the structured attribute similarity of the entity; the w_i represents entity the i th weights of the individual structured attributes; the P_{ai}, P_{bi} denote entities, respectively E_a, E_b structured attributes; the t indicates the number of public attribute sets. $SimF(E_a, E_b)$ represents the unstructured text similarity of an entity. Finally, the weighted average fused the multi information of entities to obtain the comprehensive similarity of entities, which improved the accuracy of the algorithm [11]. After the solution of the structural attribute similarity of entity pairs to be aligned is achieved by dynamic programming minimum editing distance and Doc2vec model training semantic feature vector, etc. $SimZ(E_a, E_b)$ and unstructured text similarity $SimF(E_a, E_b)$, define that w_1, w_2 are the weights occupied by structured attributes and unstructured text respectively, and finally by the following formula for the entity, the E_a, E_b, the structured attribute similarity of the $SimZ(E_a, E_b)$ and unstructured attribute similarity $SimF(E_a, E_b)$ perform weight normalization to obtain the composite similarity of entities, the $SimE(E_a, E_b)$, expressed as:

$$SimE(E_a, E_b) = w_1 \cdot SimZ(E_a, E_b) + w_2 \cdot SimF(E_a, E_b) \qquad (11)$$

For each entity E_a in data source A, the entity E_b in data source B is descending by $SimE(E_a, E_b)$ sim value to generate the optimal candidate set, generate the optimal candidate set, and select the entity with the highest similarity and greater than a set threshold E_b as an alignable entity, it realizes the in-depth integration of information on teaching and learning resources.

2.4 Instructional Resource Data Sheet Design

After the weighted fusion of multi information of teaching resources is completed, on this basis, the teaching resources data table is designed to store and manage the fused teaching resources information of aerobics courses. According to the classification of teaching resources in LOM standard, the resource data table includes text, image, audio, video, animation and other materials. There are 13 types, including test questions, test papers, courseware, cases, literature, FAQs, resource index and online courses. E-R diagram of aerobics teaching resource information is shown in Fig. 2.

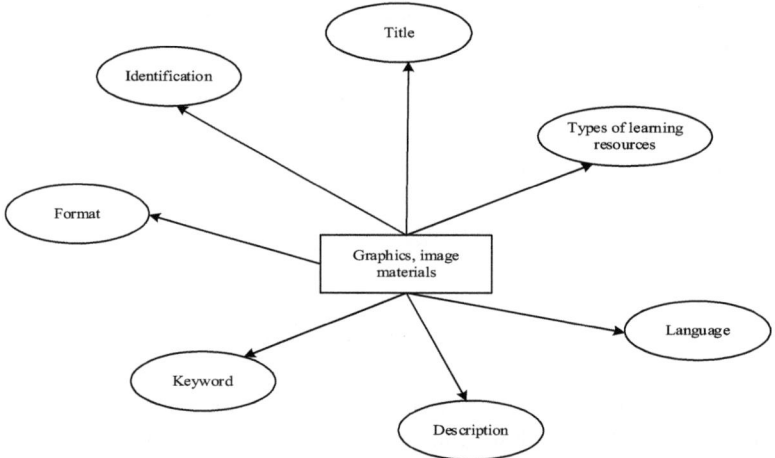

Fig. 2. E-R Diagram of Teaching Resource Information of Gymnastics Course

As shown in Fig. 2, the E-R diagram of teaching resource information of aerobics course is an embedded, disk based Java persistence engine with complete transaction characteristics, but it stores structured data on the network, referred to as a graph from a mathematical perspective, rather than in a table. E-R can also be regarded as a high-performance graph engine, encompassing all the functionalities of a well-established database. Graph Database is a special existence in the NoSql database family, which is used to store rich relational data. Neo4j is the most popular graph database at present, which supports complete transactions. In a property graph, the graph consists of vertices, edges, and attributes, both of which can be set with properties. Vertices are also called nodes, and edges are also called relationships. Each node and relationship can have one or more attributes. E-R graph creates a directed graph with vertices and edges, and its query language cypher has become the de facto standard.

In a database, a graph is comprised of nodes and relationships, where each node possesses labels and attributes. Relationships have types and attributes. Nodes represent entities. Relationships connect one-to-one nodes. Nodes can be similar to tables in mysql (different in details). Labels are equivalent to table names, and attributes are equivalent to columns in tables. The data of a node is similar to a row of data in mysql. The difference is that nodes with the same label can have user-defined attributes. They can have similar attribute structures, or they can get deep queries through the connection of relationships.

According to the resource E-R diagram shown in Fig. 2 and the specific types of teaching resources, the corresponding data table is established, as shown in Table 1.

Table 1. Aerobics Teaching Program Resource Information Data Sheet

Field name	Metadata attribute	Data type	Null or not	Explain
Id	There is no	Int	No	Unique identification of the data record
Identify	Identification	Nvarchar (1000)	Yes	A unique label for a teaching resource
Title	Title	Nvarchar (1000)	No	Teaching resource name
Language	Language	Nvarchar (100)	No	The kinds of languages used
Description	Description	Nvarchar (2000)	Yes	Text description of the resource content
Keywords	Keyword	Nvarchar (1000)	Yes	Keyword of the resource
Format	Format	Nvarchar (100)	No	The technical format of the resource
Learning resource type	Types of learning resources	Nvarchar (100)	No	Type of resource
Size	Size	Nvarchar (30)	Yes	The size of the graph/image
Location	Position	Nvarchar (1000)	Yes	The actual physical location

From Table 1, it can be seen that the designed data table structure is created in the database, including table name, column name, data type and so on. On this basis, insert test data to verify the correctness of the data table structure and data integrity. Execute various query operations to test the performance of the data table and query efficiency. Regularly monitor the performance of the data table, such as query speed, storage space, etc., to ensure the efficient operation of the data table [12]. Optimize and adjust the structure of the data table according to the actual application to meet the changing needs. Regularly back up the data table to ensure data security; in the need to quickly restore the data. Through the above process, we can design a teaching resources data table that meets the actual needs, has a clear structure and good performance. This will provide a powerful support for the efficient management and utilization of teaching resources [13] of aerobics course under the background of curriculum ideology.

3 Test Analysis

3.1 Test Preparation

In the experiment of the deep fusion algorithm of teaching resources information in aerobics curriculum, the preparation of data set is a crucial part. A reasonable and perfect dataset can provide a solid foundation for the training and testing of the algorithm, this ensures the precision and dependability of the experimental outcomes. The following are the specific steps of dataset preparation. First of all, collect aerobics course-related resources from major educational resource websites and online course platforms. These resources may include course descriptions, syllabi, video tutorials, picture materials, etc.. Grab discussions, tutorials, video sharing, etc., related to aerobics courses from social media platforms and educational forums to enrich the diversity and real-time nature of the dataset. Contact major universities, sports schools, etc., to obtain the database of their aerobics courses, including teaching materials, lesson plans, video tutorials, etc.. The obtained data results are shown in Table 2.

Table 2. Experimental dataset size and sources

No	Number of samples/PCS	Source	Scale
1	600	Major educational resource websites and online course platforms	Different styles, difficulty and duration of aerobics courses video, pictures and text introduction
2	200	Social media platforms and educational forums	Calisthenics instructional videos, pictures and discussions uploaded and shared by users
3	200	Database of aerobics courses in universities and sports colleges	Textbooks, lesson plans and expert videos

Uniformly process the above experimental data format, convert all text data to UTF-8 encoding, and convert video and picture data to a unified resolution and format. The video resolution is 720p, and the picture resolution is 1024 × 768. In the data annotation stage, 10 experts in the field of aerobics were invited to annotate the resources to ensure the accuracy and consistency of the annotation. The cleaned and pretreated data set is divided into 700 training set data, 150 validation set data and 150 test set data. In order to facilitate the processing and calling of the algorithm, the data set needs to be constructed into an appropriate data structure. For example, you can store a dataset as a CSV file or database table. Each data item contains various attributes and labels of resource information. Through the above steps, we can prepare a high-quality and diversified aerobics course teaching resource information dataset, which will provide strong support for subsequent algorithm training and testing.

Experiments need to be conducted in a stable and reliable environment. Therefore, it is necessary to build an experimental environment that contains the necessary software and hardware resources. The configuration of the experimental environment is shown in Table 3.

Table 3. Configuration of the experimental environment

No	Disposition	Argument
1	Processor	Intel Core i9-10900K (10 cores, 20 threads, 3.7GHz base frequency, up to 5.3GHz)
2	Internal memory	64GB DDR4 RAM (3200MHz)
3	Hard disk	2TB NVMe SSD (Read/write speed exceeding 3000MB/s)
4	Graphics card	NVIDIA GeForce RTX 3080 (10GB GDDR6X memory for accelerated deep learning computing)
5	Operating system	Windows 10 Pro
6	Programming language	Python 3.8
7	Machine learning framework	TensorFlow 2.5.0, PyTorch 1.8.1
8	Data processing library	Pandas 1.2.4, NumPy 1.19.5
9	Natural language processing library	NLTK 3.6.5, spaCy 3.0.6
10	Integrated development environment	PyCharm 2021.1

On this basis, the algorithm proposed in this paper is utilized to carry out an experiment on the deep integration of teaching resources information of aerobic gymnastics course under the background of curriculum ideology and politics, and the experimental results are objectively analyzed.

3.2 Analysis of Results

To test the effectiveness of the proposed method, MATLAB was used for simulation testing. The global iteration number for resource integration is set to 1200, and the sampling range for multimedia comprehensive learning teaching resources is $A_1 = A_2 = A_3 = 1$. The sampling frequency of resource information is 1004Hz, the number of resource information sources is 5000, the training sample set is 100, and the spatial sampling rate of comprehensive learning resource information is 24kHz. Based on this, resource information scheduling is carried out, and the distribution of raw resource information sampling data is shown in Fig. 3.

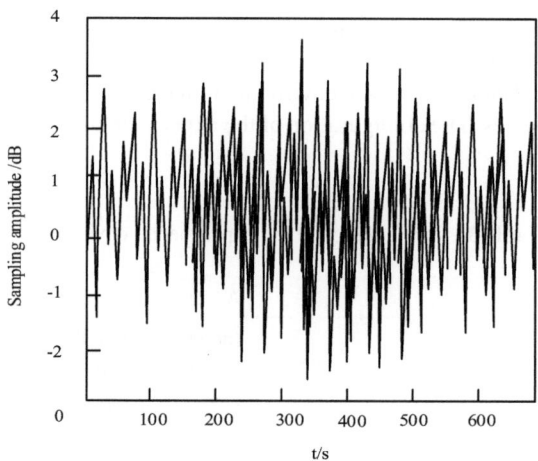

Fig. 3. Distribution of data sampled for raw resource information

Taking the resource information in Fig. 3 as a test sample, the resource integration process is carried out, and the fusion result is obtained as shown in Fig. 4.

According to the analysis of Fig. 4, the information fusion of aerobics teaching resources under the ideological and political background of the curriculum employing this method results in a superior balance. During the model training phase, the deep fusion algorithm, along with the conventional algorithms A and B proposed in this study, are trained utilizing the training set data, and the super parameters are adjusted through the verification set to optimize the algorithm performance. Test the recall of resource fusion and get the comparison results as shown in Fig. 5.

By analyzing Fig. 5, it can be seen that the method proposed in this paper has better resource information fusion effect and higher information management recall rate, it has a very good ability of fusion of teaching resources information, and the recall of resources information is higher, which has a very good value of application in fusion of teaching resources information of aerobics gymnastics course and the scheduling of information management.

To further assess the performance of various teaching resource information depth fusion algorithms, AUC-ROC (Area Under the Curve - Receiver Operating Characteristic) is selected as the main evaluation index. AUC-ROC can fully reflect the performance

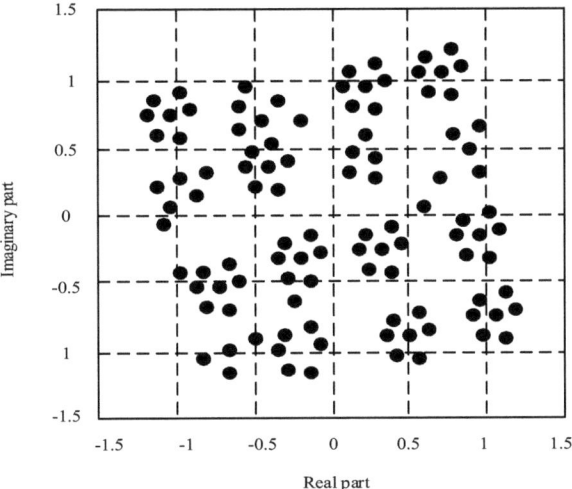

Fig. 4. Resource Fusion Output

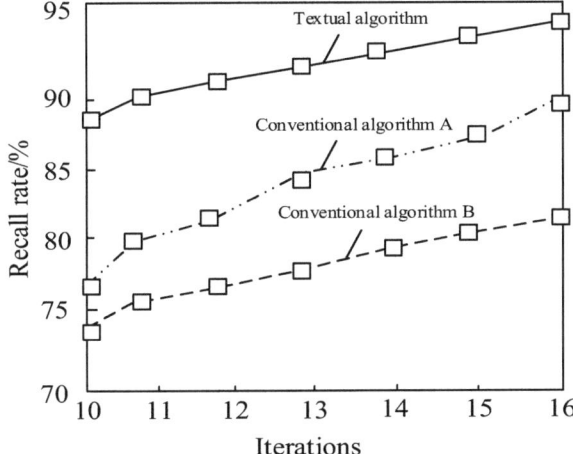

Fig. 5. Recall comparison of information fusion management

of the algorithm under different classification thresholds. The closer the value is to 1, the better the fusion performance of the algorithm will be, and vice versa. In the algorithm evaluation phase, the performance of the three algorithms was evaluated using test set data, and their AUC-ROC values were calculated and compared. The results are shown in Fig. 6.

It can be seen from the comparison results in Fig. 6 that the AUC-ROC value of the deep fusion algorithm introduced in this paper is the highest, reaching 0.95, which indicates that the algorithm has the best performance in the deep fusion task of teaching resource information, and has good generalization ability and stability in classification

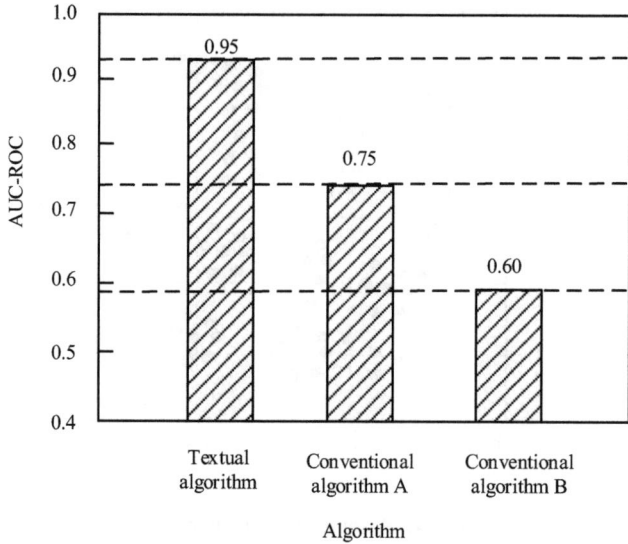

Fig. 6. Comparison Results of AUC-ROC Values of Deep Fusion Algorithm

tasks. The AUC-ROC value of conventional algorithm A is 0.75, followed by its performance. It is also competitive in the task of deep fusion of teaching resources information. However, the AUC-ROC value of conventional algorithm B is 0.60, which has poor relative performance and needs further optimization and improvement. In conclusion, the algorithm designed in this paper has the best performance in the task of deep fusion of teaching resource information, and can be used as a priority for subsequent research and application. At the same time, conventional algorithm A and conventional algorithm B also need to be further analyzed and improved to improve their performance in the task of deep fusion of teaching resources information.

In order to further test the fusion effect of different methods, F1 value was used as the test indicator, and the test results are shown in Fig. 7.

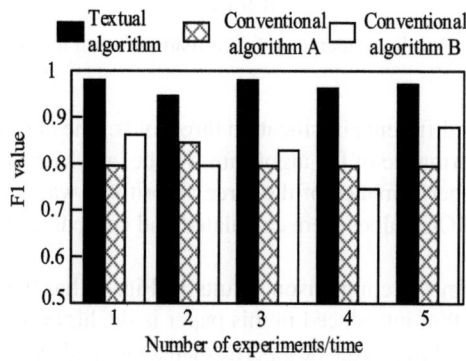

Fig. 7. Analysis of F1 value test results using different methods.

As shown in Fig. 7, the F1 value of our method is close to 1, while the F1 values of traditional methods A and B are around 0.8. This is because the deep fusion algorithm proposed in this article can more deeply explore the semantic and structural features in teaching resource information, especially through the effective extraction of unstructured text semantic information by the Doc2vec model, and the precise calculation of structured attribute similarity by dynamic programming, making the algorithm more accurate in identifying relevant and matching resources.

4 Conclusion

As information technology rapidly evolves and education undergoes a comprehensive digital transformation, the profound integration of teaching resources and information is gradually emerging as the pivotal driving force in enhancing the quality and efficiency of teaching. In this context, aerobics program, as a kind of comprehensive sport integrating fitness, beauty and entertainment, the deep integration of teaching resources and information also shows great potential and value. Aerobics program can not only shape a healthy body, but also improve the individual's physical and psychological quality, so it is loved by the majority of educators and learners. However, the traditional teaching approach in aerobics courses is frequently constrained by the scattered nature of teaching resources and limited access methods, hampering the achievement of optimal teaching outcomes. In view of this, this study is dedicated to exploring the in-depth integration algorithm of teaching resources information of aerobics course. Through the use of advanced information technology, all kinds of teaching resources are effectively integrated to realize the seamless connection and efficient use of information. This not only helps to enrich the teaching content and enhance students' learning interest, but also provides strong theoretical support and technical guarantee for the modernization and intelligence of aerobics teaching.

As we can see from the above, The deep fusion algorithm proposed in this study creatively combines dynamic programming minimum editing distance and Doc2vec model, which not only considers the structured attributes of teaching resources, but also deeply explores the semantic information of unstructured text, achieving accurate matching and recommendation. Through the weight adjustment mechanism, this study successfully integrated structured attribute similarity with unstructured text similarity, and constructed a comprehensive and three-dimensional evaluation system for teaching resources. This multi-dimensional feature fusion method not only improves the performance indicators of the algorithm, but also provides new ideas for personalized recommendation and intelligent management of educational resources. In future research, combined with the results of this study, a unified teaching resource sharing platform can be constructed based on students' learning abilities and goals, and adaptive learning path planning algorithms can be designed. This algorithm can dynamically adjust the learning order, difficulty, and progress according to the actual situation of students, ensuring that each student can learn aerobics along the most suitable path for themselves and achieve the best learning effect.

References

1. Pérez-Paredes, P.: Review of Viana (2022): teaching English with corpora: a Resource book. Int. J. Corpus Linguist. **29**(1), 116–122 (2024)
2. Huang, S.C., Fu, W.N., Zhang, Z.Y.: Global-local fusion based on adversarial sample generation for image-text matching. Inf. Fusion **103** 102084 (2024)
3. Liu, S., Huang, S.C., Wang, S.: Visual tracking in complex scenes: a location fusion mechanism based on the combination of multiple visual cognition flows. Inform. Fusion **96** 281–296 (2023)
4. Chung, C.H., Lin, Y.Y.: Online 3D gamification for teaching a human resource development course. J. Comput. Assist. Learn. **38**(3) 692–706 (2022)
5. Guyue, L.I., Aiqun, H.U.: Exploration and practice on integration of ideological, political courses into professional courses of cyberspace security specialty. Chin. J. Netw. Inf. Secur. **8**(2), 183–189 (2022)
6. Guo, W.: Streamlining applications of integration by parts in teaching applied calculus. STEM Educ. **2**(1), 73–83 (2022)
7. Gan, X., Tang, X.: Mining model of association rules for temporal data based on CNN. Comput. Simul. **38**(03), 282–285+326 (2021)
8. Biswas, S., Saha, D., Pandit, R.: A state-of-the-art association rule mining survey and its rare application, challenges, progress. Int. J. Artif. Intell. Tools **32**(06), 2350021 (2023)
9. Bruen, M., Hallouin, T., Christie, M.: A Bayesian modelling framework for integration of ecosystem services into freshwater resources management. Environ. Manage. **69**(4), 781–800 (2022)
10. Jin, J., Li, Y., Ye, L.: Integration of atmospheric stability in wind resource assessment through multi-scale coupling method. Appl. Energy **348**, 121402
11. J Lu H Gao 2022 Online teaching wireless video stream resource dynamic allocation method considering node ability[J] Sci. Program. 2022 1 8
12. Fernandes, J., Simsek, M., Kantarci, B.: TableDet: An end-to-end deep learning approach for table detection and table image classification in data sheet images. Neurocomputing **468**, 317–334 (2022)
13. Chanyawudhiwan, G., Mingsiritham, K.: Use of Smart learning resource management systems for sustainable learning. Int. J. Interact. Mob. Technol. **17**(12), 4–16 (2023)

Research on the Optimal Route Recommendation Method for Regional Tourism Based on Ant Colony Algorithm in Digital Economy Era

Ce Liu[✉] and Fang Wang

School of Management, Changchun University of Architecture and Civil Engineering, Changchun 130607, China
liuce323123@163.com

Abstract. The conventional tourism route search methods have a narrow search range and a single optimal route selection, resulting in a significant amount of time and cost waste. In response to this situation, this study designed a regional tourism optimal route recommendation method based on ant colony algorithm, based on the background of the digital economy era. Firstly, the regional tourism route planning problem is abstracted as a traveling salesman problem, with the goal of minimizing the distance traveled by the traveling salesman, and a regional tourism optimal route recommendation model is constructed; Then, the ant colony algorithm is used to solve the model. Using ant colony algorithm to simulate the swarm intelligence behavior of ants foraging in nature, guiding ants to gradually approach the optimal solution through the accumulation and updating of pheromones. By continuously iterating and updating the concentration of pheromones, it ultimately converges to the globally optimal or approximately optimal tourist route. According to the simulation experiment, it can be concluded that this method is feasible.

Keywords: Digital Economy Era · Ant Colony Algorithm · Regional Tourism · Travel Route · Optimal Route · Recommendation Method

1 Introduction

Tourism, as a comprehensive and highly related industry, has ushered in a brand new development opportunity in the wave of digital economy. In particular, the development of regional tourism not only needs to make full use of regional tourism resources, but also needs to provide tourists with personalized tourism route recommendation services to meet the growing diversified needs of tourists. Therefore, generating the most cost-effective routes for tourists is of great significance [1].

The traditional recommendation methods of optimal routes for regional tourism mainly rely on the empirical knowledge of tourism experts or simple rule-based algorithms. Although these methods can meet the basic needs of tourists to a certain extent, they have many shortcomings. First of all, they often lack comprehensive consideration of tourism resources and cannot fully integrate all kinds of resources in the region, resulting in a lack of diversity and personalization of the recommended routes. Secondly, traditional methods are difficult to cope with the dynamic changes of tourism resources, such as the impact of weather, transportation and other factors on tourism routes, resulting in a lack of real-time and accuracy of the recommended routes. In addition, traditional methods often have low efficiency in processing large-scale data and are difficult to adapt to the rapid development of modern tourism industry.

At present, more and more researchers are turning their attention to using intelligent algorithms to optimize recommendation strategies for regional tourism routes. These studies mainly focus on the use of machine learning, deep learning and other algorithms to mine and analyze tourism data to extract useful information to optimize the tourism route recommendation [2]. For example, some studies use clustering algorithms to classify tourism resources in order to provide more targeted recommendations for tourists; While others use collaborative filtering algorithms to predict the future travel preferences of tourists based on their historical behavioral data. These research results provide new ideas and methods for optimal route recommendation for regional tourism, but there are still some limitations. First of all, the existing intelligent algorithms are often difficult to take into account the global optimum and local optimum when dealing with the complex tourism route recommendation problem. On the one hand, the solution of global optimum often requires a lot of computational resources and time, and it is difficult to meet the demand of real-time recommendation; On the other hand, solving algorithms can easily lead to the dilemma of small-scale optima. How to ensure the recommendation efficiency while realizing the global optimum is an important challenge the current research is facing. Secondly, the existing research often lacks in-depth consideration of the personalized needs of tourists. Although some studies have attempted to model tourist preferences using machine learning algorithms, it is difficult to comprehensively capture the complex needs of tourists. At the same time, tourists' needs may change with time and environment.

Aiming at the limitations of the existing research results, this study introduces ant colony algorithm, hoping to recommend the optimal regional tourism routes. The innovative work of this study is as follows:

① Innovation in problem abstraction and model construction: This study cleverly abstracts the complex regional tourism route planning problem into the classic traveling salesman problem. Through this abstraction, this study provides a clear and quantifiable optimization objective for regional tourism route planning, which is to minimize the distance traveled by travelers, thus constructing a regional tourism optimal route recommendation model.

② Innovation in algorithm application: This study introduces ant colony algorithm, a heuristic optimization algorithm, to solve the problem of regional tourism route planning. Ant colony algorithm has shown unique advantages in solving large-scale and complex optimization problems due to its characteristics such as swarm intelligence, distributed computing, self-organization, and positive feedback mechanism. Applying it to the field of tourism route planning not only expands the search range, but also effectively avoids getting stuck in local optima, improving the diversity and quality of solutions.

③ Innovation in Algorithm Adaptability: When applying ant colony algorithm, this study optimized and adaptively adjusted the algorithm parameters based on the characteristics of tourism route planning problems. Reasonable parameters such as pheromone volatility coefficient and pheromone intensity were set to ensure that the algorithm can better adapt to the needs of practical problems.

④ Innovation in Practice under the Background of Digital Economy: This study closely combines the background of the digital economy era and uses modern information technology and data analysis methods to optimize tourism route planning. This is not only an innovation in traditional tourism service models, but also a positive response to the trend of intelligent and personalized development of the tourism industry in the digital economy era. Through this study, more efficient, convenient, and personalized travel route recommendation services can be provided for tourism enterprises and tourists, promoting the transformation, upgrading, and high-quality development of the tourism industry.

2 Construction of the Optimal Route Recommendation Model for Regional Tourism

In the current booming digital economy, the short-term passenger flow of regional tourism is very large, and the characteristics of distribution uniformity are not obvious, the ecological environment of the scenic area has been damaged to a certain extent. And even worse it will also lead to the occurrence of stamped. So without expanding the capacity of the scenic area, it is necessary to reasonably and efficiently plan the optimal routes for tourists to travel in the region in the digital economy era. The main objective of this chapter is to use the multi traveling salesman problem as a framework to design a model aimed at recommending the best route by minimizing the total distance traveled by the traveling salesman.

The basic mathematical model of the combinatorial optimization problem is:

$$\begin{cases} \min_{x \in R} f(x) \\ s.t. f(x) \geq 0 \\ x \in X \end{cases} \quad (1)$$

As shown in the above equation, the combinatorial optimization problem has three most important parameters (X, R, f). Among them, X is the definitional domain of decision variable x; R is a feasible domain; f is the objective function.

The traditional definition of the traveling salesman problem is that a traveling salesman must visit cities 2, 3,..., n starting from city 1, and eventually return to city 1, with only one visit to each city. In graph theory, this problem is commonly referred to as the minimum Hamiltonian loop problem. A Hamiltonian circuit is a closed path that includes all nodes in a graph and does not form branches.

Generally speaking, the planning problem of tourist routes is a complex travel salesman problem, and involves more constraints. When modeling the tourism route planning problem, it is necessary to analyze the particularity and difference between this problem and the standard TSP problem. It's as follows:

① The flexibility of tourism itinerary: Tourism routes are a core issue that tourists are extremely concerned about. Before traveling, tourists usually develop a general itinerary plan and their demand for tourism services is becoming increasingly personalized. However, in actual travel, due to unforeseeable factors such as sudden weather changes and traffic congestion, the original plan may need to be adjusted based on the actual situation, including the selection of attractions and the arrangement of travel time.
② The multidimensional complexity of resources: Tourism route planning contains numerous tourism information elements, not limited to the distance between attractions, but also including transportation modes, attraction operating hours, and expenses. The processing of this information is relatively complex and requires comprehensive consideration of multiple factors.
③ Difficulty in determining the objective function: It is difficult to describe the results of tourism route planning in terms of a quantitative and simple numerical index. Self-help tourists should consider less travel time, low travel costs, short travel routes, and rich types of attractions included in the routes and other factors when conducting route planning. Tourist route planning should meet the needs of tourists as much as possible. Therefore, how to take into account the satisfaction of tourists' needs when modeling the problem and express it with a mathematical objective function is a challenge in modeling.

To sum up, in order to meet the actual needs of optimal route recommendation for regional tourism, this paper defines the optimal route recommendation for regional tourism as a multi-travel salesman problem (mTSP) on the basis of the traveling salesman problem. It is generally described as: There are multiple traveling salesmen starting from the starting point at the same time, taking different routes respectively, and then returning to the starting point [3].

Let the decision variable satisfy the following equation:

$$x_{ij} = \begin{cases} 1, & (i,j) \text{ is on a certain circuit path} \\ 0, & \text{others} \end{cases} \quad (2)$$

In the equation, (i, j) is a city node. Then the mathematical model of the multi-travel salesman problem is:

$$\min G = \sum_{i=1}^{N} \sum_{j=0}^{n-1} D_{ij} x_{ij}, (i, j \in n) \quad (3)$$

$$s.t. \sum_{l=0}^{n-1} x_{lj} = \begin{cases} 1, j \in \{1, 2, \cdots, n-1\} \\ N, j = 0 \end{cases} \quad (4)$$

$$\sum_{l=0}^{n-1} x_{il} = \begin{cases} 1, i \in \{1, 2, \cdots, n-1\} \\ N, i = 0 \end{cases} \quad (5)$$

$$\sum_{i \in I} \sum_{j \in I}^{i=1} x_{ij} \leq n - 1 + N \quad (6)$$

Among them:

$$x_{ij} \in \{0, 1\} \tag{7}$$

In Eq. (3), min G is the objective function of the mathematical model of the multi-travel salesman problem, which requires that the sum of the paths of multiple traveling salesmen is to be the minimum; D_{ij} is the distance between urban nodes i, j; N is the number of the traveling salesmen.

Equations (4) to (6) are the feasible domains of the mathematical model of traveling salesman problem, which are often also referred to as constraints; l is the urban node; I is the set of city nodes. Equation (4) and Eq. (5) constrain every vertex except the vertex from which the traveling salesman departs to have one and only one edge in and out, while the vertex from which he departs has N edges in and out; Eq. (6) constrains the entire undirected graph to have at most $n-1+N$ edges. Compared to the $n-1$ edges in the traveling salesman problem, there will be one more edge to be added when an additional salesman is added in the multiple traveling salesman problem, thus ensuring that N loops can be formed. And similarly, this implies that there will not be any sub-loops [4]. Equation (7) is the definition domain of the mathematical model of the traveling salesman, defining the values of the decision variables x_{ij} as 0 and 1.

3 Model Solving by Ant Colony Algorithm

There are two main types of methods for solving the traveling salesman problem: Exactness algorithm and intelligent optimization algorithm. Among them:

A significant advantage of precision algorithms is their ability to ensure finding the optimal solution to the problem, but their significant disadvantage is the high time cost. As the number of cities increases, their time complexity surges, which makes it difficult to realize the solution of the exactness algorithm. The exactness algorithms include the exhaustion method, linear programming algorithm and branch and bound method.

(A) Exhaustion method

The exhaustive search method is inevitably able to obtain the optimal solution. After listing all travel options, it selects the shortest path through comparison. However, the number of schemes that traveling salesman may take increases sharply with the increase of the number of cities [5]. There are 12 possible schemes to the traveling salesman problem in 5 cities, while there are 181440 possible schemes to the traveling salesman problem in 10 cities. When the number of cities increases to 50, the possible schemes are 1.52*1064. It becomes difficult to solve the problem even with the CPU with very fast processing speed.

(B) Linear programming algorithm

The linear programming algorithm introduces more inequality constraints and generates additional cutting planes to exclude parts that do not contain integer feasible solutions, thereby gradually approaching the optimal solution. The solution efficiency of the linear programming algorithm is higher than that of the exhaustion method. But the algorithm often relies on artificial judgment when finding the cutting plane, which also means that the linear programming algorithm is not very effective in solving the traveling salesman problem with a large number of cities.

(C) Branch and bound method

The solution of the branch and bound algorithm is by effectively constraining the bounds to control the search process and make the obtained solution move towards the branch of the optimal solution. The key of branch and bound algorithm is how to select appropriate constraint limits. Different constraint limits form different branch and bound methods. Therefore, the efficiency of this algorithm is relatively low.

And the intelligent optimization algorithms for solving traveling salesman include simulated annealing algorithm, genetic algorithm, neural network algorithm, tabu search algorithm, ant colony algorithm [6], etc.

(A) Simulated annealing algorithm

The simulated annealing algorithm itself is a kind of Monte Carlo iteration, and its annealing process needs to set the initial value, constraints, iteration conditions and conditions for the algorithm to stop iteration.

(B) Genetic algorithm

Genetic algorithm is used to solve traveling salesman problems, that is to transform the solution of the traveling salesman problem into the survival process of "chromosomes". According to the evolution process of the "chromosomes" group, operations such as replication, crossover and mutation are carried out, so as to obtain an individual that is most suitable for the environment, thus obtaining the optimal solution of the traveling salesman problem.

(C) Neural network algorithm

The idea of the neural network algorithm is to map the solution of the traveling salesman problem into a permutation matrix, set a reasonable energy function, and find the optimal solution that meets the requirements of the permutation matrix according to the corresponding relationship between the optimal solution of the traveling salesman problem and the minimum value of the energy function.

(D) Tabu search algorithm

The idea of tabu search algorithm is to search from a feasible initial solution and set a tabu table to avoid the algorithm falling into the dilemma of local optimization. The use of tabu table enables the algorithm to record and select in the optimization process, thus ignoring these points in the subsequent search process [7].

Considering the practical application scenarios of the optimal route recommendation method for regional tourism in the digital economy era, this paper chooses the ant colony algorithm to solve the model, which has a better application in solving the traveling salesman problem.

3.1 Design of Ant Colony Algorithm

Ant colony algorithm is an evolutionary algorithm based on bionics. Scholars proposed ant colony algorithm. In ant colony algorithm, ants use pheromone as the medium of communication among ants. When ants are searching for food, they will leave pheromones on their walking paths as a guide for the ants behind. During the search process, ants often find the shortest path between food and ant nests. This is because ants secrete a substance called pheromones during their progress. This substance can help later ants better choose the direction to move forward [8]. Compared with other optimization algorithms in solving the multi-traveling salesman problem, the ant colony algorithm has the following special characteristics:

① Positive feedback mechanism and dynamic path selection: Ant colony algorithm uses the positive feedback mechanism of pheromone to make ants be affected by existing pheromones when selecting paths. With the operation of the algorithm, pheromones on excellent paths gradually accumulate, attracting more ants to choose these paths, so as to dynamically optimize the path selection and finally converge to the optimal solution. This dynamic property enables ant colony algorithm to flexibly cope with different situations and find the global optimal solution when solving MTSP.

② Distributed parallel computing: Ant colony algorithm is a distributed search algorithm. Each ant is independently searching for solutions and communicating indirectly through pheromones [9]. This characteristic of distributed parallel computing enables the algorithm to process a large amount of data in a short time and improve the search efficiency. In MTSP, as the scale of the problem increases, the distributed parallel computing capability of ant colony algorithm makes it more scalable and adaptive.

③ Heuristic search and global optimization: Ant colony algorithm combines heuristic information, such as distance and direction, in the search process to guide the search direction of ants.

④ Robustness and adaptability: Ant colony algorithm is not sensitive to the initial parameters and the specific form of the problem, which makes the algorithm more robust. In addition, the algorithm intelligently changes the search method for the next step, enabling it to cope with complex and ever-changing solving environments. This adaptability enables the ant colony algorithm to deal with various complex situations and improve the solution quality when solving MTSP [10].

⑤ Easy to combine with other algorithms: As an optimization algorithm with strong versatility, ant colony algorithm can be combined with other algorithms to form a hybrid algorithm to further improve the efficiency and quality of solution. For example, ant colony algorithm can be combined with genetic algorithm and simulated annealing algorithm to form complementary advantages and better solve complex problems such as MTSP [11].

Assume that there are m ants in the entire ant colony, while the number of regional tourist attractions is M; The distance between attractions u and v is $d_{uv}(u, v = 1, 2, \cdots, M)$; At a certain time t, the pheromone concentration on the path of attractions u and v is $\zeta_{uv}(t)$; Initially, set the pheromone concentration between various attractions to be the same as $\zeta_{uv}(0) = \zeta_0$. Ant $k(k = 1, 2, \cdots, m)$ can determine the next attraction that needs to be visited by the size of the concentration of pheromones between the connecting paths of individual attractions; $P_{uv}^k(t)$ indicates the probability of ant k moving from the attraction u to the attraction v at time t. The specific calculation formula is shown in the following equation:

$$P_{uv}^k(t) = \begin{cases} \dfrac{[\zeta_{uv}(t)]^\alpha [\mu_{uv}(t)]^\beta}{\sum\limits_{y \in allow_k} [\zeta_{uy}(t)]^\alpha [\mu_{uy}(t)]^\beta}, & y \in allow_k \\ 0, & y \notin allow_k \end{cases} \tag{8}$$

In the equation, $\mu_{uv}(t)$ indicates the heuristic function, which is the probability of the ant expecting to move from the attraction u to the attraction v; $allow_k$ indicates the set of the attractions that ant k has visited; α, β are the pheromone importance factor and heuristic function importance factor respectively. When ants release pheromone, the pheromone on the connecting path of the attractions will slowly disappear. So in this paper the pheromone volatilization degree is set as a parameter γ, and the value range is [0,1]. The way to update the concentration of pheromones is:

$$\begin{cases} \zeta_{uv}(t+1) = (1-\gamma)\zeta_{uv}(t) + \Delta\zeta_{uv} \\ \Delta\zeta_{uv} = \sum_{k=1}^{m} \Delta\zeta_{uv}^k \end{cases}, 0 < \gamma < 1 \qquad (9)$$

In the formula, $\Delta\zeta_{uv}^k$ represents the pheromone concentration of an individual on path $u \to v$; $\Delta\zeta_{uv}$ represents concentration and.

In particular, the formula for calculating $\Delta\zeta_{uv}^k$ is:

$$\Delta\zeta_{uv}^k = \begin{cases} \dfrac{Q}{d_{uv}}, & \text{the k - th ant visits v from u} \\ 0, & \text{others} \end{cases} \qquad (10)$$

In the equation, Q is a constant. Equation (10) gives the update rule of pheromone concentration on the path, and the ant colony algorithm can solve the traveling salesman problem according to the above equation.

3.2 Model Solving Process

The solution process is shown in Fig. 1.

Step 1: Parameter initialization. Before solving using the ant colony algorithm, the relevant coefficients are initialized, and m ants are placed on M vertex.

Step 2: Create a spatial structure for the solution and randomly assign the starting point of each ant to a set composed of the initial population solution set. Then, let these ants randomly choose and go to the next destination on their path; For each ant, calculate the probability of node transfer in accordance with the probability required by Eq. (8), and place the transfer nodes in the initialized solution set. All the ants visiting all the attractions is ended.

Step 3: Pheromone update. The length of the path traveled by each ant is calculated [12]. The optimal solution in the number of iterations is recorded and updated according to Eq. (9) and Eq. (10).

Step 4: Determination on whether to terminate or not.

Step 5: Result output. The final results of the program is output. According to the need the indicators of the process of the program optimization are output, and the trajectory route is drawn.

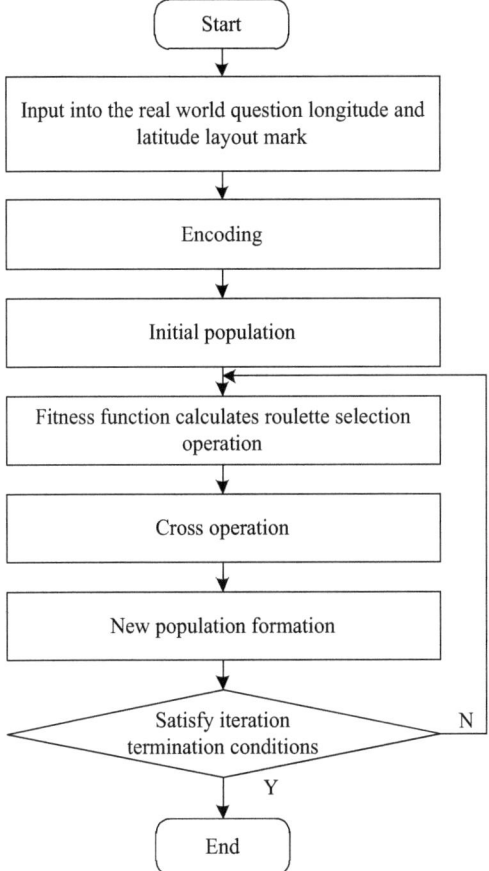

Fig. 1. Flow Chart of Ant Colony Algorithm Solving mTSP Problem

4 Simulation Experiment and Result Analysis

4.1 Design of Simulation

In order to verify the proposed tourism route planning algorithm, this paper conducts simulation experiments on the route planning algorithm proposed in the previous section in the Eclipse environment. The specific process of the experiment is shown in Fig. 2.

In Fig. 2, the input side is the basic information needed to obtain the tourism route planning. Then process the data, find feasible tourism routes using the algorithm, continuously optimize, and finally output the route planning results.

During the simulation experiment, combining the characteristics of Java programming language that is platform independent and easy to expand and the advantages of Matlab with rich function toolbox, this paper calls the functions written in Matlab in Java programs to observe the iterative process of the algorithm. The specific calling process mainly includes generating jar packages that can be called by Java projects from Matlab

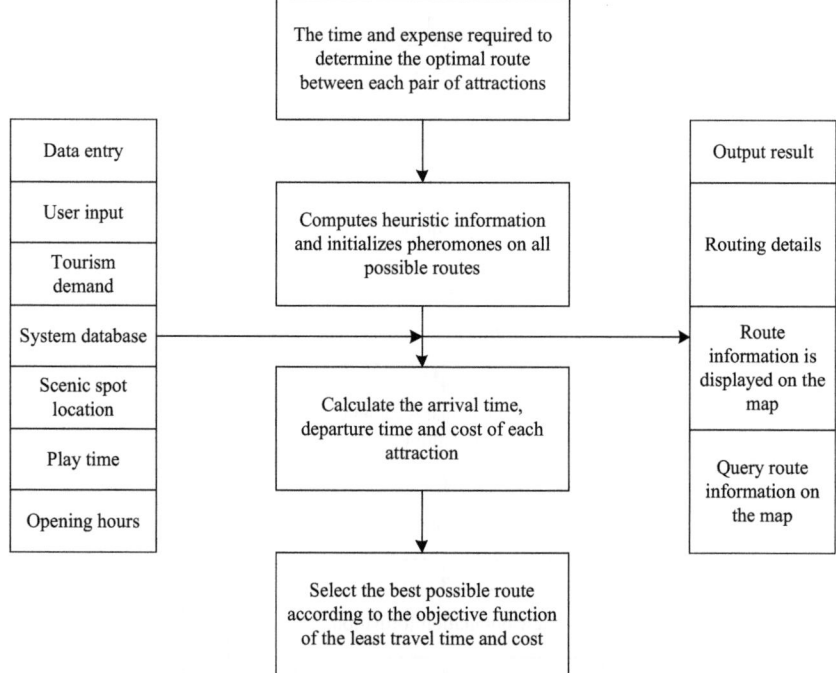

Fig. 2. Workflow Diagram of the Simulation Experiment on Optimal Route Recommendation for Regional Tourism

files, preparing to call the generated jar packages in Java programs, and using the two parts of the functions in the jar packages. The implementation method is as follows:

(A) Matlab compilation to generate jar package.

1) Enter deploytool in the programming window of Matlab, and a new window pops up.
2) In the project "Deployment Project", click the New button, select the Java package in the Target, and write the project name in the "Name", such as Quxian.prg. Then select the project storage location at the "Location" and click OK to confirm.
3) Create a new class file name at the "Add Class" where the class is added. This file name is the name of an object to be imported into Java later, that is, it is the class name placed after the new as the constructor. It is named plotquxian here. Click the "Add files" option under the class name to add the m file to be called by java.
4) In the Deployment Tool window, click the package button to generate the jar package.

(B) Java program calls Matlab function.

1) Find Matlabroot\toolbox\javabuilder\jar\javabuilder.jar and Quxian\Quxian.jar in the directory where Matlab is installed. Add these two jar packages in the directory Buildpath where external jar packages are added to the Java project.
2) Write functions in Java.

4.2 Analysis of Experimental Results

Based on the above simulation experiment platform, this paper constructs a tourism information database containing 20 points of interest and tourism routes between any two points of interest during the simulation experiment. The data information of the points of interest is shown in Table 1.

In Table 1, the point of interest types consist of two categories, where "1" represents the hotel or home, which is the starting and ending point of the tourists, and therefore a relatively small value is taken for the tour time and tour cost. "2" represents the tourists' preferred attractions, each of which has four attributes: The opening time of the attraction, the closing time, the recommended tour time, and the cost of the tour. Each tour route has two attributes: The time spent on the trip and the cost of transportation required for the trip.

Suppose that tourists randomly select 10 interest points, whose numbers are: 1, 2, 4, 7, 9, 10, 11, 14, 16, 18. The travel time is 15 h, the cost is 500 yuan, and the departure time is 7 a.m.

In order to verify the effectiveness of the proposed method, a comparison was made between the route recommendation method based on genetic algorithm and the tourism route recommendation method based on cuckoo algorithm. These three methods were used to solve the tourism optimal route recommendation model, find the optimal tourism route, and call the external jar package method according to the previous section. The drawing function generated in MATLAB was called on the Java platform, and the optimal values of the objective function for different methods were recorded. The results are shown in Fig. 3.

Due to the fact that the optimal value of the objective function (0.75) in the experiment was set in advance based on experience, the optimal search results of these three algorithms should be the same. Observing Fig. 3, it can be seen that during the entire optimization process, after applying the ant colony algorithm based regional tourism optimal route recommendation method designed in this paper, not only can the best quality solution be found before the control group in the initial iteration, but the iteration fluctuation is also relatively slow. In the control group, the two algorithms had significant iterative oscillations throughout the entire solving process. This indicates that the ant colony algorithm designed in this article has high convergence and robustness in solving the optimal route recommendation model for regional tourism, and its search solution performance is significantly better than genetic algorithm and cuckoo algorithm, further proving the effectiveness of the proposed method.

5 Conclusion

This study conducted an in-depth exploration of the optimal route recommendation method for regional tourism based on ant colony algorithm, solving the problems of narrow search range and single optimal route selection in conventional methods. In this study, ant colony algorithm has strong global search ability by simulating the collective behavior of ants in foraging. Each ant represents a potential travel route, and through parallel search, a large number of different route combinations can be explored, greatly expanding the search scope. The pheromone mechanism in ant colony algorithm enables

Table 1. Regional Tourism Points of Interest Information Sheet

NO	Name	Opening time	Closing time	Tour time (hours)	Tour cost (Yuan)	Type
1	Point of interest 1	0:00	24:00	0.00001	0.00001	1
2	Point of interest 2	6:00	15:00	0.6	20.0	2
3	Point of interest 3	7:30	19:00	0.5	0.0	2
4	Point of interest 4	7:00	20:00	0.8	30.0	2
5	Point of interest 5	8:00	21:00	0.5	50.0	2
6	Point of interest 6	9:30	17:00	0.3	55.0	2
7	Point of interest 7	6:30	17:30	0.5	10.0	2
8	Point of interest 8	8:30	16:30	0.6	20.0	2
9	Point of interest 9	8:00	18:00	1.0	80.0	2
10	Point of interest 10	9:00	21:00	1.2	40.0	2
11	Point of interest 11	9:00	21:30	0.8	30.0	2
12	Point of interest 12	9:30	22:00	0.6	25.0	2
13	Point of interest 13	8:30	22:30	1.0	5.0	2
14	Point of interest 14	8:30	20:00	0.5	30.0	2
15	Point of interest 15	7:30	19:00	0.4	40.0	2
16	Point of interest 16	6:30	19:30	0.6	60.0	2
17	Point of interest 17	7:00	20:00	0.8	50.0	2
18	Point of interest 18	8:30	20:00	0.7	15.0	2

(*continued*)

Table 1. (*continued*)

NO	Name	Opening time	Closing time	Tour time (hours)	Tour cost (Yuan)	Type
19	Point of interest 19	8:00	17:30	1.0	30.0	2
20	Point of interest 20	9:00	18:00	0.9	20.0	2

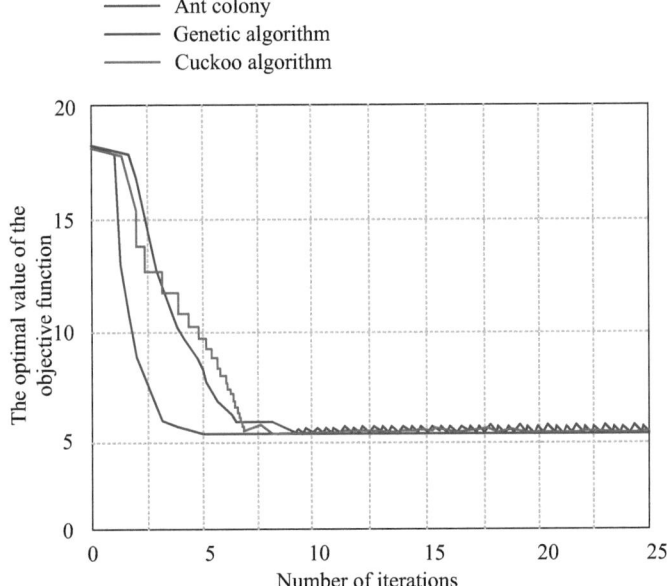

Fig. 3. The Solution Process of the Optimal Route Recommendation Model for Regional Tourism under Different Algorithms

excellent routes (i.e. shorter paths) to gradually accumulate more pheromones, thereby attracting more ants to choose the route and forming a positive feedback mechanism. By continuously iterating and updating the concentration of pheromones, the algorithm can gradually approach the globally optimal or approximately optimal tourist route.

This method is not only applicable to basic tourism route planning problems, but can also be flexibly expanded according to actual needs. For example, more constraints can be added (such as time constraints, attraction opening hours, etc.), or multi-objective optimization can be considered (such as optimizing travel time and costs). This flexibility and scalability make this method more widely applicable.

However, this study still has some shortcomings. First, the setting of algorithm parameters has a great impact on the experimental results. How to determine the optimal combination of parameters still needs further research. Secondly, this study does not fully consider the impact of the dynamic changes of tourism resources on route recommendation. In the future, real-time data can be introduced to dynamically adjust the model. In addition, this study mainly focuses on the recommendation of tourism routes in a single region, which can be expanded to cross regional tourism route planning in the future. Looking forward to the future, this paper will continue to deepen the application research of ant colony algorithm in the tourism field, and explore more efficient and accurate travel route recommendation methods. At the same time, this paper will also pay attention to the development trend of tourism and changes in tourist demand, and contribute more wisdom and strength to the development of tourism.

Acknowledgment. Research on the Development Path of Campus Tourism in Jilin Province from the Perspective of Global Tourism by Jilin Higher Education Society in 2022 (JGJX2022B50).

2022 Jilin Higher Education Association Project "Vocational Education to Empower Rural Revitalization Path Research" (JGJX2022C114).

2023 Jilin Province E-commerce Society Project "Research on Regional Brand Marketing Strategy of Featured Agricultural Products in Jilin Province under the Background of New E-commerce" (2023JLDS035).

References

1. Yu, Y.: Analysis and study on intelligent tourism route planning scheme based on weighted mining algorithm. Sci. Program. **2022**, 2022–2035 (2022)
2. Ma, Z.Q., Chen, C.C., Huang, Z.R.: Multi-objective travel-route recommendation method based on integration of user features and group-intelligence. J. Geo-inf. Sci. **24**(10), 2033–2044 (2022)
3. Sun B., Wang, C., Yang, Q., et al.: Improved evolutionary algorithm for balanced multiple traveling salesmen problem with multiple starting points. Comput. Eng. Des. **44**(07), 2030–2038 (2023)
4. Guo, H.S., Li, Z.W., Luo, C., et al.: Algorithm for solving travel salesman problems based on hybrid artificial bee colony algorithm and a-star algorithm. Sci. Technol. Eng. **23**(11), 4718–4724 (2023)
5. Dan, K., Duan, L.Z.: Dynamic programming algorithm based on optimal insertion subset for travelling salesman problem. Comput. App. Softw. **39**(12), 260–265+297 (2022)
6. Zhou, Z., Chen, E., Wu, F., et al.: Multi-satellite scheduling problem with marginal decreasing imaging duration: an improved adaptive ant colony algorithm. Comput. Ind. Eng. **176**, 108890–108923 (2023)
7. He, Q., Guan, L.H., Cui, H.H.: Hybrid variable neighborhood Tabu search algorithm for vehicle routing problem with hard time window. Computer Eng. App. 59(13), 82–91 (2023)
8. Geng, R., Zhang, Z., Niu, T.S., et al.: Research and simulation of space-based resource scheduling based on improved ant colony algorithm. J. Northeastern Univ. (Nat. Sci.) **44**(2), 168–176 (2023)
9. Song, X.B., Gao, J.W., Zhang, C.Y.: Research on off-road vehicle path planning based on improved ant colony algorithm. Comput. Simul. **40**(10), 200–204+325 (2023)
10. Liu, D., Hu, X., Jiang, Q.: Design and optimization of logistics distribution route based on improved ant colony algorithm. Optik **273**, 170405–170518 (2023)
11. Pan, H., Sun, L., Gao, S.: A hybrid algorithm of ant colony distribution estimation for the traveling salesman problem. J. Jiangsu Univ. Sci. Technol. Nat. Sci. Edn. **35**(06), 59–63 (2021)
12. Feng, Z.H., Xiao, R.B.: Extended ant colony algorithm based on mixed feedback mechanism. Control Decis. **37**(12), 3160–3170 (2022)

An Intelligent Recommendation Method for Cross-Border E-commerce with Multiple Collaborative Information Based on Deep Learning and Graph Neural Network

Liwen Zuo(✉)

School of Business, Sichuan University Jinjiang College, Meishan 620860, China
zuoliwen2021@163.com

Abstract. Due to the low matching degree and poor recommendation effect of the existing intelligent recommendation methods, we study the cross-border e-commerce multi-collaborative information intelligent recommendation method based on deep learning and graph neural network. Firstly, a heterogeneous user graph is constructed to explicitly model the interaction between users and products. The potential factor vectors of users and products are connected, and then the long vectors obtained after connection are put into the deep neural network for multilayer nonlinear transformation, and finally the predicted values are obtained directly from the output layer. By calculating the similarity, the degree of match between each product and the user's interest is obtained. According to the similarity degree, the candidate products are sorted, and the products that match the user's interest the most are put in the first place, so that they can be noticed and accepted by the user more easily. Finally, the sorted products are selected as the recommendation results. The experimental results show that the matching degree of the experimental group is more than 80%, which is the highest matching degree of the three groups, indicating that the method in this paper can accurately recommend and match the resources, and provide the users with product information recommendation services on the platform.

Keywords: Deep learning · Graph neural network · Cross-border e-commerce · Cooperative information · Recommendation

1 Introduction

Cross-border e-commerce platform developed rapidly during the epidemic, not only the quarterly online turnover grew rapidly, but also attracted a large number of domestic and foreign enterprises stationed around the cross-border trade link to build a one-stop shopping platform, from the display of commodities, marketing and promotion, payment and settlement, logistics and distribution, and so on, for both buyers and sellers to bring practical protection and convenience [1]. As the number of users rises and transaction volumes expand, the current business model of the platform gradually becomes unable to cater to the escalating demands of users, revealing certain issues that require urgent resolution. This paper hopes to explore the feasible path of its future development through the research, analysis and optimization of its business model, and to provide certain insights and references for the development of cross-border e-commerce related enterprises. The cross-border e-commerce platform business model essentially refers to the international business model in which transaction subjects belonging to different countries and regions complete the circulation of commodity information, transaction and payment by means of e-commerce platform, and finally deliver the goods and complete the transaction through cross-border logistics, based on the traditional cross-border trade, and the Internetization and platformization of the cross-border trade, which organically combines cross-border trade, e-commerce and the platform model, and possesses many characteristics in the three aspects. Characteristics of these three aspects [2]. Based on this, along with the change of transaction demand between countries and the gradual improvement of the demand of global economic integration environment, the physical boundary of Internet e-commerce has been expanding, which is not limited to a certain region or a certain country, and the scope of transaction subjects is wider. Until now, the platform's business model, characterized by a vast user base, favorable network effects, and a robust ecosystem, has rapidly emerged, aiding cross-border e-commerce in overcoming various constraints, enhancing efficiency, and gradually elevating the status of cross-border e-commerce platforms. The swift development of the Internet, the Internet of Things, communication technologies, cloud computing, and other related technologies has further accelerated this trend. "Internet+" continues to empower the industry, the concept of smart cities, smart health care, smart travel and so on continue to arise, and machine learning, big data from the software level to provide impetus for industrial development. "Internet+" for the cross-border e-commerce field empowered to give birth to intelligent shopping malls, for the traditional e-commerce field has brought about a profound change for the people's purchases to provide more and more convenient intelligent service options more abundant. E-commerce platform and the traditional e-commerce platform in the commodity content there is a certain degree of differentiation, the existence of spatial and temporal continuity between the commodity has a greater degree of relevance, browsing sequence before and after a high degree of logic, commodities in the category of a better degree of differentiation. Based on the above differences between e-commerce and traditional e-commerce, the use of sequence recommendation class method, combined with the categorization characteristics of the goods, can better capture user preferences in the field of e-commerce, and give appropriate recommendation results.

In recent years, sequence recommendation has become the main weapon of recommendation systems in capturing users' dynamic preferences. Modeling sequence interaction has achieved good results in user recommendation of social networks, cross-border e-commerce recommendation and other fields. In the work related to mining user sequence interaction, Markov chain has achieved good results in processing highly sparse data sets and obtaining short-term preference change recommendation scenarios, but it is not competent in long-term complex scenarios. Through learning a large amount of data, the recurrent neural network can model and remember the long-term preferences of users, and complete the work that is not competent based on the Markov chain model.

Today, inspired by the great success of Google Transformer model in the field of machine translation, we use the multi hop reasoning network model in TMRN for reference, and use the distribution law of e-commerce commodity hierarchy and relevance to guide the user's selection process with classified information. Multi collaborative information intelligent recommendation method not only improves the accuracy and personalization of recommendation, but also plays an important role in improving user experience and promoting commodity sales. Through this method, the cross-border e-commerce platform can better meet the needs of consumers, enhance the consumer shopping experience, thereby fostering the robust growth of the cross-border e-commerce industry. Therefore, the research on intelligent recommendation methods of cross-border e-commerce multi collaboration information is the focus of this paper.

2 Multi-collaborative Information Modeling for Cross-Border E-Commerce

2.1 Coordinated Filtering Matrix Establishment

The core idea of collaborative filtering algorithms lies in collaboration, which is a typical method of utilizing collective intelligence, manifested in mutual help and support among groups. Collaborative filtering allows users to rate a group of products, so that the system can obtain a considerable accumulation of user behavior. These user behaviors usually positively express the user's interests and preferences [3], and form a matrix of user-product interactions, as shown in Fig. 1.

	Item1	Item2	Item3	Item4
User1	1	?	1	0
User2	?	1	0	?
User3	1	?	?	?
User4	?	1	0	?
User5	?	0	0	1
User6	0	?	1	?

Fig. 1. User-commodity exchange matrix

Among them, model-based collaborative filtering method is to learn a general generalization model from the interaction matrix between users and goods, so as to fill the blank of the matrix through the model. The memory based collaborative filtering algorithm conducts mining and analysis on the basis of the relational matrix, generally only works on the relational matrix, and uses all the scores generated before the reference process. The basic idea is to remember what each user has consumed in the historical behavior, and then recommend similar things to him or recommend similar people's favorite things to users. The idea of nearest neighbor recommendation, which is often used in memory based collaborative filtering methods, needs to find similar users or similar products in high-dimensional space. Therefore, in this kind of method, the distance between two users or two goods in high-dimensional space is usually obtained by similarity measurement. Commonly employed methods for similarity measurement encompass Euclidean distance, cosine similarity, Pearson correlation coefficient, and Jaccard similarity coefficient. Euclidean distance, a frequently utilized metric, represents the authentic separation between two points in Euclidean space, or the natural length of a vector. Its equation is:

$$E(p, q) = \sqrt{\sum_{i=1}^{n} (pi - qi)^2} \tag{1}$$

In the equation: n is the spatial dimension. pi and qi respectively, in that n two vectors representing users or goods in a dimensional space. The value of Euclidean distance is non-negative and the maximum value is positive infinity. Generally, when calculating similarity, the result is expected to be in the interval [0,1] or [−1,1], therefore, the Euclidean distance usually needs a quadratic transformation when it is used. Euclidean distance measures the absolute objective value between two points in a high-dimensional space, and is biased towards analyzing the differences between user capability models, such as consumption power and contribution. Cosine similarity gauges the resemblance between two vectors by calculating the cosine of the angle they form. This cosine value indicates whether the vectors are aligned in the same direction. When the angle between the vectors is 0°, the cosine value is 1; when the angle is 180°, the cosine value is −1, which is calculated as follows.

$$\cos(p, q) = \frac{\sum piqi}{\sqrt{\sum pi^2}\sqrt{\sum qi^2}} \tag{2}$$

In the equation: pi and qi is a vector; the $\cos(p, q)$ is the cosine of the angle between the vectors. Cosine similarity is very effective in measuring the similarity of text, users, products, etc. It is characterized by the fact that the cosine similarity is independent of the length of the vectors because it is normalized by the length of the vectors. When choosing cosine similarity in collaborative filtering, after being normalized by the vector length, the recommendation results tend to be more in favor of the number of users who share the same evaluation of the two products, and there is little relationship with the size of the ratings given by the users to the products. The cosine similarity has the problem of insensitivity of absolute value size. In practical application scenarios, each user's rating criteria are inconsistent, and Pearson's correlation coefficient offers an enhanced measure by defining it as the ratio of the covariance to the standard deviation between two variables. On the basis of cosine similarity, the vector is first centralized. The vectors the p and q each subtract the mean before calculating the cosine, and the equation is:

$$R(p, q) = \frac{\sum (pi - p)(qi - q)}{\sqrt{\sum pi - p^2}\sqrt{\sum qi - q^2}} \tag{3}$$

In the equation: R is Pearson correlation coefficient. In fact, Pearson correlation coefficient measures whether two vectors are increasing and decreasing at the same time, that is, whether the change trend is consistent. Cosine similarity is applicable to explicit scoring data, while Jaccard similarity is applicable to implicit feedback data such as user favorite behavior. One of the contrast experiments involved in this paper is to mine and analyze the data based on explicit scoring. Therefore, it only follows the idea of cosine similarity in data processing and normalizes the vector length to accelerate the convergence speed of the model. In the actual recommended calculation of the prediction score, the original large matrix is no longer used, but the prediction matrix is directly obtained by multiplying the two potential factor small matrices obtained by decomposition. Due to the particularity of matrix multiplication, the blank position in the original matrix will be directly filled in the prediction matrix, thus completing the important scoring prediction process in the recommendation.

2.2 Deep Learning Behavior Prediction

A neural network, also known as a multilayer perceptual machine, is a structure of multiple neurons connected according to certain rules. Neurons in a fully connected neural network are usually laid out in layers, the leftmost layer is the input layer, which is responsible for receiving input data; the outermost layer is designated as the output layer, whence the neural network's output data is procured. The intervening layers, situated between the input and output layers, are termed hidden layers, as they remain unseen from the external perspective of the network. Within each layer, neurons lack interconnections, whereas each neuron in the present layer is interconnected with all neurons in the preceding layer. The current output subsequently serves as input for the neurons in the subsequent layer. Each connection within the neural network carries a weighted value, and often, a bias term is incorporated to enhance the network's generalization capabilities. In general, a neural network with more than two hidden layers is called a deep neural network. Deep learning is a method that uses a deep architecture. A deep neural network is actually a vector of inputs x to the output vector y function of the neural network, the output of the neural network is computed from the input, the input vector, the x which has the same dimension as the number of neurons in the input layer, needs to be firstly the input vector x, the value of each element of the neural network is assigned to the corresponding neuron in the input layer of the neural network, and then the value of each neuron in each layer is computed sequentially until the values of all neurons in the last output layer are computed, and the output vector y The dimension and the number of neurons in the output layer are the same. The computation of the output vector for each layer in a deep neural network can be expressed as the equation:

$$y(x) = f(wx + b) \tag{4}$$

In the equation: w is the weight matrix for the layer; the b is the bias vector; the f is the nonlinear converted function of each layer. The Sigmoid function with a value range of (0,1) is usually used as the activation function, and its calculation equation is:

$$sigmoid(x) = \frac{1}{1+e^{-x}} \tag{5}$$

In the equation: $sigmoid(x)$ is the activation function. Deep learning technology provides strong support for cross-border e-commerce information recommendation with its powerful feature extraction and learning ability. By building a deep neural network model, the plaorm can dig into the user's purchase history, browsing behavior, search records and other multi-dimensional data, so as to accurately capture the user's interests and needs. At the same time, the deep learning model can also automatically learn the characteristics of the product, and efficiently match the product with the user's interest. In cross-border e-commerce information recommendation, the application of deep learning models makes the recommendation results more personalized and precise [4].

2.3 Graph Neural Networks Compute Recommendation Probabilities

A graph neural network recommendation model is proposed. First, build a heterogeneous user graph. To clearly model the interaction between users and commodities. Product information can help better reflect users' interests and alleviate the sparsity of user interaction. In order to code the higher-order relationship between users and commodities, this paper uses graph neural network (GNN) to spread feature representation on the graph to learn users and representations. User embeddedness learned through the complete user click history of heterogeneous graphs can capture users' long-term interests. In addition, this paper also designs a short-term memory model based on attention, which uses the recent user reading history to model users' short-term interests. Finally, users' long-term and short-term interests are combined, and then they are spliced with candidate representations to predict the final score. Using the higher-order structure information between users, a heterogeneous map containing users and goods is established, and then the embedded information is propagated by GNN to obtain higher-order representation. GNewsRec model includes three main parts: CNN for text information extraction, GNN for long-term user interest modeling and modeling, and LSTM model based on attention for short-term user interest modeling. The first part is to extract features from titles and summaries through CNN. The second part constructs a user commodity heterogeneity map with a complete user click history, and uses GNN to encode higher-order structural information. The additional potential commodity information can alleviate the sparsity of user project, because users can aggregate more information through the commodity as a bridge by clicking fewer projects. User embedding with a complete user click history on the graph can model relatively stable long-term user interests. In the third part, we use an attention based LSTM model to encode the recent reading history to model users' short-term interests. Finally, the user's long-term and short-term interests are combined to get the final user representation, and then it is compared with the candidate representation for recommendation [5].

In this paper, we construct a heterogeneous undirected graph that $G=(V,R)$, of which V, R are node sets and edge sets. This figure contains three types of nodes, which can be mined through the topic model LDA. If the user clicks a commodity, the user commodity side is created. For each user commodity side, its commodity distribution can be obtained through LDA, and then the maximum probability commodity can be connected with the user:

$$\theta = \sum_{i=1}^{x} \theta_{d,i} = 1 \tag{6}$$

In the equation: d is user; and i is the product type. During the test, new products are inferred according to the LDA model. The attention mechanism is used to complete the task of distinguishing the importance of each item in the collection to the target item. The information formula of neighborhood node to target node is as follows, where $\pi(vi, vj)$ is the score of vi neighboring nodes vj, denoted on the global graph with the target node vi connected neighboring nodes its value of the contribution of vj; the wij is the connected edge weights of a project node vi, vj. The global item representation can be obtained by splicing the information of neighborhood nodes and target nodes. The experiment is set to 1, that is, only the first order adjacent nodes are acquired. By combining e-commerce project characterization with global project characterization through sum pooling, the final project characterization is as follows:

$$S = h_{vj} \times k \times \theta \tag{7}$$

In the equation: h_{vj} is characterization of node vj. k is the degree. The last item representation of e-commerce is taken as the basis for prediction, because the feature information obtained through three parts of the gating map neural network, self attention network and global map module has adaptively aggregated the information of other projects. The item representation is multiplied by the initial embedding of each candidate item to get all the prediction scores of the candidate item, and then the next interaction probability of the candidate item is obtained through the softmax normalization operation. In combination with the graph neural network e-commerce recommendation model GID-GNN of the self attention network [6], the model is divided into two parts as a whole: one is the representation part of the learning project, and the other is the prediction part. The project representation part also contains three important modules, namely: the gating map neural network used to learn the transfer of e-commerce projects, the self attention network used to capture the long-term dependence of e-commerce projects, and the global map module used to obtain the global level project representation. In learning project representation, the problems in the current graph neural network based e-commerce recommendation algorithm were solved in a targeted manner. Self attention network reinforcement model was used to capture the ability of long-term dependence, and the global map was used to introduce cross-border e-commerce information to enhance the model's ability to use e-commerce information, so as to obtain the probability of the final recommendation information [7].

3 Cross-Border e-commerce with Multiple Synergistic Information Recommendation

3.1 Deep Multi-Collaborative Filtering of Information

Neural collaborative filtering looks at collaborative filtering from the perspective of neural networks. In fact, it uses linear models and shallow neural networks to get the embedding vectors of users and goods. At this point, only the shallow network needs to be replaced by the deep network, and the linear model can be converted into the nonlinear activation function in the deep learning, which can be converted into the collaborative filtering recommendation model based on the deep neural network. The NCF model structure is shown in Fig. 2.

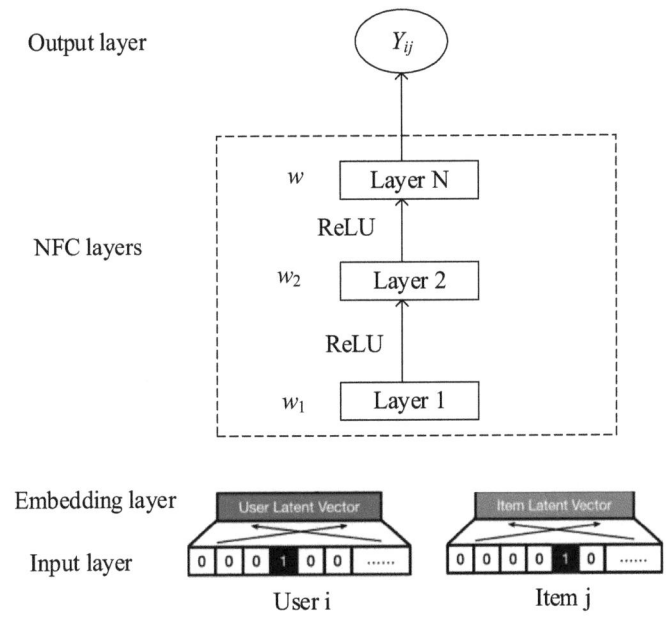

Fig. 2. Neural collaborative filtering

First, one-Hot encoded high-dimensional sparse vectors representing users and goods are utilized as inputs, and their corresponding low-dimensional dense Embedding vectors are derived through a nonlinear transformation process. Different from the traditional potential factor model method, which uses vector dot product to directly calculate the predicted value, the potential factor vectors of users and commodities are connected, and then the long vectors obtained after the connection are put into the deep neural network for multi-level nonlinear conversion, and finally the predicted value is directly obtained from the output layer. The activation function used in nonlinear conversion is ReLU function, which is considered to be the most suitable activation function for depth learning fitting, and its equation is:

$$\text{Re}LU(x) = \max(0, x) \tag{8}$$

In the equation: Re$LU(x)$ is the activation function. NCF gives two important ideas for the deep learning collaborative filtering method. The first is to connect the potential factor vectors of users and commodities before processing, which is also a common processing method in the deep learning method; The second is to directly calculate the predicted value through the deep network structure. Compared with the matrix decomposition, this processing method contains a lot of nonlinear calculations, which is conducive to mining implicit feedback information in the data. At the same time, the deep neural network here can also be replaced by other more complex deep network structures, so that the NCF model has good scalability. The scoring matrix is input into two different depth neural networks from the two directions of users and goods, and the low dimensional representation is obtained after multi-level nonlinear transformation, that is, the potential factor vector of users and goods. After obtaining the low dimensional vectors of users and commodities, DMF uses cosine calculation to normalize the predicted values, which facilitates the convergence of the model. Due to the particularity of its structure, the deep neural network can actually be regarded as the combination of multiple matrix product calculation and nonlinear activation [8], so it is very convenient and efficient to reduce the dimension of the matrix through the deep neural network. In many application scenarios, the hybrid collaborative filtering method uses many additional data to achieve better recommendation performance, especially in mitigating the cold start problem. Therefore, after considering the text information of the product, the text vector obtained by the encoder can be considered as the representation of the product text content.

3.2 Intelligent Recommendation Model Information Recommendation

Assuming that the dataset S is N dataset of cross-border e-commerce information browsed by individual users, where each browsing sequence $s = (li, ..., lm)$ is defined as an uninterrupted sequence of M texts viewed by users. Whenever the user browsing interval exceeds t days, it will start a new e-commerce. The goal of constructing this dataset is to learn that, by applying word embeddings to each browsing sequence, the li d dimension to represent. That is to browse the entire sequence data set S, learn from the idea of Skip gram model and maximize the objective function L, to learn the representation of browsing data, the objective function of the model is defined as shown in the following equation:

$$L = \log p(l_{i+j}|li) \tag{9}$$

In the equation: li is a browse sequence. That is, the probability $p(l_{i+j}|li)$ of its context domain l_{i+j} is estimated by browsing the sequence centered on li, using the soft-max function. The context of user's browsing sequence is modeled, in which the data with similar context has similar embedded representation. The Skip gram model constructed based on the browsing behavior can not only learn the user's browsing behavior habits from the click sequence, but also influence the user's use behavior learned by the recommendation model with the overall behavior trend results according to the sequence sequence generated by the single click data and the total number of clicks [9]. For example, the click sequence of a user is "cross-border commodity 1, cross-border commodity 2, cross-border commodity 3, cross-border commodity 4, cross-border commodity 5", where the click quantity of each cross-border commodity is "click quantity 1, click quantity 2, click quantity 3, click quantity 4, click quantity 5". Assuming that the target sequence is "cross-border commodity 4", the context sliding window is 1, and after optimization through negative sampling, the model only needs to calculate the probability. After improvement, the above example is changed to calculate the probability of occurrence, with additional restrictions. The improved user behavior embedding model is obtained, as shown in Fig. 3.

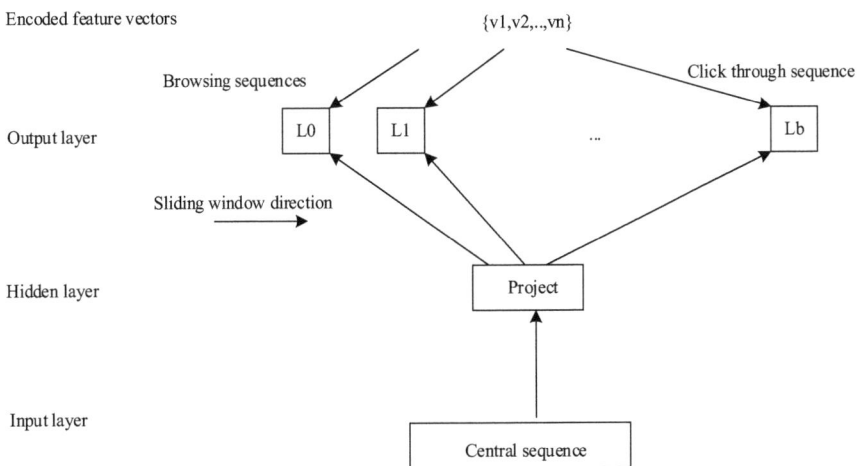

Fig. 3. Improved embedding model

The model uses a sliding window with the size of $2n + 1$ to learn the embedded vector representation. The window can slide from the first click on the list to the front of the click list. The embedded vector of the sliding center sequence is updated every time. With the sliding of the sliding window, the center sequence and context information are constantly updated, but the hit list is always kept as the global context, that is, the part pointed by the dotted line. The performance of the algorithm is that the intermediate sequence not only needs to predict its context sequence, but also needs to predict the final hits of the intermediate sequence. The reason for this design is that the total hits of e-commerce platforms are related to each user's behavior. Using the improved method as the feature extractor of the content-based recommendation algorithm, the candidate vector matrix of all users' browsing behavior can be obtained, and then the cosine similarity can be calculated according to the current browsing behavior of users and the candidate vector after model training, and the candidate text data to be recommended can be obtained. Assume there is m users, browsing candidate vector matrix can be constructed by using the improved word embedding model to train the user-browsing data. According to the user-item feature matrix, the similarity between the features is calculated as follows:

$$sim(u, v) = \cos(u, v)\frac{u \times v}{||u|| \times ||v||} \qquad (10)$$

In the equation: u and v are users; The smaller the angle between vectors, the larger the final value, and the higher the similarity. According to the similarity obtained, the top K items are selected after sorting and returned to the target user as the recommendation results. The calculation equation is as follows:

$$y = \max(sim(u, v)) \tag{11}$$

In the equation: $sim(u, v)$ is the similarity between features. When building a recommendation system, the Top-N recommendation data selection layer is a key component. The main task of this layer is to select the most suitable products from a large number of candidate products a. In order to realize this goal, it is necessary to construct a comprehensive and accurate user interest model. To accomplish this objective, it is imperative to establish a comprehensive and precise model of user interests. This is usually done by collecting users' historical behavioral data, such as browsing records, purchasing records, rating records, etc., and using machine learning algorithms for training. Through in-depth mining and analysis of user data, we can capture the user's interest preferences, consumption habits and potential needs, thus providing strong support for subsequent recommendations. After constructing the user interest model, the next task is to filter out the most compatible products with user interests from the candidate products a individual commodities. This usually involves a series of complex computation and ranking processes. First, the similarity or match between each candidate and the user interest model needs to be calculated [10]. This can be achieved by various algorithms, such as cosine similarity, Pearson's correlation coefficient, and so on. By calculating the similarity, the degree of match between each product and user interest can be obtained. Then, the candidate products need to be sorted according to the similarity. The choice of sorting algorithm can be determined according to specific scenarios and needs, and common sorting algorithms include similarity-based sorting, user feedback-based sorting and so on. Through sorting, the products that best match the user's interest can be placed in the front, so that they can be more easily noticed and accepted by the users. Finally, the sorted top is selected a products as a result of the recommendation.

4 Example Analysis

4.1 Experimental Preparation

The design of intelligent recommendation method for product information of e-commerce platform is completed from two aspects. In order to test the feasibility and reliability of this method, the following will select real registered users of e-commerce platform to participate in this experiment and recommend product information of e-commerce platform for users by using the designed method. The data set selected in this experiment is provided by a large-scale e-commerce platform in a certain area. The data set includes platform registered user data, user historical browsing data, product information, user consumption and purchase records and other data, and the data amount can reach millions. The data structure is shown in Table 1.

After mastering the relevant experimental data, 10 users were randomly selected to participate in the experiment, and two traditional methods were introduced as control

Table 1. Data sheet on e-commerce platforms

Number	Data Description	Data basis
1	Registered user id data	Sample extraction after field desensitization
2	E-commerce platform product information data	Sample extraction after field desensitization
3	Users execute product behavior data	Extract data based on recommendation requirements
4	User geolocation data	Extract longitude and latitude data
5	Classification data of e-commerce products	Sample extraction after field desensitization
6	The time of online behavior	Extract data accurate to the hour

group 1 and control group 2, respectively, to obtain the data related to the selected users as training data. When using this method to provide users with product information recommendation service of e-commerce platform, it is necessary to determine the user's preference for relevant information in this platform according to the user's historical behavior data. At the same time, the algorithm in this paper is used to calculate the matching degree between various products recommended by the platform and users' needs, and based on this, the existing information of related products of e-commerce platform is deeply mined to provide personalized products for users.

4.2 Results and Conclusions

When using traditional methods for user recommendation services, it is necessary to first establish a user behavior model, build a TPACK framework on the intelligent terminal, integrate intelligent algorithms to obtain user behavior data when browsing websites, and then complete intelligent recommendation of product information on e-commerce platforms through adaptive learning. Take the matching degree of recommendation information and user needs as an evaluation index, obtain relevant information in the experiment, and calculate the matching degree of users and recommendation information. Statistical calculation results and comparative experimental results are shown in Fig. 4.

From Fig. 4, it can be seen that the matching degree of the experimental group is always above 80% compared with the control group 1 and the control group 2, reaching the highest matching degree of the three groups, which shows that the method in this paper can accurately recommend and match resources, which proves that the method in this paper is feasible. The method in this paper is integrated into the terminal of e-commerce platform, and the traditional method is replaced by the designed method in this paper to provide product information recommendation services for users on the platform.

To sum up, by comparing the traditional recommendation method with the recommendation method in this paper, the method designed in this paper can improve the matching degree between the recommendation information and the user's needs, and

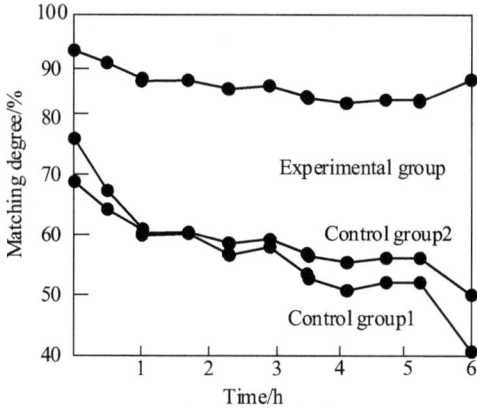

Fig. 4. Comparison results of matching degree of different methods

increase the sales of e-commerce platforms. In this way, it can create higher profits for the market operation of e-commerce platforms. After the design of this method, this paper makes a comparative experiment to compare the intelligent recommendation effects of the two methods. The test results show that the design method in this paper can improve the matching degree between recommendation information and user requirements in practical application.

5 Conclusion

The intelligent recommendation method of multi-collaborative information of cross-border e-commerce based on deep learning and graph neural network proposed in this paper realizes personalized and accurate recommendation service by deeply mining multi-dimensional information such as user behavior, commodity attributes and transaction relationship. This will not only help improve the shopping experience of users, but also promote the business growth and market expansion of cross-border e-commerce platforms.

In the future, we will continue to explore and study new technologies and methods, and constantly optimize and improve the intelligent recommendation system to better meet the needs and expectations of users. With the continuous progress of technology and the continuous accumulation of data, the cross-border e-commerce recommendation system based on deep learning and graph neural network will become more intelligent and efficient. It will be able to capture users' shopping intentions more accurately and describe the characteristics of goods more finely, thus providing users with a more intimate and satisfactory shopping experience. At the same time, it will also bring greater commercial value to cross-border e-commerce enterprises and promote the sustainable development of the whole industry.

References

1. Yan, Z., Lu, X., Chen, Y., et al.: Institutional distance, internationalization speed and cross-border e-commerce platform utilization. Manag. Decis. **6**(6), 1–5 (2023)
2. Wang, X.: Payment system of cross border e-commerce platform based on blockchain technology. Springer, Cham **28**(10), 977–984 (2022)
3. Chen, H., Huang, J., Deng, Q., et al.: Multi-granularity user interest modeling and interest drift detection. Intell. Data Anal. **27**(2), 555–577 (2022)
4. Yin, H.: The recommendation method for distance learning resources of college English under the MOOC education mode. Int. J. Continuing Eng. Educ. Life-long Learn. **2022**(2), 32–34 (2022)
5. Madhavi, A., Nagesh, A., Govardhan, A.: A study on e-learning and recommendation system. Recent Adv. Comput. Sci. Commun. **2022**(5), 15–20 (2022)
6. Liu, S., Huang, S., Weina, F., et al.: A descriptive human visual cognitive strategy using graph neural network for facial expression recognition. Int. J. Mach. Learn. Cybern. **15**(1), 19–35 (2024)
7. Papawadee, T.: The boundary between digital goods and E-services in cross-border E-commerce and implication for non-discrimination under the WTO system. Int. J. Law Inf. Technol. **31**(3), 213–230 (2023)
8. Papanicolaou, A., Fu, H., Krishnamurthy, P., et al.: A deep neural network algorithm for linear-quadratic portfolio optimization with MGARCH and small transaction costs. IEEE Access **11**(25), 16774–16792 (2023)
9. Mahesh, P.C.S., Muthumanickam, K., Vijayalakshmi, P.: Implicit spatio-temporal based hybrid recommendation model to discover malicious wireless access points. J. Intell. Fuzzy Syst. App. Eng. Technol. **44**(5), 7821–7831 (2023)
10. Brek, A., Boufaida, Z.: Enhancing information extraction process in job recommendation using semantic technology. Int. J. Performabil. Eng. **2022**(5), 18–26 (2022)

Research on English Teaching Data Location Based on Multi-agent Hierarchical Reinforcement Learning

Tingwen Wang(✉)

Shandong University of Finance and Economics, Jinan 250014, China
serenakbb123@163.com

Abstract. English teaching data often contains a amount of textual information, which poses certain difficulties for teaching due to its large volume, complex types, and difficulty in efficient localization. Therefore, a method of English teaching data location based on multi-agent hierarchical reinforcement learning is designed. Extract the challenge response delay monomer feature of English teaching data, use machine learning and data analysis to process the key information of English teaching data, and feedback the data type under the hierarchical reinforcement learning framework. A positioning model for English teaching data is formulated utilizing a multi-agent approach within a hierarchical reinforcement learning framework. Based on the characteristics and requirements of English teaching data, an appropriate multi-agent system architecture is designed, and the number, type and function of agents are determined, so as to allocate multiple subtasks to automatically locate teaching data. Generate topological adaptive levels of English teaching data positioning, dynamically adjust the hierarchical relationship, interaction mode and communication protocol of agents according to students' learning progress and feedback, so as to meet the data positioning requirements of different teaching environments. The research results demonstrate that the proposed method exhibits greater efficiency.

Keywords: Multi-agent hierarchical reinforcement learning · English teaching · Teaching data · Data location

1 Introduction

English language teaching encompasses the systematic process of imparting English knowledge and honing English skills through a diverse array of instructional techniques and strategies. This process covers all aspects of phonetics, grammar, vocabulary, listening, reading, writing and so on, aiming at enabling students to communicate and express themselves effectively in English [1]. First of all, English language teaching emphasizes the development of students' language foundation. By teaching pronunciation rules, grammar and vocabulary usage, students are helped to build a solid language foundation, which lays a solid foundation for subsequent English learning. Secondly, English teaching emphasizes the comprehensive development of listening, speaking, reading and writing skills. Through corresponding exercises, students' comprehensive English ability is improved. At the same time, it emphasizes the development of students' language perception and cross-cultural communication skills so that they can use English flexibly in different contexts [2]. In addition, English teaching also emphasizes on stimulating students' interest and motivation in learning. By designing rich and diversified teaching activities and simulations, students learn English in a relaxing and pleasant atmosphere, and cultivate their independent learning ability and lifelong learning habits. Finally, the combination of English teaching and technological development provides students with more abundant learning resources and learning methods. In conclusion, English teaching is a comprehensive, systematic and scientific process that not only cultivates students' English language abilities, but also exercises their cross-cultural communication and independent learning abilities, and provide students with a robust foundation for their future development.

Teaching data refers to various data resources used to support teaching and learning processes in the field of education. These data can include student information, course materials, learning achievements, teaching resources and other types, providing teachers and students with rich information support, and promoting the improvement of teaching effect [3]. Teaching data is the reflection of students' learning. By collecting the students' learning achievements, learning duration, learning progress and other data, teachers can understand the students' learning situation and then adjust teaching strategies to help students better grasp knowledge. Data location refers to the process of finding the required information accurately in a number of data sets. Data positioning has become more and more important, because it can help people quickly and accurately obtain the information they need to boost work efficiency and decision-making accuracy. Data location involves many aspects, including data classification, data retrieval, data mining, etc.. First, data classification is to classify data according to specific standards or attributes in order to better organize and manage data [4]. Through data classification, people can find the required data types more easily and narrow the search scope.

English teaching data location is a key link in the application of multi-agent hierarchical reinforcement learning in education. Through in-depth mining and analysis of teaching data, students' learning status, difficulties and interests can be accurately grasped, and data support can be provided for making personalized teaching plans. In addition, through real-time monitoring and feedback of teaching data, teachers can timely adjust teaching strategies to ensure the pertinence and effectiveness of teaching. Therefore, this article combines the advantages of multi-agent hierarchical reinforcement learning, and designs an English teaching data location method.

The article outlines the following key contributions:

(1) Innovatively proposed a positioning method for reinforcement learning: This article innovatively introduces the concept of multi-agent hierarchical reinforcement learning, which not only effectively extracts the challenge response delay monomer features in English teaching data, but also accurately processes key information in the data through machine learning and data analysis techniques. Through a hierarchical reinforcement learning framework, efficient feedback on data types has been achieved, significantly improving the speed and accuracy of data localization.

(2) Building a multi-agent system architecture that adapts to complex teaching needs: This article carefully designs a multi-agent system architecture based on the characteristics and requirements of English teaching data, and clarifies the number, types, and functions of intelligent agents. This architecture achieves automatic localization of teaching data by assigning multiple sub tasks to different agents, greatly enhancing the flexibility and scalability of the system.
(3) Implementing topology adaptive hierarchy to dynamically adjust agent behavior: This article further proposes the concept of locating topology adaptive hierarchy in English teaching data, dynamically adjusting the hierarchical relationship, interaction mode, and communication protocol of agents based on students' learning progress and real-time feedback. This mechanism enables the system to flexibly respond to various complex teaching environments, ensuring the accuracy and real-time performance.

2 Design of English Teaching Data Location Method Based on Multi-agent Hierarchical Reinforcement Learning

2.1 Extracting Challenge-Response Delay Monolithic Features for English Teaching Data

Challenge response delay is the total time from the landmark sending the challenge to the landmark receiving the corresponding response, which includes three parts: network transmission time, reading time, and calculation time. The network transmission time includes two parts: one is the time from L sending the challenge to P receiving the challenge, and the other is the time from P sending the response to L receiving the response. The primary determinant of network transmission time lies in the spatial relationship between the data centers of L and P. In addition, the performance of routers, network conditions and other factors also affect the network transmission time [5]. Read time refers to the time taken by P to read the file block according to the received challenge. Depending on the challenge, P may need to read multiple file blocks. The randomly selected file block necessitates that the read time hinges on the storage device's capability for random reads. Assuming a fixed number of file blocks to be read, the primary influencing factor of the read time becomes the storage device's overall performance. In this process, executing the algorithm IPScore can complete the calculation of all IP scores and obtain the IP score set S. The input of the algorithm is English data D and routing data R, and the output is the IP score of all IP addresses in the route with respect to each English data, which is recorded as S. In the fifth line of the algorithm IPScore, m represents the route the number of IP addresses in ThL.D In the seventh line, rL.D2 refers to r D. The x th IP address in. The eighth to eleventh lines calculate the IP score of the IP address IP with respect to D, which is recorded as s. If s exists, that is, in the previous iteration, the route to D contains IP, then s has a monomer feature [6]. S has been assigned and continues to accumulate. If s does not exist, initialize s. The size of s is related to the hop count of IP distance D. The smaller the hop count, the greater the value. The definition formula of monomer characteristics is:

$$s_D^{ip} = \sum_{i=1}^{n} \frac{1}{e^{m^D - x^D}} \qquad (1)$$

In formula (1), s_D^{ip} is the single body feature definition formula, indicating that n routes to D contain IP addresses; m^D is the positioning range of English data D; x^D is the x IP address of English data D. According to the IP score, the characteristics of the challenge response delay unit are obtained, as shown in Table 1.

Table 1. Challenge-response delay monolithic characterization table

Feature Name	Describe	Feature vector
Data type	Describe teaching data of types such as text, audio, video, etc	Text
Data sources	Data sources such as textbooks, the internet, and student assignments	Teaching materials
Data size	The size or length of data such as the number of words in text and the duration of audio	500bits
Delay time	The time delay from data generation to being processed or analyzed	2 h
Reason for delay	The main causes of latency include network latency, data processing complexity, and other factors	Network instability
Scope of influence	The impact range of delay on data extraction or analysis results	Single teaching unit
Remedial measures	Remedial measures taken to address delay issues	Using faster network connections
Remedial effect	Evaluation of the effectiveness of remedial measures after implementation	Reduce the delay time to 1 h

As shown in Table 1, the challenge response delay is the total time from the landmark sending the challenge to the landmark receiving the corresponding response, which includes three parts:

(1) Network transmission time. The network transmission time includes two parts: one is the time from L sending the challenge to P receiving the challenge, and the other is the time from P sending the response to L receiving the response. In addition, the performance of routers, network conditions and other factors also affect the network transmission time.
(2) Read time. Read time refers to the time taken by P to read the file block according to the received challenge. Depending on the challenge, P may need to read multiple file blocks. The randomly chosen file block, as designated by the challenge, dictates that the duration of the read operation is contingent upon the storage device's capability for random read accesses. [7].For example, mechanical hard disk and solid state disk have great differences in the read time.

(3) Calculate time. Calculation time refers to the time taken by the data center of P to generate a response after reading all the specified file blocks. When the algorithm of response generation is determined, the main factor affecting the calculation time is the CPU performance. As the proportions of reading time and computation time increase, the correlation between the overall time delay and the English teaching data diminishes. The reading duration is contingent upon the storage device's performance and the quantity of data blocks accessed, whereas the computation time is influenced by the overhead imposed by the PDP scheme and the CPU's capabilities. Hence, the design of cloud data placement strategies necessitates the consideration of these factors to minimize both reading and computation times.

2.2 Construction of English Teaching Data Positioning Model Based on Multi-agent Hierarchical Reinforcement Learning

In specific English teaching scenarios, data owners often do not deliberately perform positioning operations for server or user data. Computing time refers to the time taken by P's data center to generate responses after reading all specified file blocks. When the algorithm of response generation is determined, the main factor affecting the calculation time is the CPU performance. Therefore, according to the characteristics and requirements of English teaching data [8], design a suitable multi-agent system architecture, and determine the number, type and function of agents, so as to allocate multiple subtasks to automatically locate teaching data. The English teaching data positioning model is shown in Fig. 1.

As shown in Fig. 1, the data owner completes the initialization operation to outsource the local data to the cloud server; secondly, the data owner sends the integrity verification operation or data dynamic operation request according to the demand, specifically sends the challenge to each distributed agent, and the distributed agent also executes the protocol to the server according to the challenge content; after that, the server needs to generate the proofs or perform the operations of adding, deleting, and modifying the data in the trustworthy platform module; then, the server needs to forward each proof to all the agents immediately; in addition, each agent collects the proofs of integrity or dynamic operation from the server, and obtains the proof of addition or deletion, and modification.

After that, the server needs to generate proofs or perform data addition, deletion, modification and other operations in the trusted platform module; then, the server needs to forward each proof to all agents immediately; in addition, each agent collects integrity proofs or dynamic operation proofs from the server and obtains the corresponding network measurements, and sends them to the data owner (in case of dynamic operation of the data, the user needs to update the metadata locally); finally, the data owner performs the validation and location estimation [9]. In the initialization phase, the signature of the pedagogical data is defined with the following formula.

$$\delta_i = \left(H(M_i) \cdot \prod_{j=1}^{s} s_D^{ip} \right) \qquad (2)$$

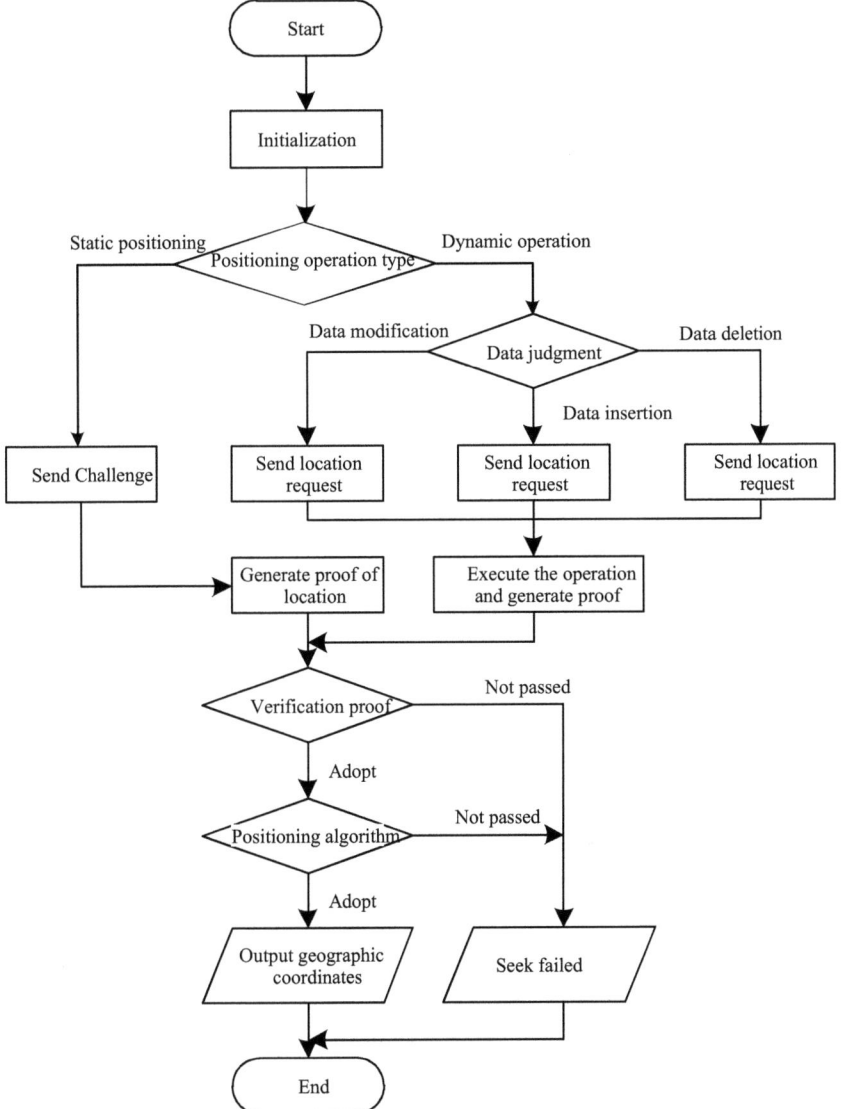

Fig. 1. Schematic of the ELT data localization model

In formula (2), δ_i is signature for pedagogical data; and M_i is a data category. H is the hash function. In the proof generation phase, the challenge-response protocol is executed with the following formula.

$$\mu_j = \sum_{(i,j)\in Q} v_i M_i \qquad (3)$$

$$\vartheta = \prod_{(i,j)\in Q} \delta_i \qquad (4)$$

In formula (3–4), μ_j is the location request received by the location model. v_i is the ith data identifier; ϑ is distributed agents. Q is the timestamp. According to the δ_i, μ_j, ϑ determine the correctness of the signature and combine it with the localization theory to improve the accuracy of localization. The theory of localization of data for teaching English language in graded intensive learning is illustrated in Table 2.

As shown in Table 2, a professional overview of the English teaching data positioning model based on multi-agent hierarchical reinforcement learning is given, covering the main components of the model, agent characteristics, learning mechanism, environmental interaction, model optimization and other aspects, which helps to understand the construction process and key elements of the model. In the validation phase, there are:

$$e(t, g) = e(\mu_j, \vartheta) \tag{5}$$

$$e(\vartheta, g) = e\left(\prod_{(i,j)\in Q} \delta_i \cdot H(M_i) \cdot \prod_{j=1}^{s} s_D^{ip}\right) \tag{6}$$

In formula (5–6), $e(t, g)$ is the signature forwarded by the proxy to the data owner g at time t; $e(\mu_j, \vartheta)$ is the localization request sent by the agent. When the formula (5) holds, perform the steps of formula (6) and summarize the results of localization of English teaching data, the $e(\vartheta, g)$, to meet data localization needs.

2.3 Generating a Topologically Adaptive Hierarchy of English Teaching Data Localization

The physical location of the data is not considered in the construction of the overlay network. Then dynamically adjust the hierarchical relationship, interaction mode and communication protocol of agents according to the learning progress and feedback of students to meet the data positioning requirements of different teaching environments. In English teaching data, nodes close to each other are organized into groups. Each group selects a node with the strongest node capability to act as a super node. The super node of each layer group is organized into the upper layer network. The maximum number of nodes in each layer is related to the dimension of ID space and the level where it is located.

The ID space of teaching data is divided among super nodes on each layer [10]. Each super node is responsible for an ID subspace belonging to the group, that is, responding to the request that the target ID value is included in the space. The size of the node subspace is related to the level of the node. The ID subspace of the super node of the group is the sum of the subspaces of all nodes in the group. The lowest common node is not responsible for the subspace, and the nodes are organized into a hierarchy through the hierarchical relationship of the subspace. The English teaching data positioning topology adaptive hierarchy code is generated, as shown in Fig. 2.

As shown in Fig. 2, (1) is to build a Q network to estimate the value of state action pairs; (2) select actions for using Q network; (3) choose the action that possesses the highest Q value; (4) to arbitrarily select a batch of samples from experience playback for learning; (5) initialize the state space, action space and other environments; (6) to

Table 2. Theoretical table of data localization for English language teaching in tiered intensive learning

Concept	Assembly	Describe	Association method
Model Overview	Multi Agent system	Multiple agents collaborate to complete tasks, with each agent responsible for different data localization tasks	Multi Agent Reinforcement Learning
	Hierarchical structure	Decompose complex localization tasks into multiple levels, with each level handling tasks of varying difficulty	Layered reinforcement learning
Agent characteristics	Agent type	Agents for different data types, such as text agents, audio agents, etc.	Customized agent design
	State space	A feature set that describes the current state of the Agent, such as data features, location information, etc.	Feature engineering
	Action space	The set of actions that an Agent can take, such as data filtering, classification, etc.	Action Definition
Learning mechanism	Reinforcement learning algorithms	Strategies and value functions used to train agents, such as Q-learning, Actor Critic, etc..	Reinforcement learning algorithms
	Reward function	Define the pros and cons of Agent behavior and guide Agent learning	Reward design
	Exploration - Utilizing balance	Exploring new actions for balancing agents and utilizing known optimal actions	ε- greedy strategy
Environmental interaction	Environmental status	The status of English teaching data, such as data type, quantity, etc.	Environmental modeling

(continued)

Table 2. (*continued*)

Concept	Assembly	Describe	Association method
	Action execution	Agent performs actions in the environment, such as data retrieval, filtering, etc.	Environmental interface
	Feedback signal	Environmental feedback on Agent actions, such as rewards, punishments, etc.	feedback mechanism

```
# Defining Agent Classes
class Agent:
    def __init__(self, env):
        self.env = env
    def build_q_network(self):
        # (1)


model.add(tf.keras.layers.Dense(self.env.action_spac
e.n, activation='linear'))
    # (2)
    if np.random.rand() < 0.1:  # Random
exploration
        minibatch = self.memory.sample(batch_size)
            # Define Environment Class
    def __init__(self):
    # (3)
        Self.observation_space = spaces.Box(low=0,
high=1, shape=(10,))  # (4)
        self.action_space = spaces.Discrete 5
    # (5)   def reset(self):
    # (6)    self.topology = self.initialize_topology()
        return np.random.rand(10)  # (7)  def
step(self, action):
    # (8)    reward = 0
    # Update status return self.state, reward, done,
next_state
    def update_state(self, action):
    # (9)
```

Fig. 2. English teaching data localization topology adaptive hierarchy partial code

initialize the adaptive topology; (7) to reset the environment, return to the initial state; (8) is to return to random initial state; (9) to update the topology according to actions and current status, calculate rewards, and adjust connections and levels in the topology.

At present, the hierarchical topology structure is mainly used to improve the accuracy of positioning by taking advantage of the difference of data, and a two-layer structure is established on Brocade, CoCO (Chord over Chord), and Fast Track. Some data is selected from a specific structure to serve as super data and a super data layer is established. The hierarchy of topology structure is fixed, and the difference of data is fully utilized, the overlay network constructed by using the physical location information of data can obtain specific DHT data location, thus ensuring the accuracy of data location.

3 Experimental Analysis

To validate the suitability of the method proposed in this paper for English teaching data localization, an experimental analysis is conducted on the aforementioned approaches. The conclusive experimental outcomes are presented in a comparative format, juxtaposing them against the traditional cloud-based approach for English teaching data localization, the conventional machine learning based The final experimental results and the specific experimental preparation process are shown below.

3.1 Experimental Preparation

This experiment consists of two parts of SFSChord simulator, one part is traffic_gen, the other part is sim. The codes of the two parts are independent. The function of traffic_gen is to generate a detailed simulation description, and the generated code is interpreted and executed by the sim program to complete the simulation. When a node joins the BIRChord ring, first build two data sets FingerTable and ReFingerTable for each data node. FingerTable and ReFingerTable respectively save the information of some subsequent nodes of English teaching data in the clockwise and counterclockwise orientations of this node within the ring. The structure of FingerTable and ReFingerTable datasets is shown in Table 3.

Table 3. Structure of the ELT dataset

Dataset name	Domain name	Definition
Finger table	Finger[k].start	$(n + 2k\text{-}1) \mod 2m \ 1 \leq k \leq m$
	Finger[k].node[j]	finger[k].start Subsequent nodes of (quantity j)
	Successor	Subsequent to this node
	Predecessor	Previous to this node
ReFinger table	REfinger[k].start	$(n + 2k\text{-}1) \mod 2m$ or $(n + N\text{-}2k\text{-}1) \mod 2m \ 1 \leq k \leq m$ N: Maximum value of data location node ID
	REfinger[k].node[j]	REfinger[k].start The predecessor nodes of (quantity is j)
	Successor	Subsequent to this node
	Prede cessor	Previous to this node

As shown in Table 3, traffic_ gen: traffic_ gen has two parameters: input.sc and seed. Input.sc contains a brief description of the experimental scenario, including three commands: events, num, avg, wjoin, sleep, wfail, winsert, wfind. This command uses

Poisson distribution to generate num events, and the average arrival time of each event is num, where num is in ms. Wjoin, weep, wfail, winsert, and wfind represent five basic events: data join, data leave, data access failure, document insert, and document query. Here, we define wtotal = wjoin + wireless + wfail + winsert + wfind. Therefore, the probability of data location events can be expressed as wjoin/wtotal, the probability of leaving events can be expressed as wireless/wtotal, and so on. The data positioning principle is generated, as illustrated in Fig. 3.

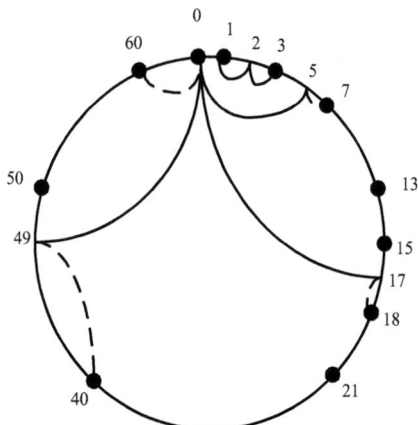

Fig. 3. Schematic diagram of data localization for English teaching

As shown in Fig. 3, 0 ~ 60 indicates that there are 60 nodes that need to be localized in the English teaching data set, each data node locates two types of teaching data, respectively, to save part of the node information in the clockwise and counterclockwise directions in the ring, and each dataset maintains a successor chain table to save a certain number of available successor nodes. The nodes are located from the initial node of the query in the ring from the clockwise and counterclockwise directions for bidirectional location search, from both directions at the same time to locate the target node.

3.2 Experimental Results

This article randomly selects coverage, location accuracy, F1 value, response time and other indicators, and judges the data location effect under different data set sizes and number of agents. Among them, coverage refers to the proportion of the data covered by the model in the whole dataset. High coverage means that the model can process more data, thus providing more comprehensive positioning services. The response time represents the time required from the input query to the location result. Short response time means that the model can provide results faster, which is particularly important for applications with high real-time requirements. The data location accuracy rate represents the proportion of the model correctly predicting the data location in the data location process. In other words, the extent to which the model's predicted outcomes align with the actual data location represents its accuracy.

Additionally, the F1 value, which is the harmonic mean of precision and recall, serves as a comprehensive metric to assess the model's overall performance. A higher F1 value indicates superior performance of the model in terms of accuracy and recall. When other conditions are known, this paper compares the performance of the conventional cloud based English teaching data location method, the performance of the conventional machine learning based English teaching data location method, and the performance of the multi-agent hierarchical reinforcement learning based English teaching data location method designed in this paper. The results are illustrated in Table 4.

Table 4. Experimental results

Method	Dataset size	Number of intelligent agents	Coverage rate /%	response time /ms	Positioning accuracy /%	F1 value
Traditional method	1000	5	85.26	2.50	86.21	0.88
	2000	10	84.23	3.30	85.36	0.87
	3000	15	90.12	6.50	91.27	0.92
	4000	20	76.27	4.00	77.33	0.78
Machine learning driven approach	1000	5	92.36	0.50	93.42	0.94
	2000	10	91.15	0.70	92.58	0.93
	3000	15	90.27	0.90	91.36	0.92
	4000	20	93.34	1.50	94.27	0.95
The method of this paper	1000	5	99.26	0.05	99.99	0.99
	2000	10	97.24	0.05	98.96	0.99
	3000	15	98.37	0.05	99.27	0.99
	4000	20	99.21	0.05	99.99	0.99

As shown in Table 3, when the data set size, the number of agents and other conditions are consistent, and after using the conventional cloud based English teaching data location method, the coverage varies from 76% to 91%, the location accuracy varies from 77% to 92%, the F1 value varies from 0.75 to 0.95, and the response time varies from 2ms to 7ms. It can observe that the data positioning effect of this method is not good, which affects the effectiveness of English teaching. After using the conventional machine learning based English teaching data location method, the coverage, location accuracy, F1 value, response time and other indicators are lower than those of the conventional cloud based method.

However, on the whole, there is still some room for improvement in the location effect, which needs further optimization. However, after using the English teaching data location method, the coverage varies from 97% to 100%, the location accuracy varies from 98% to 100%, the F1 value is stable at 0.99, and the response time is stable at 0.05ms. The proposed method can quickly locate English teaching data with higher accuracy.

4 Conclusion

Artifical intelligence and machine learning have made significant breakthroughs in many fields. As an important machine learning method, reinforcement learning has been widely used in decision-making and control problems. In recent years, multi-agent reinforcement learning has gradually become a research hotspot, which shows great potential in dealing with multi-agent cooperation and competition. Especially in the field of education, the introduction of multi-agent hierarchical reinforcement learning provides a new perspective for personalized and intelligent teaching. English teaching is of great significance in cultivating students' cross-cultural communication ability and global vision. However, the traditional English teaching methods often have the problems of low efficiency, lack of personalization and intelligence. English teaching has produced a amount of teaching data. Utilizing these data efficiently to improve the standard of teaching has emerged as a pressing issue.

Therefore, this paper uses multi-agent to strengthen learning and designs a method of locating english teaching data. From the aspects of features, positioning models, topological levels, etc., the teaching content, student performance, teaching strategies, etc. are regarded as different agents. By constructing a multi-agent system, accurate data localization has been achieved, truly meeting the needs of English teaching.

References

1. Byun, M., Evans, S., Herman, B., et al.: Game development as speculative design: teaching data science ethics using decentralized research groups. Proc. Assoc. Inf. Sci. Technol. **59**(1), 630–632 (2022)
2. Macgillivray, H.: Statistical meaningfulness, teaching craft and writing about teaching statistics and data science. Teach. Stat. **44**(1), 1–4 (2022)
3. Taub, M., Azevedo, R.: Teachers as self-regulated learners: the role of multimodal data analytics for instructional decision making. New Dir. Teach. Learn. **2023**(174), 25–32 (2023)
4. Sun, X., Hu, G.: Direct and indirect data-driven learning: an experimental study of hedging in an EFL writing class. Lang. Teach. Res. **27**(3), 660–688 (2023)
5. Anumudu, C.: Practical implications from teaching adult English as a second language in faith-based programs. New Direct. Adult Continuing Educ. **2022**(175–176), 95–103 (2022). https://doi.org/10.1002/ace.20471
6. Sun, C., Wei, L., Young, R.F.: Measuring teacher cognition: comparing Chinese EFL teachers' implicit and explicit attitudes toward English language teaching methods. Lang. Teach. Res. **26**(3), 382–410 (2022)
7. Mckinley, J., Rose, H.: English language teaching and English-medium instruction: putting research into practice. J. Engl. Med. Instruct. **1**(1), 85–104 (2022)
8. Dimova, S., Kling, J.: Certifying lecturers' English language skills for teaching in English-medium instruction programs in higher education. J. Engl. Med. Instruct. **1**(2), 137–152 (2022)
9. Ja, Z.P., Zheng, S., Ren, J.J., et al.: Hierarchical collaborative edge cache architecture based on reinforcement learning. Comput. Simul. **40**(7), 296–299 (2023)

Optimal Allocation Method of College Students' Ideological and Political Education Resources Based on PSO Algorithm

He Kong[1(✉)], Zhengfang Lu[1], and Xiaodi Li[2]

[1] Student Affairs Department, Jilin Communications Polytechnic, Changchun 130000, China
konghe0412aa@163.com
[2] Jilin Communications Polytechnic, Changchun 130012, China

Abstract. Aiming at the problem of uneven distribution and limited coverage of ideological and political education resources for college students, a PSO algorithm based method for optimizing the allocation of ideological and political education resources for college students is proposed. By using the ID3 algorithm to deeply explore and classify the multidimensional attributes of ideological and political education resources for college students, including resource types, content depth, audience preferences, and expected effects, a comprehensive resource feature library is constructed, providing a solid data foundation for subsequent allocation decisions. Designed and implemented a customized PSO algorithm framework that fully considers multiple factors such as resource scarcity, diversity of student needs, and maximization of educational outcomes. By simulating the search behavior of particle swarm optimization in the solution space, the resource allocation scheme is iteratively optimized until the optimal solution that satisfies all preset conditions or approaches the optimal solution is found, achieving fast convergence. Test results demonstrate that this method effectively and rationally distributes these resources, ensuring minimal disparities across varying request conditions, thus exhibiting a noteworthy advantage in terms of coverage.

Keywords: PSO Algorithm · ID3 Algorithm · College Students · Ideological and Political Education · Optimal Allocation of Resources

1 Introduction

With the continuous deepening of educational reform, ideological and political education for college students is also receiving increasing attention. However, in today's university education, there is a problem of uneven allocation of ideological and political education resources [1]. Therefore, exploring scientifically effective methods for allocating educational resources is of great significance. The distribution of resources for ideological and political education among college students ought to take into account various considerations, including the genuine requirements of the students, the environmentally friendly utilization of educational resources, as well as other pertinent factors.

Traditional resource allocation methods often rely on manual experience and subjective judgment, and lack scientific and uniform standards and quantitative indicators [2]. As a result, the process of resource allocation is often not objective and fair, and making it difficult to allocate resources reasonably and effectively. The traditional method of resource allocation is often static and lacks consideration of the dynamic changes in the demand for educational resources and individual differences [3]. As the times progress and the demands for students' ideological education evolve, this static allocation method can no longer meet the actual needs. Due to the lack of scientific allocation basis and criteria, the traditional method often leads to waste of resources and duplication of investment. Some important resources for ideological and political education may not be sufficiently supported, while some relatively minor resources may be over-invested [4]. Traditional resource allocation methods often focus on short-term resource allocation without long-term planning and sustainability considerations. This may lead to short-term excess or long-term shortage of resources, ultimately impeding the fulfillment of students' long-term ideological and political educational requirements. Furthermore, the resource allocation under traditional methods is very hasty, thus ignoring the individual differences and needs of different students (such as ethnic minority students, poor students, etc.) not being able to meet their needs for ideological and political education.

The PSO algorithm can transform the resource allocation problem into an optimization problem through mathematical modeling of educational resources, and find the optimal resource allocation scheme [5] through the adaptive search process of the algorithm. Particle Swarm Optimization (PSO) has garnered widespread application across numerous fields because of its simplicity, fast convergence and other advantages. PSO algorithm regards each solution as a "particle" in the search space by simulating the group cooperation behavior of birds in the process of foraging, and constantly updates the speed and position of particles through information sharing between particles and accumulation of individual experience, so as to find the optimal solution of the problem [6]. Applying PSO algorithm to the optimal allocation of college students' ideological and political education resources can make full use of its characteristics of global search and rapid convergence, and improve the rationality and effectiveness of resource allocation. In addition, PSO algorithm has good scalability and flexibility, which can be adjusted and optimized according to the actual situation to adapt to different educational environments and needs. For example, the search speed and accuracy of the algorithm can be controlled by adjusting the number, speed, acceleration and other parameters of particles; different constraints and optimization objectives can be introduced to adapt to different educational resource allocation problems. Therefore, the method proposed in this paper based on PSO algorithm is not only theoretically feasible, but also has practical application value and prospect.

2 The Design of the Optimal Allocation Method for the Resources of Students' Ideological and Political Education

2.1 Evaluating the Resources of Students' Ideological and Political Education

The key link in optimizing the allocation of resources for student ideological and political education is resource evaluation, and its aim is to attain a comprehensive and in-depth understanding of the existing resources, therefore serving as a foundation for informed decision-making in devising subsequent allocation strategies. First of all, we use interviews, questionnaires, and group discussions, collect students' needs and expectations for ideological and political education. We analyze students' age, gender, professional background, interests and other factors in order to better customize educational resources. List all the available ideological and political education resources as shown in Table 1.

Table 1. Types of Ideological and Political Education Resources for University Students

Types of educational resources	Specifically cover
Textbooks and reference materials	Ideological and political theory textbooks, supplementary reading materials, electronic resources, etc.
Online courses and videos	Ideological and political courses, special lectures, documentaries and other video resources on the online platform
Practical activities and bases	Take stock of ideological and political practice opportunities provided by the school or community, such as volunteer service, field trips, simulation exercises, etc.
Teaching staff	Statistics of ideological and political education of full-time teachers, part-time teachers, outside experts and other human resources

For each resource, a quality assessment is made. Assessment criteria could include.

(1) Content accuracy: whether the information conveyed by the resource is accurate and authoritative, and whether it is in line with the Party's education policy.
(2) Timeliness: whether the content of the resources is updated in a timely manner and whether it can reflect the latest political theories and the results of social practice.
(3) Applicability: whether the resources are suitable for the age, psychological characteristics and cognitive level of college students, and whether they can meet their learning needs [7].
(4) Attractiveness: whether the resources are vivid, interesting and able to attract students' attention and interest.

Questionnaires and interviews were used to find out how students and teachers use the available resources, as shown in Table 2.

We predicted the demand trend for ideological and political education resources among students based on their age, gender, and professional background. They are as follows:

Table 2. Use of Ideological and Political Education Resources

Item	Usage situation
Usage frequency	The number and frequency of the resource being used
User feedback	Student and faculty satisfaction with resources, comments and suggestions
Problems	Identify the problems and challenges encountered in the use of resources, such as difficulty in obtaining resources and outdated content

(1) Content preferences: students in different majors may have different preferences and focuses for ideological and political content.
(2) Form demand: modern college students may prefer digital and interactive learning methods.
(3) Practical opportunities: students may want more practical opportunities to enhance their understanding and application of ideological and political theory.

Match students' needs with available resources and analyze the gap between supply and demand of resources. Through the above detailed resource assessment, we can have a comprehensive understanding of the situation of the existing civic education resources and provide powerful data support for the subsequent optimization of resource allocation. This helps to ensure that the distribution of educational resources is more accurate and effective, and meet the diversified needs of college students.

2.2 Classification of Educational Resource Attributes

To guarantee the equilibrium in the allocation of educational resources, this paper utilizes the ID3 algorithm to meticulously dissect the particular attributes of ideological and political education resources, categorizing them in accordance with the insights derived from the analysis. In the specific calculation process, this paper uses the gain of data information as the index parameter. The calculation method of mixed ideological and political education resources can be expressed as:

$$X(x_1, x_2, ..., x_n) = -\sum_{i=1}^{n} px_i(k) \qquad (1)$$

Among them, X represents a collection of ideological and political education resources for college students; x_n represents educational resources characterized by multiple attributes; n represents the total amount of ideological and political education resources for college students; $x_i(k)$ represents the attributes of k in x_i resources; $px_i(k)$ represents the likelihood of the occurrence of attributes within resources. However, the attribute composition of ideological and political education resources for college students exhibits a more pronounced degree of autonomy, which leads to the fact that for any resource, the attributes it contains are not fixed, and the corresponding range of $px_i(k)$ values is also relatively dynamic. In response to this feature, this paper sets that the upper limit of the threshold of $px_i(k)$ result is "1" and the lower threshold of the value is "0" [8]. When $px_i(k) = 1$, it indicates that no resources contain the k attribute, when $px_i(k) = 0$, it indicates that no resources do not contain the k attribute. According to the above formula, there is an entropy of the attribute category of any resource corresponding to the value of the central attribute. The calculation method of this parameter can be expressed as follows:

$$e(x_i) = -x_i \sum_{j=1}^{m} \frac{W_j}{W} \qquad (2)$$

Among them, $e(x_i)$ indicates the entropy of attribute category and central attribute value of ideological and political education resources x_i; W_j serves as an indicator of the weight coefficient assigned to the j attribute of ideological and political education resources, represented by x_i; W indicates the weight coefficient of educational resources x_i in its category of the attribute; m indicates the total number of attributes contained in the educational resources x_i [9]. As evident from the aforementioned formula, the magnitude of the entropy value for any mixed ideological and political education resources is intricately linked to the composition of its attributes and the value of the central attribute within the classification it belongs to. The calculation of entropy expectation value can be expressed as follows:

$$E(x_i) = -\sum_{i=1}^{n} \sum_{j=1}^{m} p_{ij} x_{ij}(k) \qquad (3)$$

Among them, $E(x_i)$ indicates the expected entropy value; p_{ij} indicates the probability of a multi-attribute resource x_i appearing in a category with the j attribute as the core. In this paper, the use of resource data gain to classify resources can not only effectively improve the classification accuracy of resource attributes, but also maximize the role of multiple attributes for the subsequent stage of merit allocation. In this paper, the calculation of educational resource data gain can be expressed as follows:

$$G(x_i) = E(x_i) - e(x_i) \qquad (4)$$

Among them, $G(x_i)$ it signifies the entropy gain parameter associated with each attribute of multi-attribute ideological and political education resources x_i. On the basis of the calculation result of the above formula, the attribute with the largest gain is classified as the final attribute of the resource. The calculation result has been completed according to the above method, which provides the basis for the subsequent resource allocation tables.

2.3 Design PSO Algorithm Process

The optimization allocation method for ideological and political education resources among college students, utilizing the Particle Swarm Optimization (PSO) algorithm, strives to determine the most effective resource allocation strategy by simulating the information sharing mechanism observed in the predatory behavior of birds. Has a significant impact on promoting education. The operation is shown in Fig. 1.

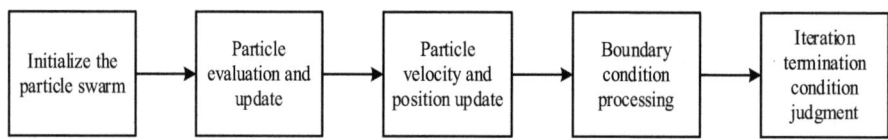

Fig. 1. Operation Flow of PSO Algorithm

As depicted in Fig. 1, the initial step involves setting the number of particles within the particle swarm, which is determined based on the problem's scale and available computational resources. Randomly initialize its position and velocity in the solution space for each particle. Define a fitness function for evaluating the advantages and disadvantages of the ideological and political education resource allocation scheme represented by each particle. The design of the fitness function should be based on the goal and effect of ideological and political education, and also consider the utilization efficiency of resources, student satisfaction and other factors [10]. This function can incorporate diverse indicators. For each particle, use the fitness function to calculate its fitness value. It is assumed that there are n different types of ideological and political education resources that need to be allocated to m different educational sessions or programs. For the number of each resource i is R_i, the quantity demand for resources i of each session or course j is D_{ij}. The resource allocation program can be represented by using a matrix A of $n \times m$, where A_{ij} represents the number of resources i assigned to sessions or courses j. The formula for the fitness function is shown below:

$$F = \alpha \cdot S + \beta \cdot E + \gamma \cdot E_f \tag{5}$$

$$S = 1 - \frac{\sum_{i=1}^{n} \sum_{j=1}^{m} |D_{ij} - A_{ij}|}{\sum_{i=1}^{n} \sum_{j=1}^{m} D_{ij}} \tag{6}$$

$$E = \frac{\sum_{i=1}^{n} \sum_{j=1}^{m} \min(D_{ij}, A_{ij})}{\sum_{i=1}^{n} R_i} \tag{7}$$

$$E_f = \frac{\sum_{j=1}^{m} S_j}{m} \tag{8}$$

Among them, α, β, γ all represent weighting coefficients, which are used to balance the importance of different components; S represents the evaluation value of resource satisfaction, which can be calculated on the basis of the difference between actual allocations and requirements; the smaller the difference is, the higher the level of satisfaction is; E indicates an evaluation value of the resources utilization, which may involve factors such as resources utilization, the balance of distribution, etc., the higher the ratio is, the higher the efficiency is. E_f represents the evaluation value of the education effects can be calculated based on students' feedback and statistical data of educational results. In practical applications, the design of the fitness function needs to be adjusted and optimized according to the specific problems and data characteristics to ensure that the algorithm can find the best resource allocation scheme that meets the actual needs.

If the fitness value of the current particle surpasses its previously recorded optimal value, its individual optimal position and corresponding fitness value should be updated. By comparing the individual optimal fitness values across all particles, the most favorable one is chosen as the global optimal solution, and the global optimal position is subsequently updated. Fine tune the velocity of each particle based on individual and global optimal positions. According to the updated velocity, update the position of each particle, i.e., update the resource allocation scheme. Check whether the updated particle position exceeds the boundary of the solution space. If it does, it is necessary to perform boundary processing, such as taking the boundary value, random jump, etc.. Based on this, set the boundary value of the particle. On this basis, set a maximum number of iterations and stop iteration when the number is reached. Set a fitness threshold, when the global optimal fitness value is not significantly improved for many iterations, the iteration can be stopped early. If the found global optimal solution already meets certain quality requirements, the iteration can be stopped early. The visualization of each particle's position and fitness value throughout the iterative process provides valuable insights into the algorithm's operational dynamics. This visualization aids in comprehension and analysis of the algorithm's execution. A thorough analysis of the optimization results reveals pertinent information and recommendations that inform subsequent resource allocation decisions for ideological and political education.

2.4 Optimal Allocation of Ideological and Political Education Resources Based on PSO Algorithm

On this basis, this paper aims to optimize the allocation of ideological and political education resources for college students, establish a constraint objective function, and solve the problem of educational resource heterogeneity in the cloud computing environment. The algorithm formula can be expressed as follows:

$$R(x_i) = \frac{aT(x_i) + bD(x_i)}{cB(x_i)} \quad (9)$$

Among them, $R(x_i)$ represents the objective function for allocating mixed ideological and political education resources in a cloud computing environment; a denotes the weighting factor for the execution time constraint; while b ignifies the weighting factor for the delay constraint; c represents the weighting factor for the bandwidth constraint; $T(x_i)$ represents the time cost in implementing the allocation of these resources x_i; $D(x_i)$ ignifies the latency encountered in executing the allocation of diverse ideological and political education resource x_i; and $B(x_i)$ denotes the distribution of bandwidth during the allocation of mixed ideological and political education resources x_i. The corresponding constraint is to guarantee that the allocation result can guarantee that the load of resources is in a balanced state. Therefore, the corresponding constraints can be expressed as follows:

$$s.t. \begin{cases} T(x_i) \prec T_{max} \\ D(x_i) \prec D_{max} \\ B(x_i) \prec B_{max} \end{cases} \quad (10)$$

Among them, T_{max} represents the maximum allowed time overhead when the resource perform the corresponding assigned task; D_{max} represents the maximum delay allowed when the resource perform the corresponding assigned task; B_{max} represents the maximum bandwidth available for utilization during the execution of the assigned task by the resource. The specific allocation process of this educational resource is depicted in Fig. 2.

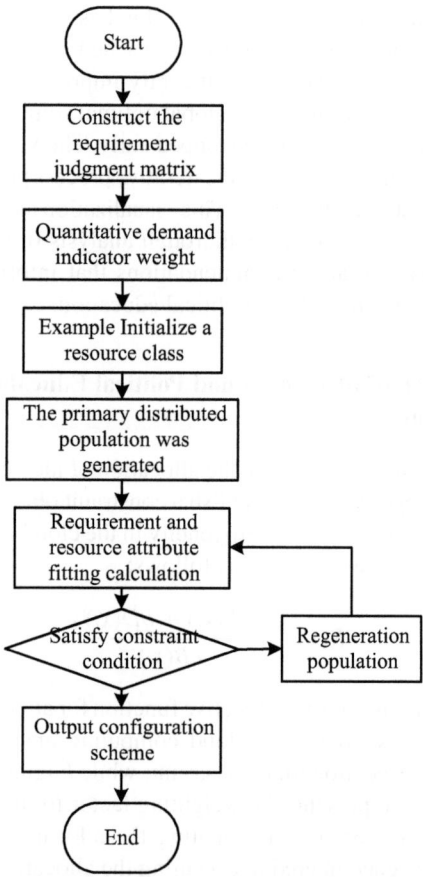

Fig. 2. Schematic Diagram of the Process of Allocating Resources for Ideological and Political Education

In this way, educational resources are fully allocated, ensuring a balanced load distribution among the resources during the practical application phase.

3 Test Analysis

3.1 Test Preparation

To validate the efficacy of the optimized allocation approach for ideological and political education resources for college students, utilizing the PSO algorithm, the statistical data of education resources in S city districts and counties are collected, as shown in Table 3.

Table 3. Statistical Data on Educational Resources

No	Number of students	Number of teachers	Student-teacher ratio	Number of books	Number of books per capita
1	11779	1881	6.262	1021446	86.717
2	20185	2995	6.740	1914772	94.861
3	10925	1792	6.093	880434	80.589
4	6546	1894	3.456	908681	138.815
5	12678	2524	5.023	1072874	84.625
6	17008	3578	4.753	1776209	104.434
7	4179	462	9.045	303686	72.670
8	12850	1157	11.106	559515	43.542
9	17665	2135	8.274	1230236	69.643
10	12158	1052	11.557	363090	29.864
11	13288	1142	11.636	594900	44.770
12	12988	1300	9.990	442549	34.074
13	3830	842	4.549	269879	70.646

The data presented in Table 3 reveals a significant disparity in terms of the student-teacher ratio and the number of books available per student in some districts and counties. To improve this issue the educational resource allocation model will, according to the objective function, appropriately decrease the student-teacher ratio while simultaneously increasing the number of books available to each pupil in the process of resource allocation, so that the student-teacher ratio and the number of books available per pupil between districts and counties converge to a balanced state as much as possible.

The ideological and political education of college students from a particular institution is chosen as the experimental sample object for this study. The university has approximately 10000 students and multiple colleges in disciplines such as humanities, science, and engineering. The overview of ideological and political education resources is as follows:

(1) Classroom resources: there are a total of 50 classrooms that can be used for ideological and political education, with a maximum capacity of 100 students per classroom.

(2) Teachers' resources: there are 100 ideological and political teachers, with an average of 2 h/week per teacher.
(3) Textbook resources: there are five different types of ideological and political textbooks, with 2,000 copies of each.

In order to ensure the smooth running of the experiment, the experimental environment was built as shown in Table 4.

Table 4. Configuration of the Experimental Environment

No	Disposition	Detailed parameter
1	Server	Intel Xeon E5–2620 v4 processor, 32GB memory, 1TB hard disk, and Windows Server 2019 operating system
2	Client	It has an Intel Core i7 processor, 16GB of RAM, 512GB of solid-state drive, and runs Windows 10
3	Development language	Use Python 3.8 as the development language
4	Data processing and analysis tools	Pandas was used for data processing and Matplotlib and Seaborn were used for data visualization analysis
5	Local area network	The experimental environment is connected through Gigabit Ethernet to ensure the high efficiency and stability of data transmission
6	External network	Connect to the external network through the campus network to facilitate access to external data and resources

Collect the use of ideological and political education resources of the university in the past year, including classroom utilization rate, teacher teaching hours, textbook borrowing and other data. Simulated data is generated according to historical data to simulate the demand for ideological and political education resources in different scenarios. Through the above experimental sample objects and environment configuration, we can conduct a comprehensive experimental verification of the method proposed in this paper, evaluate its performance in different scenarios, and provide strong support for practical applications. First of all, it is assumed that the goal is to maximize the coverage and effect of ideological and political education, taking into account the constraints of classrooms, teachers, textbooks and other resources. This problem can be transformed into a mathematical optimization problem, such as multi-objective optimization problem or constrained optimization problem. Next, use Python programming language to implement the PSO algorithm. Use the above prepared data to initialize particle swarm and evaluate particle fitness. During the execution phase of the algorithm, the particle swarm undergoes initialization, and an iterative search is conducted to find the optimal solution. Update both individual and global optimal solutions simultaneously.

3.2 Analysis of Results

In order to further analyze the application effect of the design allocation method in this paper, targeted comparative tests were carried out. Among them, the allocation methods used by the control group are the conventional allocation method based on differential evolution and the allocation method based on deep reinforcement learning. This paper employs simulation techniques to evaluate the performance of various allocation methods.The TSPLB database categorizes mixed ideological and political education resources into 10 categories according to their corresponding attributes. The total number of resources in each category is 120. In the test phase, set the scale of the number of concurrent resource requests, and separately count the load balance of a single resource under different allocation methods. Considering the impact of contingency on the test results, this paper uses three methods to conduct 50 independent tests respectively, and takes the average value of the test results as the final evaluation basis. The allocation of any resource is based on resource requests. Therefore, this paper takes the difference coefficient of resource allocation under each request condition as an assessment criterion, and the final experimental data is shown in Table 5.

Table 5. Statistical Results of the Coefficient of Variation in the Distribution of Resources for Ideological and Political Education

No	Number of concurrent requests per item	Optimal allocation method based on PSO algorithm	Distribution method based on differential evolution	Allocation method based on deep reinforcement learning
1	50	0.12	0.26	0.21
2	100	0.13	0.30	0.25
3	150	0.12	0.34	0.26
4	200	0.15	0.35	0.28
5	250	0.14	0.29	0.31
6	300	0.13	0.38	0.35
7	400	0.13	0.40	0.37
8	500	0.14	0.42	0.40
9	600	0.15	0.46	0.43
10	700	0.16	0.52	0.45
11	800	0.16	0.54	0.46
12	900	0.17	0.58	0.48
13	1000	0.17	0.64	0.57

Through the analysis of the data in Table 5, it can be seen that under the three allocation methods, the maximum difference coefficient of the corresponding resource allocation has certain differences. In the test results of the allocation method based on differential evolution, the maximum difference coefficient of educational resource allocation is 0.64. In the test results of the allocation method based on deep reinforcement learning, the maximum difference coefficient of educational resource allocation is 0.57. In contrast, in the test results of the design method in this paper, the maximum difference coefficient of educational resource allocation is only 0.17, and the corresponding difference coefficient is not affected by the scale of concurrent requests. The test results show that the method designed in this paper can achieve reasonable allocation of resources, to ensure a uniform distribution of resources with minimal disparities across various request conditions, we further validate the optimal allocation effectiveness of our proposed method by employing the coverage rate of ideological and political education resources as a comparative metric. This metric quantifies the proportion of students covered by these educational resources. And the calculation formula for this coverage rate is as follows:

$$C = \frac{M_F}{M} \times 100\% \qquad (11)$$

Among them, set the total number of students to M; M_F is set for the number of students allocated to educational resources. Organize and analyze the collected data, evaluate the effectiveness of its results of the resource allocation scheme by comparing the changes in the number of students covered. The calculated comparison results are shown in Fig. 3.

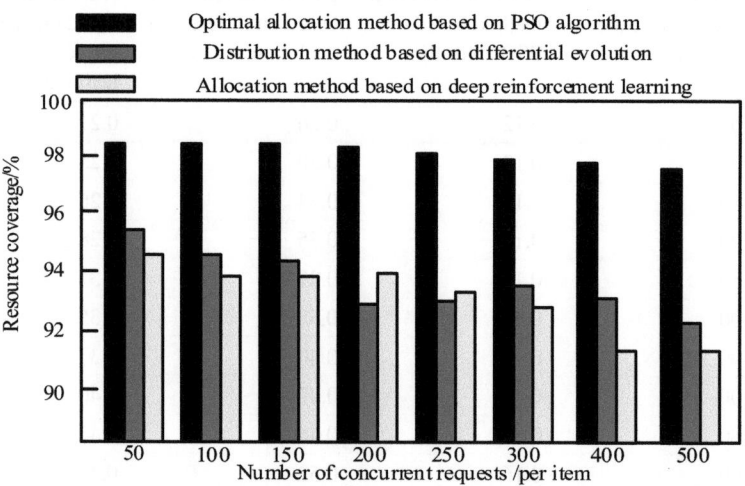

Fig. 3. Comparative Results of the Coverage of Ideological and Political Education Resources

It can be seen from the comparison results in Fig. 3 that compared to the coverage calculated by traditional methods, the method proposed in this paper has better accuracy, which shows that the algorithm can allocate ideological and political education resources more effectively, it holds the potential to play a more significant role in practical applications and thus possesses valuable potential for promotion and widespread application.

On this basis, we analyze the time complexity of three methods of optimal allocation of resources: n represents the number of populations; D represents the dimension; the number of iterations of the algorithm is set to m; the number of optimized allocation attempts is set to t. It is calculated that the allocation method based on differential evolution has a higher time compared to the allocation method grounded in deep reinforcement learning, is $O(m \cdot n \cdot D \cdot (t + n))$; the proposed method in this paper exhibits a time complexity of $O(m \cdot n \cdot D)$. When compared to the other two methods, achieving the optimal solution requires fewer iterations with the proposed approach. The average running time of the three methods in the process of allocating educational resources for students' ideology and politics when the adaptability value converges, and the average running time of the educational resource allocation methods are shown in Table 6.

Table 6. Average Running Time of the Educational Resource Allocation Methodology

No	Allocation method	Teacher resources	Book resources
1	Optimal allocation method based on PSO algorithm	0.054	0.103
2	Distribution method based on differential evolution	0.526	1.249
3	Allocation method based on deep reinforcement learning	0.671	1.374

It can be seen from Table 6 that the average running time of the optimized allocation of educational resources based on the PSO algorithm proposed in this paper is the shortest, while the average running time of the other two conventional educational resource allocation methods is relatively long. In each iteration of the algorithm, the calculation process is relatively complex, so this method has a long running time and low allocation efficiency in allocating educational resources. Through the above comparison and analysis, the efficacy of the resource optimization allocation approach introduced in this paper has been thoroughly validated.

In order to further test the effectiveness of different methods in allocating educational resources, the accuracy of allocation was used as the testing indicator, and the test results are shown in Fig. 4.

As shown in Fig. 4, with the increase of concurrent requests, the allocation accuracy of each method shows a downward trend. However, the allocation accuracy of our method is always above 95%, while the allocation accuracy of the comparative method is around 90% and 85%, which is lower than that of our method. This proves that the resource allocation effect of the proposed method is better.

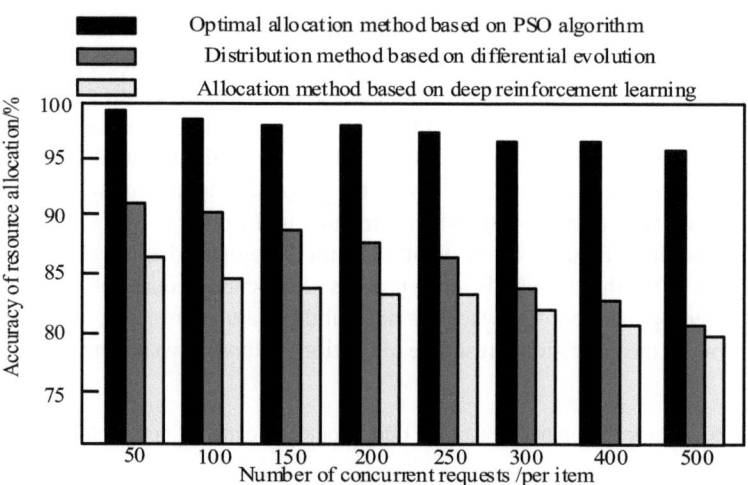

Fig. 4. Analysis of the Accuracy of Educational Resource Allocation Using Different Methods

4 Conclusion

This paper offers a thorough investigation into the optimal allocation approach for ideological and political education resources, leveraging the Particle Swarm Optimization (PSO) algorithm. Conduct a comprehensive evaluation and in-depth analysis of the current status of ideological and political education resources for college students, in order to clearly grasp the detailed information of the current resource library, including resource types, quantities, quality, and distribution. Based on the ID3 decision tree algorithm, conduct a detailed analysis of the specific attributes of ideological and political education resources. By calculating indicators such as information gain or gain rate, identify the most influential attributes for resource classification, and scientifically and reasonably classify ideological and political education resources into different categories based on this, in order to facilitate subsequent management and utilization. Based on the Particle Swarm Optimization (PSO) algorithm, a fitness function is defined to quantitatively evaluate the effectiveness of different resource allocation schemes. At the same time, the initial state and motion rules of the particle swarm are set to simulate the optimization process of particles in the solution space. Optimize the allocation of ideological and political education resources for college students. Through algorithm iteration, continuously adjust the resource allocation plan until the solution with the highest fitness is found, that is, achieve the optimal allocation of resources.

In the future research, we will further explore the combination of PSO algorithm and other intelligent optimization algorithms, as well as the application expansion in other aspects of education. Through continuous research and practice, optimize this educational resource allocation system and make greater contributions to higher education and talent cultivation.

References

1. Peng, C., Zhou, X., Liu, S.: An introduction to artificial intelligence and machine learning for online education. Mobile Networks Appl. **27**(3), 1147–1150 (2022)
2. An, B., Gao, L.: Construction of an inquiry-based teaching model for ideological and political education in colleges and universities from the perspective of deep learning. Mobile Inf. Syst., (Pt.30) (2022)
3. Li, X., Dong, Y., Jiang, Y., et al.: Analysis of the teaching quality of college ideological and political education based on deep learning. J. Interconnection Networks **22**(Supp02) (2022)
4. Sun, Y., Zheng, H.: Research on improving the accuracy of ideological and political education in colleges under artificial intelligence technology in the era of big data. Mobile Inf. Syst., (Pt.17) (2022)
5. Han, S., Wang, W., Lu, D., et al.: Design optimization and simulation of seedling picking mechanism based on PSO algorithm. Comput. Simul., **40**(03), 338–345+394 (2023)
6. Jin, C., Li, J., Xi, Y., et al.: Problems and solutions in ideological and political education under the training mode of excellent and innovative talents in colleges and universities. Asian Agric. Res. **14**(10), 51–57 (2022)
7. Cheng, P., Yang, L., Niu, T., et al.: On the ideological and political education of material specialty courses under the background of the internet. J. Higher Educ. Res. **3**(1), 79–82 (2022)
8. Liu, S., Peng, C., Srivastava, G.: What influences computational thinking? A theoretical and empirical study based on the influence of learning engagement on computational thinking in higher education. Comput. Appl. Eng. Educ. **31**(6), 1690–1704 (2023)
9. Jing, Z., Chen, C.: High-quality development of curriculum ideological and political education and talent cultivation: practice and thinking of school of information management in Sun Yat-sen university. Library Inf. Serv. **66**(1), 39–45 (2022)
10. Sanhong, D., Ping, L., Suyang, L., et al.: Pattern construction and implementation path of "ideological and political education" of library, information and archives management: taking Nanjing university as an example. Library Inf. Serv. **66**(1), 22–29 (2022)

A Study of Collaborative Filtering-Based Recommendation Algorithms for University Aesthetic Education Teaching Resources

Xiaodi Li[1(✉)] and He Kong[2]

[1] Jilin Communications Polytechnic, Changchun 130012, China
Mylove1024D@126.com
[2] Student Affairs Department, Jilin Communications Polytechnic, Changchun 130000, China

Abstract. Aiming at the shortcomings of current recommendation algorithms in capturing changes in user interests, an innovative university aesthetic education teaching resource recommendation algorithm based on collaborative filtering is proposed. In the process of developing collaborative filtering recommendation algorithms, the core features of university aesthetic education teaching resources were deeply explored, and data clustering was successfully implemented to optimize the recommendation effect of educational resources. Designed an exclusive recommendation algorithm for aesthetic education teaching resources in universities. By constructing an interest model for college students, we can accurately calculate the similarity between individuals and achieve personalized recommendation of aesthetic education teaching resources to meet the needs of aesthetic education teaching in universities The experimental data shows that the algorithm achieved a low MAE value, demonstrating excellent accuracy, and high user feedback satisfaction, proving its significant advantage in recommendation performance.

Keywords: Resource Recommendation · Education and Teaching · Teaching Resource Recommendation · University Aesthetics · Collaborative Filtering Algorithm

1 Introduction

Aesthetic education is an important way to cultivate students' aesthetic concept, aesthetic interest and aesthetic ability. People's pursuit and appreciation of beauty are constantly improving with social development and cultural progress, therefore, university aesthetic education is particularly important. At present, university aesthetic education teaching resources are becoming more and more abundant, covering a wide range of fields from classical art to modern creativity, including teaching materials, multimedia resources, art exhibitions, cultural activities and so on. However, the vast amount and diversity of these resources make it challenging for learners to find teaching resources that meet their interests and needs. In addition, access to and utilization of these resources are often limited by factors such as geography, funding, and teachers' strength, this situation has resulted in uneven distribution or low utilization efficiency of aesthetic education teaching resources in some universities.. With the deepening application of information technology in education and the innovation of student learning models, With the development of the times, students' demand for aesthetic education is gradually showing a trend of personalization and diversification. The traditional teaching model of aesthetic education is inadequate in responding to the diverse needs of students. Therefore, how to accurately screen high-quality and personalized teaching content from massive resources has become an urgent challenge for aesthetic education in universities. Therefore, it is necessary to explore more flexible and personalized teaching resources recommendation methods.

As a classic recommendation algorithm, collaborative filtering has achieved remarkable success in e-commerce, social media and other fields. The core strategy is to use user historical behavioral data to identify user groups with similar interests, or explore other content that is similar to the user's past preferences, and based on this, present recommendations that may attract their attention. Collaborative filtering algorithms stand out among numerous algorithms due to their excellent recommendation accuracy, strong scalability, and efficient ability to meet the needs of recommending aesthetic education teaching resources in universities.[1] In the recommendation system for aesthetic education teaching resources in universities, collaborative filtering algorithms can deeply analyze the learning process, browsing trajectory, and rating feedback of students, thereby accurately capturing their personalized learning preferences and actual needs. At the same time, by calculating the similarity between resources, the algorithm can find potential teaching resource associations and provide students with more abundant and diversified learning options. In addition, the collaborative filtering algorithm can also find hidden popular trends and hot topics according to the behavior patterns of user groups, providing data support for the update and optimization of teaching resources.

In view of this, this study innovatively designed a university aesthetic education teaching resource recommendation algorithm based on collaborative filtering. Through in-depth analysis of the core principles and key technologies of collaborative filtering algorithms, combined with the specificity of aesthetic education teaching resources in universities and the actual learning needs of students, this study aims to build an efficient and accurate recommendation model. At the same time, we will pay attention to the dynamic update of the algorithm and privacy protection to ensure that the recommendation system can continuously adapt to the changes of students' interests and provide personalized recommendation services under the premise of protecting students' privacy. The aim of this study is to explore new ways of recommending aesthetic education teaching resources in universities, contribute new ideas and methods to the development of educational informatization and personalized learning, and promote the effective utilization of educational resources and the personalized development of learning. While simultaneously anticipating that the research outcomes will serve as valuable references and guidance for further inquiries in this domain and practice in related fields, and promote the optimization and utilization of university aesthetic teaching resources. In conclusion, the research on collaborative filtering-based recommendation algorithm for university aesthetic education teaching resources has important theoretical value and practical significance. Through in-depth research and exploration, we will provide powerful technical support and innovative power for the recommendation of university aesthetic teaching resources, and contribute significantly towards enhancing the quality of education and fostering the cultivation of innovative talents.

2 Collaborative Recommendation Algorithm

The research on collaborative filtering based recommendation algorithms for aesthetic education teaching resources in universities is a comprehensive exploration that spans multiple disciplines such as data mining and machine learning. In-depth understanding of users' behavioral patterns and preferences on university aesthetic education teaching platforms is the basis for designing effective recommendation algorithms [2]. Therefore, it is crucial to study the methods and tools of user behavior analysis to improve the recommendation effect. By analyzing diverse data such as user browsing history, liking records, favorite lists, and comment content, it is possible to accurately extract user interests and learning needs, thereby providing strong support for personalized recommendations.

2.1 Extracting the Characteristics of Aesthetic Teaching Resources

University aesthetic education teaching resources cover a rich variety of contents and forms, aiming to enhance students' aesthetic literacy and creativity. The following are some common university aesthetic education resources.

Textbooks and course materials: These are the basic resources for teaching aesthetic education, including textbooks and course materials related to the theory of aesthetic education, art history, principles of aesthetics, etc.. These materials are usually written or recommended by professional teachers to ensure the accuracy and authority of the contents. These materials are usually written or recommended by professional teachers to ensure the accuracy and authority of the contents.

Works of art and exhibitions: artwork serves as a vital medium for teaching aesthetic education, encompassing a diverse array of forms such as paintings, sculptures, photographs, films, and numerous other artistic expressions. Schools usually organize art exhibitions to give students the opportunity to appreciate and learn from these works up close.

Online resources and platforms: with the development of the Internet, more and more teaching resources for aesthetic education have been brought online. These resources include online courses, e-books, online exhibitions, etc.. Students can learn anytime and anywhere.

Practical teaching bases: some schools will establish practical teaching bases, such as art studios, design laboratories, etc., to provide students with opportunities for practical operation, so that they can improve their aesthetic ability and creativity in practice.

Teachers: excellent teachers are the key resources for teaching aesthetic education. They possess profound professional knowledge and possess the capability to not only guide students in discovering, appreciating, and creating beauty, but also fostering a deeper understanding and appreciation of the aesthetic realm.

In order to maximize the utilization of these resources, schools often adopt comprehensive strategies, such as strengthening teacher training, adjusting curriculum design, and optimizing practical teaching systems, to ensure the quality and effectiveness of aesthetic education teaching. In addition, the school actively encourages students to participate in various aesthetic activities, such as art competitions, cultural salons, etc., aiming to comprehensively enhance students' comprehensive literacy.

Collaborative filtering algorithm, as a classic recommendation method, has a long and extensive application history. It is based on deep mining of user historical behavior data, analyzing user characteristics in detail, and identifying user groups with similar interests. Based on the preferences of these similar user groups, the algorithm can calculate a set of recommended content for specific users. Collaborative filtering algorithms are mainly divided into two categories: user oriented collaborative filtering and project oriented collaborative filtering. Given the unique learning characteristics of college students, this article chooses a user oriented collaborative filtering algorithm to recommend suitable aesthetic education teaching resources for college students. This method is based on a core idea that users with similar interests often have similar evaluations of the same resource. By analyzing the user's historical operation records, we identify other users with similar interests as the target user, and use their preferred content as a recommendation basis to provide personalized recommendations for aesthetic education teaching resources for the target user. As shown in Fig. 1, this process clearly demonstrates the recommendation mechanism based on user collaborative filtering.

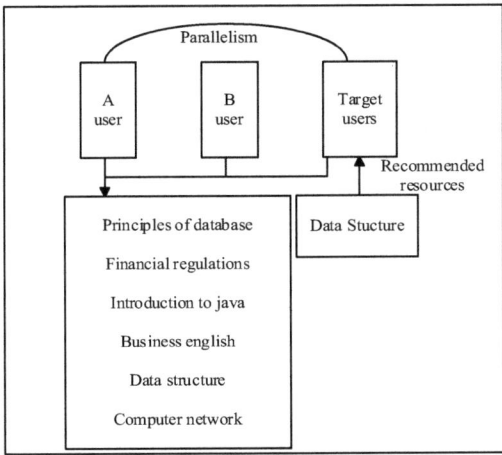

Fig. 1. Diagram or schematic diagram of collaborative filtering algorithm

From the user history operation analysis shown in Fig. 1, we found that the similarity between user A and the target user is particularly significant, and they are considered the most similar users. Therefore, the "data structure" in aesthetic education teaching resources is accurately recommended to target users. Furthermore, as an advanced stage of aesthetic education, the core goal of universities is to cultivate talents with high quality and innovative spirit. Aesthetic education plays a unique and irreplaceable role in stimulating imagination, promoting innovative consciousness, and creativity. With the assistance of aesthetic education, professional education is expected to achieve greater breakthroughs and leaps. Therefore, taking moral education and talent cultivation as the fundamental mission of higher education, comprehensively implementing the strategy of "five educations simultaneously", and building a comprehensive and fully inclusive aesthetic education system are essential tasks of higher education [3, 4].

Based on the above analysis, the primary task is to gather aesthetic teaching resources. The data collection process focuses on obtaining information closely related to user characteristics, preferences, and activities, which provides a crucial data foundation for constructing user interest models. Usually, this process can be achieved through both explicit and implicit pathways. Explicit data collection requires that college users personally provide data that indicate their interests and preferences, such as their ratings or scores on a program, and explicit goal statements for a sample set. This simple and clear method helps to accelerate the speed of learning algorithms.

Secondly, feature extraction of aesthetic teaching resources is carried out. Extracting features of aesthetic teaching resources usually involves analyzing the content of the resources in order to transform them into numerical forms that can be processed by recommendation algorithms. These features may include text features, image features, audio features, etc., depending on the type of resource.

Taking text features as an example, for text resources, such as teaching articles, descriptions, comments, etc., the following features can be extracted: word frequency inverse document frequency, as a weighting technique, is often used in information retrieval and data mining, playing an important role in evaluating the importance of words in documents or corpora. TF-IDF consists of two core components: Term Frequency and Inverse Document Frequency.

Word frequency (TF) is a quantitative indicator that measures the number of times a word appears in a specific document. Generally speaking, the higher the frequency of a word appearing in a document, the greater its importance to the content of the document. However, in order to avoid words in long documents receiving excessive weight due to their document length, word frequency is usually standardized. The specific calculation formula is as follows.

$$TF(t, d) = frac[f\_(t, d)(F\_d)] \qquad (1)$$

In formula (1), $f\_(t, d)$ represents the frequency of occurrence of word t in document d, while (F_d) represents the total number of words in document d.

Inverse Document Frequency (IDF) is an indicator of the universality or uniqueness of a word across all document collections. Specifically, if a word frequently appears in multiple documents, the IDF value will be relatively low, indicating that the word's role in distinguishing specific documents is weak. On the contrary, if a word only appears in a few documents, its IDF value will be higher because such words may have greater importance in distinguishing these documents. The following is its calculation formula:

$$IDF(t) = \log frac \cdot N \cdot df(t) \qquad (2)$$

In formula (2), N represents the total number of documents in the document set, while $df(t)$ represents the number of documents containing a specific word t.

TF-IDF is obtained by multiplying the word frequency (TF) by the inverse document frequency (IDF), which combines the frequency of a word appearing in a single document and its importance in the entire document set. Its expression formula is as follows:

$$[TF - IDF(t, d) = TF(t, d) \times IDF(t)] \quad (3)$$

The TF-IDF algorithm usually tends to downplay the saliency of common words to achieve its filtering effect, while retaining words that can distinguish documents. Therefore, it is an effective feature weighting method, especially suitable for text mining and information retrieval tasks.

Based on the above steps, the collection and processing of data on aesthetic education teaching resources were completed.

2.2 Clustering of Aesthetic Education Teaching Resource Data

Cluster analysis is a pivotal data analysis technique that boasts a diverse array of applications, encompassing pattern recognition, image processing, document classification, and numerous other fields. This process involves grouping sets of physical or abstract objects to form multiple categories composed of similar objects. Its core goal is to divide data into different clusters based on the similarity between objects. Ideally, the similarity within each cluster should be maximized, while the similarity between objects belonging to different clusters should be minimized. The data clustering steps of the aesthetic teaching resources are shown in Fig. 2.

The similarity measure between objects is the key to clustering, and its accuracy directly affects the effect of clustering. There are two kinds of similarity measures: distance and similarity coefficient, the fundamental principle underlying both is that as the distance between objects decreases, the similarity between them increases and the greater the similarity coefficient is, the greater the similarity between the objects is. Let an object have n attribute, then it is possible to use a n dimensional vectors to represent, e.g., objects X_i and X_j can be expressed as $X_i = (x_{i1}, x_{i2}, ..., x_{in})^T$, $X_j = (x_{j1}, x_{j2}, ..., x_{jn})^T$ the formula for their Euclidean distance can be expressed as follows:

$$d_{ij} = \left(\sum_{k=1}^{n} |x_{ik} - x_{jk}|^2 \right)^{\frac{1}{2}} \quad (4)$$

The result of the similarity coefficient between two objects can be found by using the correlation coefficient, which is calculated as follows:

$$c_{ij} = \frac{\sum_{k=1}^{n} (x_{ik} - \overline{x}_i)(x_{ik} - \overline{x}_j)}{\sqrt{\sum_{k=1}^{n} (x_{ik} - \overline{x}_i)^2} \sqrt{\sum_{k=1}^{n} (x_{ik} - \overline{x}_j)^2}} \quad (5)$$

By clustering aesthetic teaching resource data, these resources can be more effectively organized and managed, while revealing potential structures and correlations, thereby providing users with more accurate and personalized recommendation services.

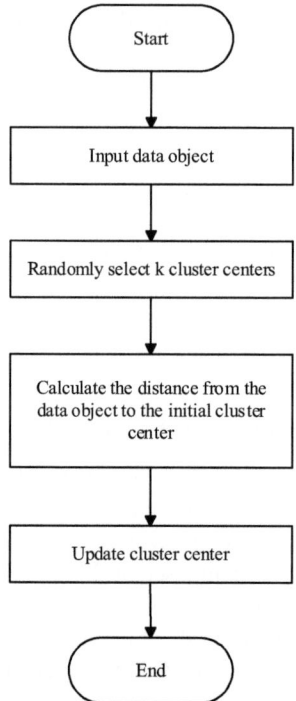

Fig. 2. Clustering steps for data clustering of aesthetic education teaching resources

3 The Development or Construction of Personalized Recommendation Algorithms for Aesthetic Education Teaching Resources in Universities

The design of university aesthetic teaching resources recommendation algorithm needs to consider multiple factors, including user's behavioral data, characteristics of resources, and the evaluation of recommendation effect. Based on the preferences and behavioral data of neighboring users, similar aesthetic teaching resources are recommended for target users. Weighted average, voting and other methods can be used to synthesize the preferences of neighboring users to get the final recommendation results. In addition, content-based recommendation technology and hybrid recommendation strategies are integrated, which, combined with collaborative filtering algorithms, further improve the accuracy and personalization level of recommendations, thereby optimizing the overall user experience.

3.1 Constructing a Model of College Students' Interests

Building a user interest model for college students is an indispensable part of the university art education resource recommendation system. In order to provide personalized information recommendations, it is necessary to accurately understand user preferences and construct a model that can represent user interest characteristics based on this. This model is like an intelligent filter that filters and recommends resources based on the user's interests and tendencies. Of course, as user interests change, this filter also needs to have corresponding adjustment capabilities to ensure the continuity and accuracy of recommendations [5].

Colleges and universities have different groups of students with different interests and hobbies, based on the unique characteristics of school teaching explored in this paper, the student population is organized and classified according to four distinct levels: the type of discipline, faculty, specialty, and grade, and numbering of each student numbering method as shown in Fig. 3 [6].

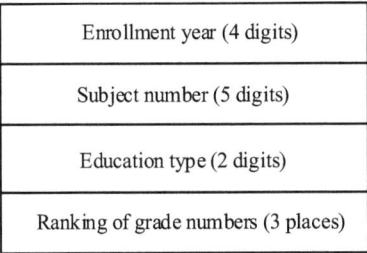

Fig. 3. Student numbering rules

Among them, Discipline numbering can be standardized and uniformly numbered in accordance with the national standard "Discipline Classification and Code", for example, a student studying computer science , after the number in the query standard , it can be specified that the subject type is D520, and the classification of user groups is helpful for recommendation , because the student group at the bottom of the classification tree must have the highest similarity. Education type , disciplines, majors and year of enrollment all correspond to some attributes of educational resources in the educational resources storage table , so as to form a certain connection between students and resources. Since a student can be interested in the information resources of his/her major, and most of the students will have other hobbies and pay attention to other information, the student groups show a cross state. Therefore, for the target users to be recommended, only consider the two groups as a whole , that is, when acquiring similar groups of target users, it is divided into two categories as a whole , carry out similarity calculation for two types of processing , then make recommendations. The calculation of similarity is introduced in the next chapter [7].

The construction of student interest models and resource models should be closely linked in order to effectively manage student information in the database. The relevant data of students can be divided into two main parts: the basic information table of students and the interest table of students. The former stores basic information of students, such as name, gender, and major; The latter is used to track and record the dynamic changes in students' interests after visiting the website. The keyword storage for student interests is shown in Table 1 [8].

Table 1. Student Interest Keyword Storage Table

Attribute	Type	Notes
SID	Character type	Student ID
Keywords	Character type	Keywords accessed by users
Response ratio	Integer	Response ratio of keywords visited by users
Professional type	Boolean type	0-Major; 1-Non Major

Table 1 is mainly in the form of interest keywords to reflect the data in the table is also a user interest filter template, according to the table is constantly updated to change the interest of the keywords to dynamically recommend to the students. In addition, in the consideration of user interest must also take into account a time problem, In short, the recent visit records of users within a certain time range have the highest reference value for understanding their current interests and needs, if too long ago to access things may not be the user is interested in the content of course, recently accessed some of the content does not mean that the user is interested in the resource. Therefore, it is necessary to set a time-related coefficient that can express the user's interests [9, 10].

Based on the above analysis, the following expression formula has been designed for the interest model of college students:

$$s = (SID, INTEREST)$$
$$INTEREST = (INTERESTWORD, XYB) \qquad (6)$$

In formula (6): *SID* indicates the student number, which is numbered in the manner shown in Fig. 3. *INTEREST* indicates keywords of interest to the user, the *XYB* indicates the time response ratio for the keywords that the user is interested in.

3.2 Calculating Similarity

Similarity calculation is a key step in collaborative filtering recommendation technology, helping us identify user groups with similar interests to achieve more accurate recommendations [11].

In order to overcome the difficulty of traditional similarity algorithms accurately measuring the similarity between users and projects in extremely sparse user project evaluation matrices, this paper proposes an innovative method. When calculating the similarity between project i and project j, we first determine the union of the user sets that have rated both project i and project j $U_{ij} = U_i \cup U_j$, rather than just considering the intersection. Here, U_i re represents the user group that has evaluated project i, while U_j re represents the user group that has evaluated project j.

In set U_{ij}, for users who have not rated item i or item j, we can predict their potential ratings by evaluating their similarity with other users, and calculate the similarity between item i and item j. This method not only cleverly addresses the problem of scarce user commonly used rating data in similarity algorithms, but also effectively solves the cold start problem caused by the cosine similarity algorithm and Pearson similarity algorithm processing all unrated items due to the same score. Through this method, we can obtain more accurate similarity of target items, which is significant for improving the recommendation quality of recommendation algorithms [12]. The following is the specific calculation process:

Step 1: enter the User-Program Evaluation Matrix.

Given a user project evaluation matrix $R(m, n)$, where rows represent m users and columns represent n projects. The elements in the i-th row and j-th column of matrix $R_{i,j}$ represent the rating of user i on item j. If user i does not rate item j, the element will be marked as a null or missing value.

Step 2: The construction of a collection of similar projects.

Using an improved similarity calculation method, calculate the similarity between any two items i and j in System $Sim(i, j)$. Given that the relationships between projects are relatively stable over a period of time, the similarity between projects can be calculated offline in advance and stored in a dedicated database table, while maintaining regular updates. For any project k, by searching the entire project set, we select the first project with the highest similarity to project k and make it a member SI_k of the similar project set for project k N.

Step 3: formation of user's nearest neighbor.

For the union $I_{ui} = I_u \cup I_i$ of the item sets rated by target user u and user i respectively, in the case where user u has not rated item k but user i has already rated it, we can use the similar item set SI_k of item k to predict the potential rating of user u on item k. The specific prediction calculation formula is as follows:

$$P_{u,k} = \overline{R}_k + \frac{\sum_{n \in SI_k} Sim(k, n) \cdot (R_{u,n} - \overline{R}_n)}{\sum_{n \in SI_k} Sim(k, n)} \quad (7)$$

By iterating through the above method, the predicted scores can be filled into the rating matrix, ensuring that users u and i have a corresponding rating record for each item in item set I_{ui}.

Step 4: Assess the degree of similarity between users.

The similarity calculation formula between target user u and user i is as follows:

$$Sim(u, i) = \frac{\sum_{k \in I_{ui}} (R_{i,k} - \overline{R}_i)(R_{u,k} - \overline{R}_u)}{\sqrt{\sum_{k \in I_{ui}} (R_{i,k} - \overline{R}_i)^2} \sqrt{\sum_{k \in I_{ui}} (R_{u,k} - \overline{R}_u)^2}} \quad (8)$$

After the similarity calculation between the target user and other users is completed, a similarity threshold can be set as the filtering criterion, or the top n users with the highest similarity can be directly selected as candidate neighbors for the target user. In order to accurately determine the nearest neighbor set for the target user, we will conduct a comprehensive search of the entire user group and select the top N users with the highest similarity to user u to form the nearest neighbor list for user u.

3.3 Personalized Aesthetic Education Teaching Resource Recommendation System

Based on the above analysis, the personalized recommendation process designed in this article is summarized as follows:

Step 1: the basis of personalized recommendation lies in the deep mining and analysis of student data. This includes students' learning history, browsing records, clicking behavior, etc.. Through these data, a personalized learning portrait of each student can be drawn. In addition, students' learning preferences, professional background, learning goals and other information is also an important basis for building personalized recommendation models.

Step 2: in terms of algorithm selection, collaborative filtering algorithms have become a common method in the field of personalized recommendation because of their advantages in user behavior prediction. By analyzing the behaviors of similar users, we can predict the teaching resources of aesthetic education that the target users may be interested in.

Step 3: Meanwhile, by combining content recommendation strategies and accurately matching specific content features of resources, the accuracy of recommendations can be effectively improved. In the process of implementing personalized recommendations, the diversity and balance of resources are also important considerations that cannot be ignored. Avoid over-recommendation of a certain type of resources, resulting in students' limited horizons. In addition, attention should be paid to the timeliness and quality of resources to ensure that the recommended resources not only meet the needs of students, but also have practical application value and significance.

Step 4: the personalized recommendation system should also have good interactivity and customizability. Students can adjust the recommendation strategy according to their own needs, such as setting the type, difficulty and duration of recommended resources. The system should also provide a feedback mechanism so that students can evaluate and give feedback on the recommended resources in order to optimize the recommendation algorithm.

Step 5: personalized recommendation is not only a technical problem, but also an educational problem. In the recommendation process, it is necessary to fully consider the essence and purpose of education, ensuring that the recommended resources not only meet the personalized preferences of students, but also contribute to their comprehensive growth and development.

The use of collaborative filtering algorithms can significantly enhance the effectiveness of resource recommendations and provide personalized recommendation services that are continuously updated for university users. The core goal of recommendation algorithms is to recommend aesthetic education teaching resources that may be more interesting to target users. These resources usually come from the user group with the closest interests to the target user, that is, resources with higher ratings and those that the target user has not yet rated. By using the following formula, the predicted rating of the target user on these resources can be calculated, which is the most likely average rating:

$$Rec(u_i, i_j) = \frac{\sum_{u_y \in N(u_i) \cap l(i_j)} P_{yj}}{|N(u_i) \cap l(i_j)|} \qquad (9)$$

In formula (9), $Rec(u_i, i_j)$ represents the predicted rating of the target user's preference for a certain resource., the $|N(u_i) \cap l(i_j)|$ denotes the collection of the nearest users who have non empty ratings for the item at the same time. After completing the calculation of all ratings, the "Top N" strategy is used to sort the candidate items and select the top ranked items in the list to construct the final recommendation list, in order to provide personalized recommendations for the target users.

To sum up, the design of university aesthetic teaching resources recommendation algorithm needs to comprehensively consider multiple aspects such as data collection and processing, user profile construction, similarity calculation, neighbor user selection, resource recommendation, feedback and optimization, as well as effect evaluation and adjustment. By continuously optimizing and upgrading algorithms, provide users with more personalized and accurate recommendation services for aesthetic education teaching resources.

4 Experimental Verification and Result Analysis

4.1 Preparation Before the Experiment

To verify the effectiveness of the collaborative filtering based recommendation algorithm for college aesthetic education teaching resources proposed in this study, corresponding experimental tests were conducted. To ensure the accuracy and reproducibility of the experimental results, the hardware environment was set to an Intel Core i7-9700K processor, 32 GB of memory, and an SSD solid state drive; Set the software environment to Windows 10 system, Python 3.8+ programming language, and Pandas data processing library.

The aim of this experiment is to construct a personalized recommendation system by collecting 694 rating data from 50 users on 200 aesthetic education teaching resources. The rating range is 1–5, and the higher the score, the more popular the resource is. Considering the sparsity of the dataset, its sparsity reaches 93.06% (i.e. 1 minus the proportion of scoring data to all possible scoring combinations). Divide the dataset into training and testing sets in a 4:1 ratio to simulate actual recommendation scenarios. The specific resource composition of the dataset is shown in Table 2.

Table 2. Resource composition of the data set

Resource category	Number of resources
Computer network	30
Graphic image recognition	27
Operating system	24
Database	35
Programming language	63
Software tool	21

To evaluate the accuracy of recommended predictions for testing, mean absolute error (MAE) is used as the evaluation metric. MAE is calculated based on formula (10) and measures the average absolute deviation between predicted scores and actual scores. In recommendation systems, MAE is a widely used evaluation method. Usually, the smaller the MAE value, the higher the accuracy of the recommendation.

Among them, $\{a_1, a_2..., a_n\}$ represents the set of ratings predicted by the recommendation system, while $\{b_1, b_2..., b_n\}$ represents the set of real user ratings for each resource, therefore, the specific formula for calculating MAE is as follows:

$$MAE = \frac{1}{n}\sum_{i=1}^{n}|a_i - b_i| \tag{10}$$

4.2 Implementation Steps and Operational Procedures

The following are the specific implementation steps of this experiment:

Step 1: based on the collected data, a user-resource matrix is constructed, with rows representing individual users and columns denoting various resources, and the values in the matrix represent user ratings or interaction information for the resources.

Step 2: extract user features from the user-resource matrix, which may include user's interests and preferences, behavioral patterns, and so on. Use statistical methods, machine learning algorithms, etc., to extract user features.

Step 3: Based on the extracted user features, the collaborative filtering recommendation algorithm proposed in this paper has been implemented and applied, and optimized to improve traditional resource recommendation techniques.

Step 4: In order to evaluate the performance of the recommendation algorithm, an appropriate evaluation metric was selected, which is the Mean Absolute Error (MAE).

Step 5: select the initial number of neighbors K as 2, increase 2 each time, 2, 4, 6, 8, 10, 12, 14, and evaluate the absolute deviation data according to different tests of K value.

4.3 Results and Conclusions

First, according to the above experimental preparation, Table 3 presents in detail the test results of Mean Absolute Error (MAE).

According to the test results in Table 3, the collaborative filtering algorithm generates recommendation results by comprehensively considering user historical behavior data and similarity between users, which more effectively captures user interests and preferences. Compared with traditional methods, this method significantly reduces the MAE index, indicating superior accuracy. Therefore, the collaborative filtering based algorithm for recommending aesthetic education teaching resources in universities has shown good performance in accuracy evaluation and achieved high user satisfaction. Compared to other comparative algorithms, collaborative filtering algorithms do have certain advantages in recommendation performance.

Secondly, Table 4 presents in detail the test results of resource recommendation efficiency.

From the test results in Table 4, it can be seen that compared to content-based recommendation algorithms and hybrid recommendation algorithms, collaborative filtering algorithms have improved resource recommendation efficiency. This improvement is

Table 3. Summary of MAE evaluation results

Number of nearest neighbors	Proposed method	Traditional method
2	0.458	0.945
4	0.645	0.821
6	0.481	0.764
8	0.539	0.813
10	0.620	0.757
12	0.387	0.659
14	0.531	0.948

Table 4. Test results of resource recommendation time (unit: s)

Resource category	Proposed method	Traditional method
Computer network	13.89	21.69
Graphic image recognition	15.64	19.48
Operating system	7.62	15.32
Database	10.38	17.60
Programming language	5.87	9.43
Software tool	6.30	12.76

mainly due to the collaborative filtering algorithm's ability to utilize user behavior data to explore their potential interests, rather than relying solely on the content features of resources.

Finally, a survey was conducted on user satisfaction with resource recommendations, and the specific results are shown in Fig. 4.

According to the test results in Fig. 4, the collaborative filtering algorithm successfully recommended aesthetic teaching resources that meet the personalized needs of users based on their historical behavioral data, thus achieving high user satisfaction. Through the questionnaire survey and interviews with users, it is found that users are very satisfied with the collaborative filtering-based aesthetic education teaching resources recommendation algorithm. The general feedback from users is that the resources recommended by the algorithm are highly compatible with learning needs and interests, which helps to more effectively discover and utilize high-quality teaching resources. At the same time, users also said that the resources recommended by the algorithm are diversified and novel, which can stimulate their learning interest and motivation.

Overall, the collaborative filtering, based recommendation algorithm for aesthetic education teaching resources in universities has demonstrated outstanding performance and effectiveness in the experiment. This algorithm not only enhances the precision and effectiveness of the recommendation process, but also enhances user satisfaction and

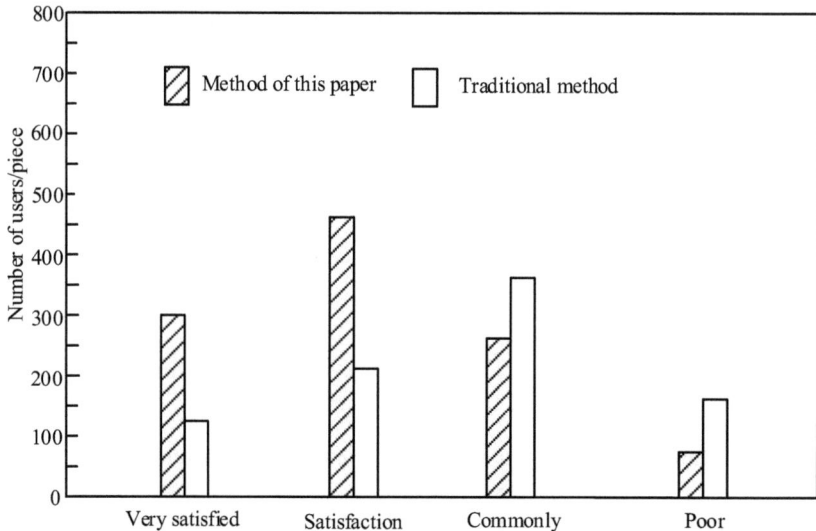

Fig. 4. User satisfaction survey results

optimizes resource utilization. Looking ahead to the future, will continue to explore the in-depth application of collaborative filtering algorithms in the field of aesthetic education teaching resource recommendation, committed to further improving the performance of algorithms, in order to provide users with higher quality and more personalized recommendation services.

5 Conclusion

This study focuses on aesthetic teaching resources in universities and conducts in-depth research on recommendation algorithms based on collaborative filtering, aiming to improve the accuracy and personalization level of teaching resource recommendations to meet the needs of teachers and students. Looking back at the entire process of this study, a detailed theoretical exploration was first conducted on the collaborative filtering algorithm. Collaborative filtering, as a widely used strategy in recommendation systems, focuses on analyzing user behavior, comparing it with the behavior patterns of other users, identifying similar users or items, and generating personalized recommendations for users based on these similarities. Collaborative filtering algorithms have shown significant potential in recommending aesthetic education teaching resources in universities. After in-depth analysis of its applicability and unique advantages in this field, a collaborative filtering based recommendation model was designed and constructed to meet the needs of practical teaching scenarios. Through experimental verification, the model is not only effective but also easy to operate, providing a practical and feasible solution for personalized recommendation of aesthetic education teaching resources in universities, and has important practical application value. The results demonstrate that the algorithm is capable of accurately effectively alleviate the problem of information overload, and enhance users' learning experience.

However, there are still some shortcomings in this study. For example, in terms of data sources, the experimental data mainly come from the aesthetic education courses of a university, and the data scale is relatively small, which may affect the extensiveness and universality of the recommendation results. Looking ahead, we plan to further expand our data sources and include more universities and art education courses to enhance the universality and practicality of recommendation algorithms. In addition to collaborative filtering algorithms, we will also explore the application of other recommendation strategies (such as content-based recommendation, mixed recommendation, etc.) in the recommendation of aesthetic education teaching resources, in order to continuously optimize the recommendation effect and improve user experience. In addition, by combining user profiling and emotion analysis techniques, it is expected to achieve more refined and personalized recommendation services.

Looking ahead to the future, we will continue to deepen our research on collaborative filtering algorithms, actively seek more innovative optimization and improvement strategies, and strive to provide users with higher quality and efficient recommendation services for aesthetic education teaching resources. At the same time, eagerly look forward to working together with more researchers and practitioners to promote continuous progress and innovation in the field of recommending aesthetic education teaching resources.

In summary, the research on collaborative filtering based recommendation algorithms for aesthetic education teaching resources in universities has profound significance and practical application value. Through this research, not only can it provide an innovative and effective method for recommending aesthetic education teaching resources, but it can also provide valuable references for exploration and practice in related fields.

Acknowledgment. Project of Jilin Provincial Department of Education: Research on Collaborative Education Strategies of Ideological and Political Education and Aesthetic Education in Universities (JJKH20231024SZ).

References

1. Liu, S., Xiyu, X., Zhang, Y.: A reliable sample selection strategy for weakly-supervised visual tracking. IEEE Trans. Reliab. **72**(1) 15–26 (2023)
2. Vagt, C.: Design as aesthetic education: on the politics and aesthetics of learning environments Hist. Hum. Sci. **33**(1), 175–187 (2020)
3. Todorova, V., Sosina, V., Vartovnyk, V.: Features of the individual style of coach and teacher work of the choreographic team. Sci. Educ. **2020**(3) 149–155 (2020)
4. Fu, D.: Research on the practice path of aesthetic education for normal university students. Adult Higher Educ. **5**(12), 66–70 (2023)
5. Wu, L.: Collaborative filtering recommendation algorithm for MOOC resources based on deep learning. Complexity **2021**(46), 1–11 (2021)
6. Zhu, W.: Topic recommendation system using personalized fuzzy logic interest set. J. Intell. Fuzzy Syst. **40**(2), 1–11 (2020)
7. Li, M., Wang, M., Liu, F.: Clustering simulation of big data access trace based on similarity calculation. Comput. Simul. **40**(3), 485–489 (2023)

8. Huang, Z., Stakhiyevich, P.: A time-aware hybrid approach for intelligent recommendation systems for individual and group users. Complexity **2021**(2), 1–19 (2021)
9. Wang, H., Ding, S., Li, Y., et al.: Hierarchical physician recommendation via diversity-enhanced matrix factorization. ACM Trans. Knowl. Disc. Data. **15**(1), 1–17 (2020)
10. Li, R.: The construction of college aesthetic education information management system under the background of big data. Int. Conf. Cyber Secur. Intell. Anal. **18**, 557–563 (2022)
11. Zha, P., Mahat, G., Qureshi, R., et al.: Utilising a WeChat intervention to improve HIV and AIDS education among college students in China. Health Educ. J. **80**(8), 1002–1013 (2021)
12. Newton, A.C.I.: Race, sports, and education: improving opportunities and outcomes for black male college athletes. J. Sport Manage. **34**(5), 504–505 (2020)

Data Topic Mining Method of Online English Teaching in Higher Vocational Colleges Based on LDA Model

Yuanyuan Zhang[✉]

Sanya Aviation and Tourism College, Sanya 572000, China
zhangyuanyuan1030@yeah.net

Abstract. In order to optimize the effect of topic mining of online English teaching data in higher vocational colleges and improve the consistency of topics, a topic mining method of online English teaching data in higher vocational colleges based on LDA model was proposed. Relevant data were collected from online English language teaching in higher education institutions and pre-processed to provide a reliable database. An LDA model was created and a document generation process for the LDA model was designed. On this basis, the document term matrix is constructed, and the LDA model is combined to deeply mine the data theme of online English teaching in higher vocational colleges. The test results reveal that upon the implementation of the proposed method, the theme consistency score consistently surpasses that of the traditional approach. This enhanced methodology is more effective in capturing and representing the underlying themes within the online English teaching data of higher vocational colleges, thereby exhibiting significant performance advantages.

Keywords: LDA model · Vocational English · Online teaching · Data topic mining

1 Introduction

Online teaching has become an important component of higher education [1, 2]. Especially in higher vocational English education, its flexibility and convenience provide more learning opportunities and ways for learners. Higher vocational English online teaching data usually contain a large amount of textual information, such as students' discussions, assignments, comments, etc., which contain rich teaching feedback and learning behavior data. By mining these data for themes, we can reveal students' concerns, difficulties and learning preferences in the learning process. This information is an important reference value for teachers, which can assist them in fine-tuning their teaching strategies and refining the content of their instruction, thus improving the effectiveness of online teaching.

However, the current traditional online teaching data theme mining method for higher vocational English still has certain problems in the process of practical application, mainly including the following aspects: (1) Limitations of data collection: the traditional method often relies on specific data sources, such as learning management systems, online course platforms, etc., which may only be able to provide limited, structured data. The collection and processing of unstructured data (e.g., textual information such as student discussions and teacher feedback) is weak, resulting in a large amount of valuable information being ignored. (2) Complexity of data processing and analysis: online teaching data usually contain a lot of noise and irrelevant information, and traditional methods may face difficulties in processing these data. In addition, due to the diversity and heterogeneity of data, traditional methods may require complex data preprocessing and feature engineering, which increases the complexity and workload of data analysis. (3) Insufficient depth and accuracy of theme mining: traditional methods may not be able to deeply mine hidden themes and patterns in online teaching data. For example, they may not be able to accurately identify hot topics in student discussions, learning difficulties, or teachers' teaching strategies [3]. This may lead to an incomplete understanding of the online teaching environment, which in turn affects the effectiveness of teaching improvement and decision making. (4) Lack of real-time and dynamics: Traditional methods usually analyze historical data in batch, which cannot reflect the changes in the online teaching environment in real time. However, online teaching is a dynamic process, students' learning behavior, teachers' teaching strategies and course content may change.

As an unsupervised machine learning technology, LDA model is widely used in topic modeling and text mining. The LDA model can find hidden topic information from a large number of document sets, and show the relationship between each document and topic, as well as the distribution of each word under each topic in the form of probability distribution [4]. Therefore, employing the LDA model to extract the themes from the online English teaching data in higher vocational colleges not only enables a profound comprehension of students' learning requirements and interests, but also find out the hot issues and potential teaching improvement directions in the teaching process. The purpose of this study is to explore the themes in online English teaching data of vocational colleges using the LDA model, reveal the learning needs and interests of students, and contribute new ideas and methods to expand the online English teaching practice of these institutions.

2 Design of Data Theme Mining Methods for Online Teaching of English in Higher Education

2.1 Data Collection Related to Online Teaching of English in Higher Education

In order to effectively mine the theme of online English teaching in higher vocational colleges, relevant data should be collected first. These data include students' learning behavior data, teaching resource use data, student interaction data, etc.. The following is a detailed introduction to the collection process of data related to online English teaching in higher vocational colleges. First of all, it is necessary to clarify the type and objectives of data collected, such as students' learning duration, learning progress, homework completion, forum participation, etc. Select appropriate acquisition tools according to the acquisition target, such as learning management system (LMS), online teaching platform, student assignment management system, etc. Secondly, set the frequency of data acquisition [5]. When setting the acquisition frequency, it is necessary to comprehensively consider several factors, such as teaching cycle, data size, storage and processing capacity, and analysis needs [6]. The frequency of data collection related to online English teaching in higher vocational colleges set in this paper is shown in Table 1.

Table 1. Frequency of data collection related to online teaching of English in higher education

Acquisition frequency	Specify
Daily collection	It is collected every hour to record students' online learning hours per hour. Collected at the end of each day, counted the number of learning tasks completed by students on that day, recorded the use of teaching resources by students on that day, including the number of videos watched and the number of materials downloaded.
Weekly collection	Collected every Sunday evening, the number of homework submitted by students that week and the number of posts and replies in the forum that week were counted.
Monthly collection	At the end of each month, collect and summarize students' learning data for the whole month, including learning time, learning progress, homework completion, etc., to form a learning summary report.
Other specific acquisition	At the beginning and end of the course, the basic information of the student (such as student number, name, class, etc.) and the comprehensive score at the end of the course are collected.

The collection frequency is adjusted according to the teaching cycle (e.g. semester, course, etc.) to ensure the continuity and integrity of the data. Through the above settings, we can ensure that we can not only obtain the students' learning dynamics and behavioral data in a timely manner, but also store and process them within a reasonable range of data volume, so as to provide strong support for the subsequent data analysis and mining.

2.2 Pre-processing of Online Instructional Data

In the process of data mining, it is necessary to preprocess the data to ensure its quality and applicability. Data preprocessing is also crucial in subject mining of online teaching data for higher vocational English [7]. First, using sorting methods, duplicate data records were identified and deleted. The utilization of mean interpolation for replacing missing values is contingent upon the specific type of data and the extent of missingness present. The formula employed for mean interpolation is exhibited as follows:

$$\overline{X} = \frac{1}{n}\sum_{i=1}^{n} X_i \tag{1}$$

Among them, X_i denotes a non-missing value. $\overline{X}$ denotes the mean value. n indicates the number of non-missing values. If a value is missing, the mean value of the variable use $\overline{X}$ to replace. Secondly, the irrelevant characters, punctuation marks and stop words are removed for stemming extraction. Normalize the processed text and convert the numerical data to the same scale to eliminate the influence of dimensionality. The formula is as follows:

$$x' = \frac{x - \min(x)}{\max(x) - \min(x)} \tag{2}$$

Among them, x signifies the original data, while $\min(x)$, $\max(x)$ respectively represent the minimum and maximum values within the dataset. After normalizing the numerical data, it is converted to the same proportion, eliminating the influence of amplitude.

Through the above data preprocessing steps, a high-quality and standardized online teaching dataset of higher vocational English can be obtained, which provides a reliable data base for the subsequent theme mining.

2.3 Build LDA Model

LDA stands as a probabilistic generation model specifically tailored for topic modeling, enjoying widespread application in the fields of text mining and document classification. The LDA model built in this paper is shown in Fig. 1.

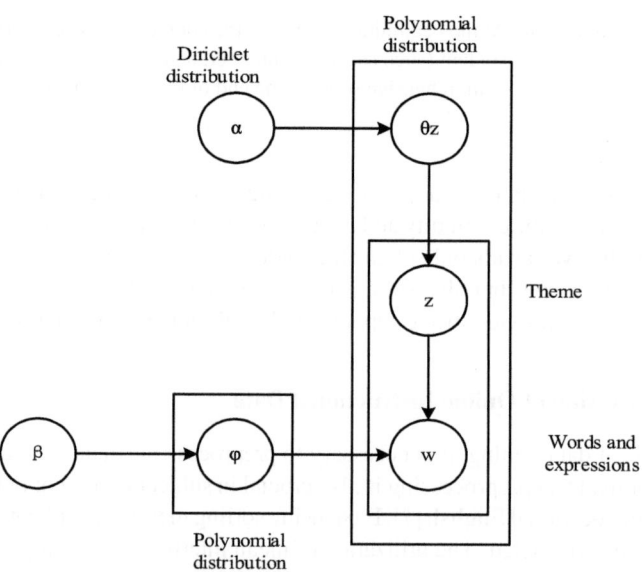

Fig. 1. LDA Model Diagram

In Fig. 1, nodes represent random variables, arrows represent probability dependencies, and rectangles represent iterations of the contents. The meanings of characters used in LDA model diagram are shown in Table 2.

Table 2. Character Meaning in LDA Model

Character	Implication
α	Subject Dirichlet prior probability distribution, θ_z hyperparameters.
β	Dirichlet prior probability distribution of words, hyperparameters of φ.
φ	Polynomial distribution of topic corresponding words.
θ	Polynomial distribution of documents corresponding to topics.
z	θ randomly generated a certain topic.
w	Words produced by combining z and φ.
K	The number of topics in the data set.
M	The number of words in a document.
N	Number of documents in the data set.

The process of LDA model document generation is as follows:

(1) α polynomial distributions of randomly generated documents corresponding to topics θ.
(2) θ randomly generate a theme z.
(3) β polynomial distribution of words corresponding to randomly generated topics φ[8].
(4) Synthesizing themes z and the distribution of theme pairs of words φ generating words w.
(5) Cycle like this, to generate a document containing M words.
(6) Final generation K under the theme N documentation.

2.4 Topic Mining of Online Teaching Data Based on LDA Model

After the completion of the LDA model, on this basis, the topic mining of online English teaching data in higher vocational colleges is conducted. The topic mining process of online teaching data based on LDA model designed in this paper is shown in Fig. 2.

Following the process shown in Fig. 2, after completing the above steps, start to mine online teaching data topics. First, convert the online teaching related data after the above preprocessing into a document term matrix, where each row represents a document and each column represents a term, and construct a document term matrix [9]. The numerical values within this matrix typically reflect the frequency of a term's occurrence within a document or are derived from the TF-IDF metric, which combines both word frequency and inverse document frequency. The TF-IDF calculation formula is as follows:

$$TF - IDF(t, d, D) = TF(t, d) \times IDF(t, D) \qquad (3)$$

Among them, $TF(t, d)$ indicates the frequency of words for lexical items t in the document d; the $IDF(t, D)$ indicates the inverse document frequency for lexical items t in the document set D. Use the gensim library to initialize the LDA model, and set the number of topics, iterations and other parameters. Determine the parameters of the

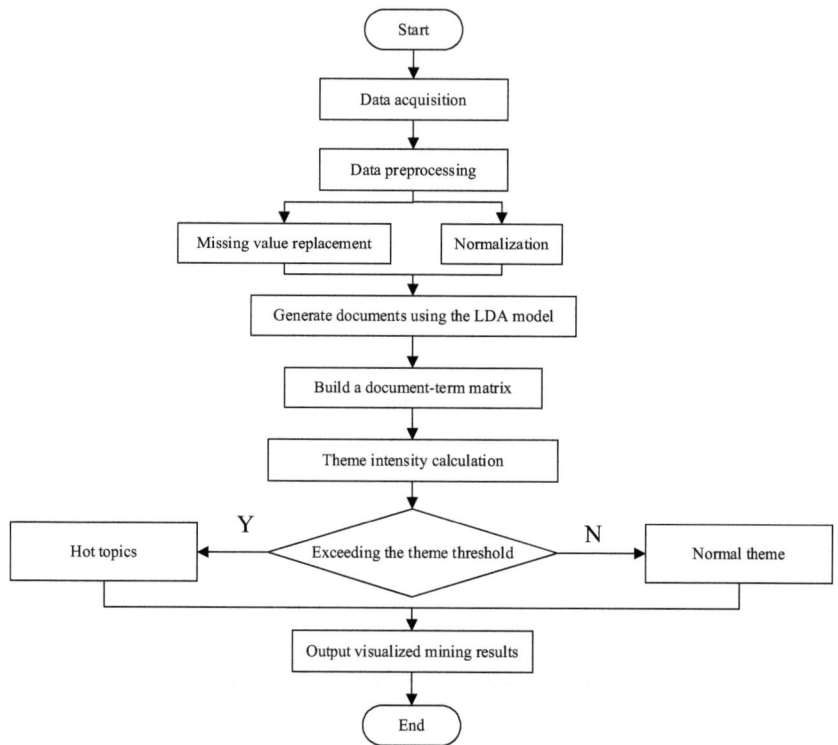

Fig. 2. Online teaching data topic mining process based on LDA model

LDA model, read the time information of the text and divide the time window, disperse the topic to each time window, and then conduct topic mining and analysis according to the intensity of the topic [10]. Theme intensity is a key indicator to determine whether a theme is popular. The formula for calculating theme intensity is:

$$\delta_k^t = \frac{\sum_{d=1}^{D_t} \vartheta_k^d}{D_t} \tag{4}$$

Among them, δ_k^t represents the topic intensity of topic k in time window t; D_t represents the number of documents on time window t; ϑ_k^d represents the posterior probability distribution of topic k in document d. Through this formula, the topic intensity of topic k of the time window t, the evolution trend of the themes in the online teaching data of higher vocational English can be analyzed by drawing a line graph of the evolution of each theme over time. On this basis, a theme threshold is set to select popular themes for analysis. The formula of the theme threshold is as follows:

$$T = \frac{\sum_{d=1}^{D_t} \sum_{k=1}^{K} \vartheta_k^d}{D_t \cdot K} = \frac{1}{K} = \frac{\sum_{d=1}^{D} \sum_{k=1}^{K} \vartheta_k^d}{D \cdot K} \tag{5}$$

Among them, *K* indicates the number of topics. *D* indicates the number of documents. *T* represents the topic strength threshold. Through calculation, the topic threshold of online English teaching data in higher vocational colleges is obtained, and topics higher than the topic intensity threshold can be identified as hot topics. The document term matrix and hot topics are used as input to train the LDA model. The objective of the LDA model is to identify a comprehensive set of topics, enabling each document to be expressed as a composite of these topics, thereby capturing the essence of its content. The distribution of terms under each topic was viewed to understand the meaning of the topic. The distribution of each document over topics was analysed to understand the thematic tendencies of the document. Relationships between topics and terms were visualised using tools such as pyLDAvis.

Through the above process, the goal of topic mining of online English teaching data in higher vocational colleges based on LDA model can be achieved, and the association and evolution between different topics can be analyzed, which can further reveal the potential laws and trends in online teaching data.

3 Test Analysis

3.1 Test Preparation

In order to more effectively validate the effectiveness of the LDA model-based data topic mining method for vocational English online teaching, New Oriental Online, which is a representative and fully functional platform for vocational English online teaching, was chosen as the experimental sample source. New Oriental Online is known for its cutting-edge online education technology and diverse interactive features, which tailor-make an efficient and user-friendly online learning environment for students and teachers. In this platform, the student discussion area is a particularly active and important part, where students can freely express their views, exchange learning experiences, ask questions and answer doubts. In addition, the platform also has functional modules such as homework submission area and teacher feedback area to provide students with all-round learning support.

In this experiment, the posts in the students' discussion forum were especially chosen as the sample objects of this experiment. These posts reflect students' real thoughts and feelings in the learning process, and also contain a lot of information related to the content of higher vocational English teaching. By mining and analyzing the topics of these posts, we can gain a deeper understanding of students' learning needs, learning difficulties, and their understanding and mastery of the teaching content. The details of the experimental dataset are shown in Table 3.

In the dataset, key information such as the title, content, publishing time, and publisher of each post is recorded in detail for further text analysis and topic mining. Through careful study of these data, it is expected to find valuable information hidden in them and furnish robust data support that significantly enhances and refines the online English teaching practices within higher vocational colleges. Preprocessing the experimental data in Table 3 is an indispensable step in the process of topic mining. The purpose of preprocessing is to remove characters, punctuation marks and stop words irrelevant to topic

Table 3. Experimental data set

Item	Instructions
Data source	Posts in student discussion boards over the course of a semester (e.g. Spring Semester 2023) that cover different course topics, learning difficulties, student interactions, and more.
Time frame	From March 1, 2023 to July 1, 2023, make sure to cover the entire semester of discussion.
Data volume	Total posts: 500 posts. Total word count: about 100,000 words.
Data content	Course theme: It covers a number of vocational English course modules, such as reading comprehension, grammar analysis, listening training, writing guidance, etc. Learning difficulties: Students mentioned various problems they encountered in the process of English learning, such as word memory, grammar rules, listening comprehension, etc. Interactive content: including students' questions about the course content, answers to other students' questions, sharing of learning experiences, etc.

mining from the data, so as to purify the text content and improve the accuracy and efficiency of subsequent analysis. The professional Chinese word segmentation tool - jieba word segmentation tool is used to process Chinese text word segmentation. The Jieba word segmentation tool has gained widespread popularity in the realm of Chinese text processing, attributed to its exceptional efficiency and precision in word segmentation tasks. Through jieba word segmentation, continuous chinese text can be divided into separate words, which provides a basis for subsequent topic mining.

After word segmentation, further stemming operations were performed. Stem extraction is a technique in natural language processing, which is used to reduce words to their basic forms or roots, in order to minimize word redundancy and enhance the precision of topic mining. Through stemming, the key information in the text can be better recognized and understood, which provides the basis for the subsequent topic classification and naming.

After completing the above preprocessing steps, the processed data is divided into training set and test set. The training set contains 400 posts, about 80000 words, for training LDA models and mining topics. The test set contains 100 posts, about 20000 words, to verify the performance of the model and the accuracy of the topic. Such division not only ensures sufficient data for model training, but also can test the generalization ability of the model.

In order to ensure the smooth implementation of the experiment of topic mining method of online English teaching data in higher vocational colleges based on LDA model, the experimental environment was configured in detail. The experimental environment includes high-performance computing equipment, sufficient storage space and necessary software tools. The specific configuration information is shown in Table 4.

Table 4. Configuration of the experimental environment

Disposition	Argument
Processor	Intel Core i7-11700K, with 8 cores and 16 threads
Internal memory	16 GB
Programming language	Python 3.8
Operating system	Ubuntu 20.04 LTS
Dependency library	jieba (Chinese word segmentation tool), numpy (numerical calculation library), pandas (data analysis library), gensim (topic model library).
Integrated development environment	PyCharm, which provides functions such as code editing, debugging, and version control.

By reasonably configuring the experimental environment, the stability and reliability of the experiment can be ensured, which provides a strong guarantee for the final accurate and objective conclusions. Through the configuration of the above experimental environment, a stable and efficient experimental environment is provided for the topic mining method in this paper. This not only ensures the smooth progress of the experiment, but also provides a strong guarantee for the reliability of the experimental results.

Load the preprocessed training set and test set data into the Python environment, and use the gensim library to train the LDA model. First, the data needs to be converted to a format acceptable to the gensim library for subsequent model training. Then, the model parameters are carefully set. Set the number of topics to 10, which means that we hope to mine 10 main topics from the data to reflect the key areas or topics of online English teaching in higher vocational colleges. To fully train the model and capture the subject information in the data, set the number of iterations to 10. This means that the model will traverse the entire training set 10 times to optimize the estimation of topic distribution. The learning rate controls the learning speed of the model in the training process. Set it to 0.5, which is a moderate value, which can not only ensure the learning efficiency of the model, but also avoid over fitting or under fitting. Use the LdaModel class of the gensim library to train the LDA model. In the training process, the model will automatically learn the hidden topics in the document set, and calculate the probability distribution between each post and topic, topic and words. This will provide strong support for subsequent thematic analysis and interpretation.

3.2 Analysis of Results

Extract the first five topics from the trained LDA model, and check the distribution and weight of keywords in each topic. Analyze the frequency of each topic in different posts to understand students' learning needs and interests. The LDA model is evaluated using test set data. In order to evaluate the performance of LDA model in topic mining of online English teaching data in higher vocational colleges, the Coherence Score is calculated as an evaluation index. The calculation formula of subject consistency is as follows:

$$C = \frac{1}{|T|} \sum_{t \in T} \frac{1}{|W_t|} \sum_{w_i, w_j \in W_t} \log \frac{P(w_i, w_j)}{P(w_i)P(w_j)} \qquad (6)$$

Among them, T denotes the set of topics; the symbol W_t epresents the collection of words pertaining to the subject matter t; whereas $P(w_i, w_j)$ denotes the likelihood of words w_i and w_j occurring together within the corpus.; the $P(w_i)$, $P(w_j)$ denote the probability of occurrence in the corpus words w_i, w_j. This metric assesses the semantic consistency of words within a topic by calculating the logarithmic mean of the ratio of the probability of co-occurrence to the probability of independence for pairs of words within the topic, the higher the value C, the more related the words within the topic are, the better the online teaching data topic mining effect is, and vice versa, the same reason. To bolster the credibility of the experimental test results, the mining approach introduced in this paper is benchmarked against two conventionally established mining techniques, and the number of topics in the online teaching data of higher vocational English is set to be 5, 10, 15, 20, 25, and 30, respectively, and the experiments are carried out in the case of gradual increase of the number of topics, and the topic consistency of the topics mined by each method is calculated. Compare the scores of the themes mined by different methods on the theme consistency index, as shown in Fig. 3.

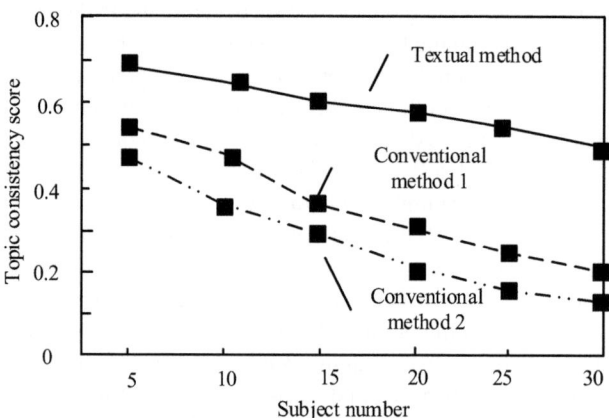

Fig. 3. Comparative results of thematic coherence scores

It can be clearly seen from Fig. 3 that the LDA model has advantages in subject consistency scores. This advantage is not only reflected in the overall trend, but also in each specific scoring point. The topic consistency score of LDA model is always higher than the other two methods. This is because LDA models can automatically discover potential topics in documents, and each topic has a representative set of vocabulary that can better summarize the content of that topic. Therefore, the LDA model has excellent performance in capturing and representing potential themes in online teaching data of vocational English.

LDAvis using the R language has been developed, so pyLDA vis is made on the basis of the R language version of LDAvis. In this study, the R language version of LDAvis is selected between the two, mainly because compared with pyLDAvis, the R language version uses fewer load packages and is more convenient to use. Use LDA model to analyze the theme of text data through LDAvis in R software and get visual results. Analyze the first and second types of text data. First, analyze the first type of Chinese data and then visualize the results with LDAvis. The result of sorting out the subject words in the LDA visualization results of the first and second types of data in the order of occurrence probability from high to low. The probability here refers to the probability of words appearing without forming a subject. The analysis results of LDA topic mining are shown in Table 5.

Table 5. LDA Topic Mining Analysis Results

First type data (in Chinese)				First type data (in English)			
Theme1	Theme2	Theme3	Theme4	Theme1	Theme2	Theme3	Theme4
Enterprise (0.49)	Cooperation 10.47)	Economy (0.35)	Culture (0.35)	United States (0.35)	Enterprise (0.27)	North Korea (0.23)	Region (0.20)
Development (0.48)	ASEAN (0.48)	Global (0.48)	Tourism (0.48)	Economy (0.48)	Market (0.48)	Korea (0.48)	India (0.48)
Market (0.41)	Europe (0.37)	People (0.35)	Media (0.29)	World (0.28)	Investment (0.23)	Cooperation (0.20)	Central Asia (0.17)
Country (0.38)	Countries (0.34)	Society (0.31)	Communication (0.26)	Government (0.20)	USD (0.21)	Russia (0.18)	Career (0.15)
Industry (0.33)	Relationship (0.31)	Reform (0.24)	Movie (0.23)	Collaboration (0.16)	Physics (0.18)	Access (0.17)	City (0. 14)
Investment (0.24)	Deepening (0.24)	Era (0.44)	Spread (0.21)	International (0.11)	Yuan (0.17)	Asia (0.15)	High speed rail (0.12)
Innovation (0–18)	Initiative (0.22)	Implementation (0.17)	News (0.16)	Society (0.06)	Income (0.15)	Relationship (0.13)	Mongolia (0.10)
Taizuo (0.17)	Project (0.15)	Concept (0.12)	Art (0.11)	China and the United States (0.05)	Overseas (0.11)	Strengthening (0.11)	Connection (0.08)
International (0.16)	Consensus (0.13)	19th National Congress (0.11)	The Silk Road (0.09)	Diplomacy (0.01)	Bank (0.09)	Summit (0.08)	History (0.0G)
Project (0.13)	Co construction (0.09)	Future (0.10)	Asia (0.07)	Battle Road (0.04)	Eurasia (0.08)	Forum (0.08)	Kazakhstan (0.06)

In text analysis, the processing of high-frequency words has always been an important link. In order to ensure the accuracy and objectivity of the analysis, it is necessary to conduct in-depth screening and selection without arbitrarily excluding these high-frequency words. In each theme, the most representative 30 words will be selected from the numerous words that appear. These words often reflect the core content and characteristics of the topic. The proposed method can convert textual data into numerical form by constructing a reasonable document term matrix, which is convenient for computer processing and analysis. In this article, a term matrix that accurately reflects the theme content of the document is constructed through reasonable word segmentation, term selection, and weight calculation. This matrix not only preserves the association information between documents and terms, but also highlights the role of important terms through weight adjustment, further improving the topic recognition ability of the LDA model. Therefore, the subject analysis content obtained through the LDA model can combine the actual background and situation to draw meaningful and relatively objective and accurate conclusions. These conclusions not only help to understand the theme and content of the text, but also have important reference value.

4 Conclusion

As an unsupervised machine learning technique, the LDA model excels at automatically uncovering latent topics from document collections. It effectively represents each document as a probability distribution across these topics, offering a comprehensive and meaningful representation of the document's content. In the topic mining of online English teaching data in higher vocational colleges, the application of LDA model can not only help identify key topics in teaching data, but also reveal the relationship and evolution trend between topics, providing strong support for teaching improvement. In this experimental study, we undertake a comparative analysis of the LDA model's performance against other traditional methods in the context of topic mining within online English teaching data in higher vocational colleges. The experimental results show that the LDA model shows superior performance in topic consistency. This fully proves the validity and reliability of LDA model in the topic mining of online English teaching data in higher vocational colleges. By utilizing the LDA model, we can have a deeper understanding of the basic themes in online English teaching data in vocational colleges, facilitating educators to obtain more targeted teaching suggestions.

Of course, there are still some limitations and challenges in the application of LDA model in the topic mining of online English teaching data in higher vocational colleges. For example, LDA model has limited ability to process short text data. How to process a large number of short text data while ensuring the quality of topic mining is also a problem worth discussing. In the future, we will continue to improve and optimize model algorithms, combine more teaching data and practical application scenarios, offer enhanced and precise support for the advancement of online English teaching in higher vocational colleges, thereby fostering the sustained growth and progress of English education within these institutions.

References

1. Liu, S., He, T.H., Li, J.Y., et al.: An Effective learning evaluation method based on text data with real-time attribution - a case study for mathematical class with students of junior middle school in China. ACM Trans. Asian and Low-Resour. Lang. Inf. Process. **22**(3), 63 (2023)
2. Peng, C.L., Zhou, X.Y., Liu, S.: An introduction to artificial intelligence and machine learning for online education. Mob. Networks Appl. **27**(3), 1147–1150 (2022)
3. Arslan, O., Xing, W., Inan, F.A., et al.: Understanding topic duration in Twitter learning communities using data mining. J. Comput. Assist. Learn. **38**(2), 513–525 (2022)
4. Zhang, Y., Zhao, J.: Personalized recommendation method of network information integrating LDA and attention. Comput. Simul. **39**(12), 528–532 (2022)
5. Gurcan, F., Cagiltay, N.E.: Exploratory analysis of topic interests and their evolution in bioinformatics research using semantic text mining and probabilistic topic modeling. IEEE Access **10**, 31480–31493 (2022)
6. Qiu, Z., He, B.: Research on the evolution of public opinion and topic recognition based on multi-source data mining. Int. J. Comput. Appl. Technol. **69**(3), 219–227 (2022)
7. Abidar, L., Asri, I.E., Zaidouni, D., et al.: a data mining system for enhancing profit growth based on RFM and CLV. In: 2022 9th International Conference on Future Internet of Things and Cloud (FiCloud), pp. 247–253 (2022)
8. Shi, L., Di, X.: A recognition method of learning behaviour in English online classroom based on feature data mining. Int. J. Reason. Based Intell. Syst. **15**(1), 8–14 (2023)
9. Zhong, Q.: Employment distribution modeling analysis and MATLAB simulation based on data mining and meta-analysis model. In: 2022 Second International Conference on Artificial Intelligence and Smart Energy (ICAIS), pp. 206–209 (2022)
10. Huang, L.: Psychology of adolescents' internet life based on computer technology and online communication data mining. In: 2022 Second International Conference on Artificial Intelligence and Smart Energy (ICAIS), pp. 1132–1135 (2022)

A Study on the Clustering Method of Digital English Teaching Resources Based on Deep Learning

Hui Xu[✉] and Xiaorong Zhu

Department of Public Foundation Courses, Wuhan Institute of Design and Sciences, Wuhan 430000, China
uniqueariel@163.com

Abstract. In order to improve the quality of clustering of digital English teaching resources, this paper introduces deep learning and conducts a research on the clustering method of digital English teaching resources. This method designs a cluster quality evaluation index of digital English teaching resources from multiple dimensions to evaluate the level of digital English resources. Collect English resources and establish a digital English resource bank. On this basis, we use deep learning to design resource clustering model. Finally, it realizes the accurate calculation of resource distribution density and designs the clustering process of digital English teaching resources. The findings from the experiment reveal that this technique has the capability to reach the maximum F-value measurement while maintaining a rapid pace, thereby leading to a substantial enhancement in the clustering quality of digital English instructional materials.

Keywords: Deep Learning · Digitization · English Language Teaching · Resource Clustering · Distributional Density

1 Introduction

As an important part of the education system, the effective clustering and utilization of English teaching resources are of great significance for improving the quality and efficiency of teaching. In the digital era, there is an explosive growth in the number of English teaching resources, including electronic textbooks, multimedia courseware, online video courses and other forms [1]. The disarray and fragmented dispersion of these resources have posed considerable inconveniences for both educators and learners. Addressing the efficient clustering and organization of these materials, expeditiously locating necessary resources, and enhancing the efficacy of teaching and learning have become pressing imperatives.

Presently, researchers both domestically and internationally have undertaken studies on the clustering of digital teaching resources. Traditional clustering methods predominantly encompass techniques such as K-means, hierarchical clustering, and so forth. While these methods have achieved a certain level of classification and organization for teaching resources, their reliance on artificially designed feature extraction methods often leads to less than ideal clustering outcomes for complex and adaptable English teaching resources. Among them, the clustering method proposed in literature [2] usually relies on manually designed feature extraction methods. This approach not only requires a lot of time and manpower, but also often fails to comprehensively and accurately reflect the internal characteristics and semantic information of resources. For English teaching resources, their content is rich and diverse, including text, image, audio, video and other forms. It is difficult to fully capture the complexity and diversity of these resources by manually designed feature extraction methods. The clustering method proposed in literature [3] lacks sufficient intelligence and adaptability. In the face of massive English teaching resources, these methods are often difficult to effectively deal with data noise and outliers, which affects the accuracy and stability of clustering results. At the same time, these methods can not adapt to the individual needs of users, and it is difficult to meet the diversified needs in the teaching and learning process. The clustering method proposed in literature [4] usually faces the challenges of computational efficiency and scalability when dealing with large-scale data. With the continuous increase of digital English teaching resources, traditional clustering methods may consume a lot of computing resources and time to process, which is often unrealistic in practical applications. The clustering method proposed in literature [5] lacks in-depth interpretation and visual display of clustering results. This makes it difficult for teachers and learners to understand the meaning and logic behind the clustering results, and it is difficult to effectively apply the clustering results to the actual teaching and learning process.

In response to the problems existing in the above methods, this study proposes a clustering method for digital English teaching resources based on deep learning. Deep learning technology can automatically extract high-level features from multimodal data, achieving effective clustering of complex resources such as text, images, audio, and video. In addition, this method can dynamically adjust according to different resources and learner needs by introducing adaptive learning mechanisms and personalized learning path recommendations, improving learning efficiency and meeting personalized needs. At the same time, this method also has adaptability across languages and cultures, improving the model's generalization ability and practicality. By enhancing the interpretability and transparency of the model, as well as achieving real-time updates and dynamic clustering, this method not only improves the quality and timeliness of clustering results, but also promotes the digitalization and intelligence process of English teaching, providing educators and learners with more abundant, efficient, and personalized learning resources.

2 Design of Clustering Method for Digital English Teaching Resources

2.1 Evaluating Digital ELT Resources

In this paper, we designed the quality assessment indexes for clustering digital English teaching resources from multiple dimensions, as shown in Fig. 1.

The five primary indicators in Fig. 1 include many secondary indicators. The A1 semantic relevance includes two secondary indicators, namely, the relevance of I1

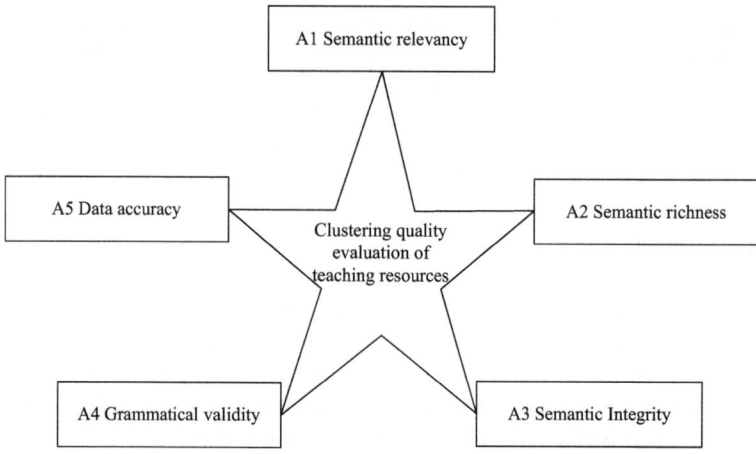

Fig. 1. Indicator map of quality assessment of clustering of digital English teaching resources

retrieval results and the relevance of I2 recommendation results; A2 semantic richness includes two secondary indicators, namely, the number of I3 triples and the number of I4 entities; A3 semantic integrity includes three secondary indicators, namely, I5 thesaurus data integrity, I6 link integrity, and I7 presentation entity integrity; A4 syntax validity includes two secondary indicators, including I8 type and attribute validity and 9 data type matching; A5 data accuracy includes two secondary indicators, namely I10 content accuracy and I11 metadata accuracy.

(1) The calculation formula for A1 semantic relevance is:

$Semantic\ Relevance = \alpha \times R_r + \beta \times R_e$, among them, α and β are weight parameters, R_r is the relevance of search results, and R_e is the relevance of recommendation results.

(2) The calculation formula for A2 semantic richness is:

$Semantic\ Relevance = \alpha \times N_{Triples} + \beta \times N_{Entities}$, among them, $N_{Triples}$ represents the number of triples and $N_{Entities}$ represents the number of entities.

(3) The calculation formula for A3 semantic integrity is:

$Semantic\ Integrity = \alpha \times V_t + \beta \times L_t + \gamma \times E_t$, among them, V_t represents lexical data integrity, L_t represents link integrity, and E_t represents entity integrity.

(4) The calculation formula for A4 grammar validity is:

$Grammatical\ Validity = \alpha \times C + \beta \times D$, Where C represents class and attribute validity, and D represents data type matching.

(5) The calculation formula for A5 data accuracy is:

$Data\ Accuracy = \alpha \times C_A + \beta \times M_A$, among them, C_A represents content accuracy and M_A represents original data accuracy.

The set of factors affecting the quality of resource clustering U can be categorized into 3 levels, the factor set of the first level is $E = \{A_1, A_2, A_3, A_4, A_5\}$, of which, at the first level of influence A_i is made up of second level factors I_i, which can be expressed as $A_1 = \{I_1, I_2\}$, $A_2 = \{I_3, I_4\}$, $A_3 = \{I_5, I_6, I_7\}$, $A_4 = \{I_8, I_9\}$, $A_5 = \{I_{10}, I_{11}\}$. The results of the assessment are categorized into five grades. The results of the assessment were categorized into five levels, indicating the success of $V = \{V_1, V_2, V_3, V_4, V_5\} =$

{Poor, Medium, Good, Excellent, Extremely}. Using hierarchical analysis to determine the weight vectors of the evaluation factors, and then calculating the rank ratings of each single factor to obtain the fuzzy relationship matrix R, expressed as:

$$R = \begin{bmatrix} r_{11}, r_{11}, ..., r_{1n} \\ r_{21}, r_{11}, ..., r_{2n} \\ ... \\ r_{m1}, r_{m1}, ..., r_{mn} \end{bmatrix} \quad (1)$$

According to the fuzzy analysis method of value conversion, and based on the results of Eq. (1) to determine the final quality assessment of resource clustering. This completes the evaluation of digital English resources in the early stage.

2.2 Establishment of a Digital English Language Resource Bank

After the evaluation of digital English teaching resources is completed, a digital English resource database is established. First, collect English resources from various channels, including textbooks, exercise books, listening materials, video tutorials, e-books, academic papers, etc. Ensure the legitimacy and effectiveness of resources. Classify the collected resources, such as by theme, difficulty, format, etc. This helps users quickly find the required resources [6–8]. Select resources to ensure the accuracy and practicability of resources. Eliminate duplicate, outdated or low quality resources. Scan paper resources and convert them into digital formats (such as PDF, JPG, etc.). For audio and video resources, perform format conversion and compression to adapt to different playback devices. OCR recognition is carried out for e-books and academic papers to extract text content. Proofread and edit the text to ensure the accuracy of the content. Add descriptive metadata for each resource, such as title, author, publication date, keywords, etc. This helps users quickly find relevant resources through search engines [9, 10]. Select appropriate digital platforms according to needs, such as self built websites, cloud storage services, etc. Ensure that the platform has good stability and scalability. The user friendly interface is designed to make the resource library easy to browse and use. Clear navigation menu and search function are set to facilitate users to find resources. Upload the digitized resources to the platform. Ensure the integrity and security of data during uploading. The structure of the digital English teaching resource database established in this paper is shown in Fig. 2.

Drawing from the blueprint of the resource database architecture, the decision was made to tailor the DSpace framework to cater to the unique clustering demands of the available resources. The design scheme for the resource data table in the database is outlined in detail in Table 1, offering a clear depiction of the structural layout.

Regularly check the operation status of the plaorm to ensure normal access and download of resources. Handle and repair any problems in a timely manner. Continuously update the content of the resource library according to user needs and market changes. Add new resources and delete outdated or invalid resources [11, 12]. Set up user feedback channels to collect users' opinions and suggestions on the resource library, and improve and optimize according to the feedback.

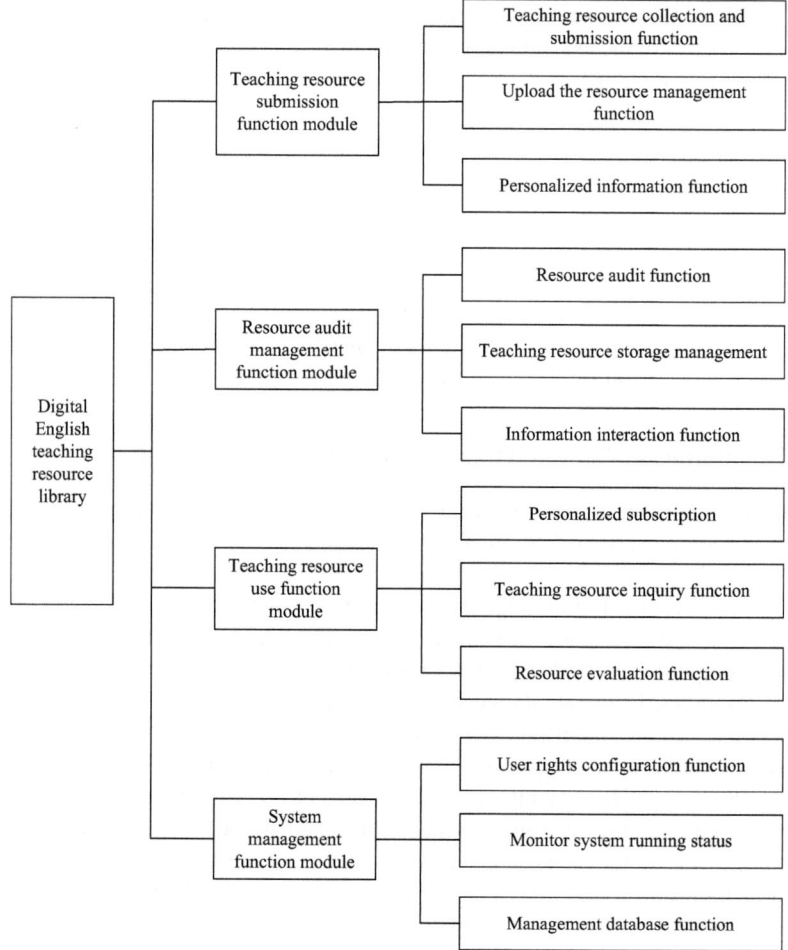

Fig. 2. Architecture of the digital English repository

Table 1. Data sheet on digitized English language resources

Field name	Field type	Field length	Remark
ID	int	4	Resource number, primary key
ResTypeID	int	4	Resource category, foreign key
ResName	nchar	12	Resource name
UpDate	datetime	8	Release time
OwnerID	int	8	ID of the organization to which the resource belongs
ResContent	text	8	Resource introduction

2.3 Designing a Resource Clustering Model Based on Deep Learning

After the establishment of the digital English teaching resource database, we will use in-depth learning to design resource clustering models, which will lay a good foundation for subsequent resource clustering. Set resource clustering keywords, summarize teaching resources according to keywords, and form a matrix of teaching resources to be processed. Use Hadoop technology to design modular teaching resource cluster structure, as shown in Fig. 3.

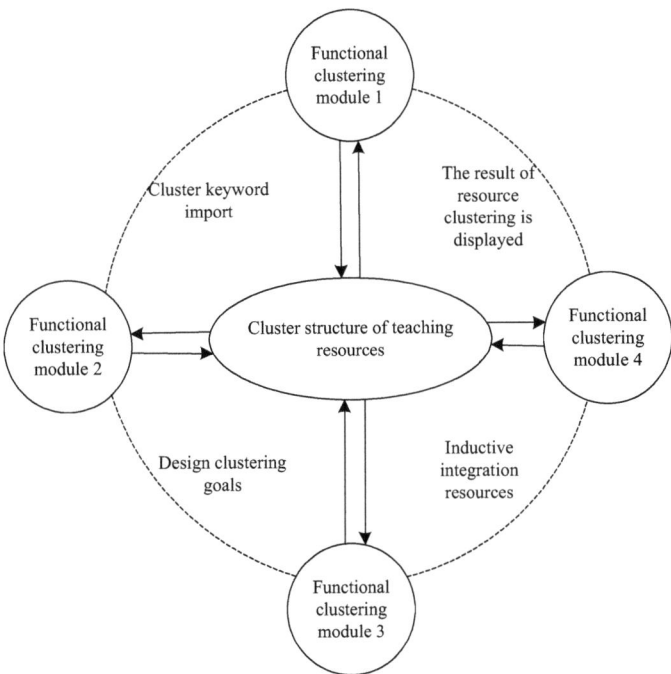

Fig. 3. Clustering structure of teaching resources

Use Hadoop Distributed File System (HDFS) to establish hierarchical teaching resource programs, and deploy a certain number of monitoring nodes in the teaching platform. The formula for calculating the cluster similarity of teaching resources is:

$$Q = \sum_{S=1} FS - \frac{(1+L)}{\psi \rho} \times L \qquad (2)$$

Among them, Q denotes the resource clustering similarity; the F denotes the value of the unit. S indicates the number of matches. ψ denotes the real-time conjugate function; the ρ denotes the covariance. L indicates the controllable cluster range. Combined with Hadoop technology, the teaching resources of various professional courses are differentiated and summarized, and converted into resource blocks and data blocks for storage for future use. The clustering model of teaching resources based on deep learning mainly includes five layers, including two layers of convolution layer and three layers of full connection layer. The output layer is constituted by the final connection layer. ReLU serves as the activation function in the convolutional layers. Following each convolutional layer is an immediate pooling layer for local normalization and pooling of convolutional outputs. The last layer encompasses the fully connected layer of the output layer, employing the Softmax regression model [13–15]. This model serves as the cornerstone for training, with subsequent phases adjusting the network model based on experimental findings.

During the training procedure of the model, the preprocessed image dataset is fed into the network as the foundational data input. The network then segments this dataset into training, validation, and testing datasets based on predefined proportions. Subsequently, the convolutional layer processes the input images to extract distinct features and produce diverse feature maps. The human neural system is very complex, the convolutional neural network model in order to simulate the neuron's work process in the human visual neural system, before outputting the features of the image, let each output feature through the activation function calculation and add a bias value, and then output. After passing through 2 convolutional layers and 2 pooling layers, the data is then passed into a fully connected layer, which completely transforms the features into a one-dimensional vector output.

The convolutional neural network (CNN) model uses a loss function to optimize its parameters and assess its performance. This function calculates the difference between the actual output and expected output, which is then used in the back-propagation algorithm to adjust the network parameters. High values of the loss function suggest suboptimal network parameters, necessitating further training. This process repeats until the loss function value converges to a small value indicating optimal performance. The number of iterations during training is significant, as it governs when the model stops training. If the maximum number of iterations is reached before the model reaches optimal performance, it stops training, and the output model is taken. The final CNN model contains all image types and feature information from each layer, organized hierarchically, with each layer having a specific role in the image analysis process. Through continuous optimization, the CNN model efficiently extracts feature information and produces the optimal model.

2.4 Clustering of Digital English Teaching Resources

To start with, a dense clustering technique is utilized for resource classification, taking into account the interconnections among them. In the course of its implementation, the weighted grid span is defined as follows.

$$f_{(\lambda)} = \{N(G_i) | \forall s 1 \leq i \leq m\} \tag{3}$$

Among them, $N(G_i)$ denotes the weighted grid range of action. i denotes the mesh within the range of action of the weighted mesh. m indicates the total number of meshes constructed.

For the setting of the specific weight of the weighted grid, the grid boundary expansion principle is adopted. Assuming that there is English education resource a in G, if there is aeG and GEN (G), it is considered that the location of English education resource a is the boundary of two grid objects, and there is also a corresponding association between the English education resources of the corresponding grid. In the subsequent clustering phase, merge processing 4 is required. At this point, set the weight value of the mesh to 1. Otherwise, there is no corresponding association between English education resources of the corresponding grid, and the weight value of the grid is set to 0. On this basis, the density of any grid is:

$$\rho(i) = P(density(i) = t) = \frac{count(t)}{count(n)} \quad (4)$$

In the formula, the $\rho(i)$ denotes the total amount of English education resources in the corresponding weighted grid after English education resources are gridded. $density(i)$ denotes the size of the grid cells involved in the statistics. $count(t)$, $count(n)$ denote the number of grid cells and the number of non-empty grid cells densities of t, respectively. According to the above way to realize the accurate calculation of the distribution density of English education resources, to provide a reference basis for the subsequent resource clustering. On this basis, the clustering process of digital English teaching resources is designed, as shown in Table 2 below.

Table 2. Clustering process of digital English teaching resources clustering

No	Flow	Procedure
1	Input	English teaching resource classification data set
2	Initiate	Create a data set point buffer with radius R
3		The query generates the target contained in the bullet buffer
4		Determine the similarity between the source classification data set and the buffer target
5		Select the target with the highest similarity and match it
6	Judgment	Matching result
7	No	If no, go to step 2
8	Yes	Match, direct output
9	Exportation	Integrated result

According to the process of clustering history teaching resources as shown in Table 2, assume that the target data set in its m categories is B, i.e. $B = \{b_1, b_2, b_3, ..., b_n\}$, assuming that the set of all categories of digitized ELT resources is $A = \{a_1, a_2, a_3, ..., a_m\}$, based on the set of all categories A category document in the a_1 established to a_1 as the radius buffer, with the target dataset within the radius buffer B and the matching process is shown in Fig. 4.

Figure 4 shows in detail the clustering process of digital English teaching resources, which is achieved by calculating the similarity (S value) between texts. Firstly, the system receives all pending English teaching resource document datasets as initial input.

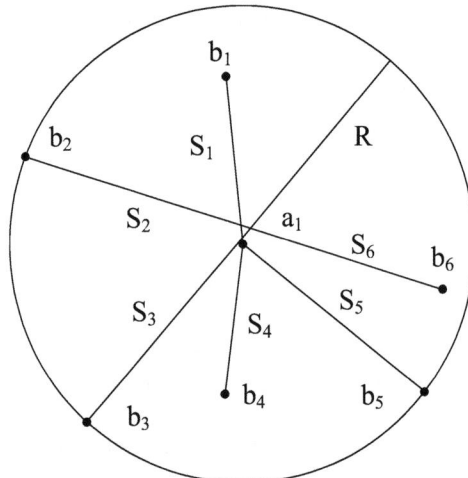

Fig. 4. Matching process of digitized English teaching resources

Subsequently, the system calculates the similarity S between every two datasets to form a similarity matrix. During the initialization phase of clustering, each dataset is considered as a separate cluster. Enter the merge step, search for the two clusters with the highest similarity in the similarity matrix, and merge these two clusters according to predefined criteria (such as maximum similarity threshold or expected number of clusters). After merging, update the similarity matrix and recalculate the similarity between the newly merged cluster and other clusters. This merging step will be repeated iteratively until all clusters that meet the requirements are completed. After each merge, it is necessary to recalculate and update the similarity matrix to ensure dynamic adjustment of clustering.

3 Experiment

To verify the clustering method of digital English teaching resources in this study, a JavaScript operating platform was selected, along with three digital English teaching resource datasets: Read Write Think, Many Things, and Reading Rocks.

Read Write Think: This dataset provides many free K-12 educational resources, including course plans, reading materials, student activities, audio and video teaching resources, teaching PPTs, and task sheets.

Many Things: This dataset has many interesting learning resources, such as vocabulary games such as crossword puzzles and puzzles suitable for the classroom, tips on proverbs and colloquial usage in American, popular English songs, classic English jokes with audio, novels and stories, and other reading materials.

Reading Rockets: This dataset contains many concepts, methods, techniques, and related resources related to English reading and reading teaching, such as recommended reading lists by age or topic, expert interview videos, and parent-child reading tips. The website also provides many free downloadable resources related to vocabulary, pronunciation, spelling, reading comprehension, reading fluency, writing, and teaching evaluation.

Record the deep learning based clustering method in this study as the experimental group, and record the two teaching resource clustering methods mentioned in references [2] and [3] as the control group A and control group B, respectively.

According to the characteristics of the clustering method of digital English teaching resources, and taking digital English teaching resources clustering as the experimental research direction, this paper determines the clustering parameters of digital English teaching resources, compares three groups of digital English teaching resources clustering methods, and aims at the data clustering quality and clustering time of three data sets.

3.1 Experimental Preparation

This time, we verified the clustering method of digital English teaching resources, and selected four directions of source data, including relevant teaching data, online teaching data, paper teaching data, and external teaching data, as the clustering method of this digital English teaching resources, and the digital English learning resources needed to be clustered. The JavaScript platform was selected as the data collection platform for this experiment, and Google Clomme was used as the operating environment of the JavaScript platform. The network environment configuration of its JavaScript platform is shown in Table 3.

Table 3. Platform network environment configuration

No	Disposition	Argument
1	Processor	Intel core i5
2	Internal memory	8.0 GB
3	Hard disk	256 GB solid state
4	Search environment	Google Chrome v71
5	Operating environment	macOS High Sierra x64
6	Operating platform version	Java SE vl.8.0–151

Based on the digital English teaching resource data collection environment set above, three digital English teaching resource datasets, Read Write Think, Many Things, and Reading Rocks, were selected. In this experiment, the selected digital English teaching resource datasets include file format, number of elements, average depth, number of fuzzy DTDs, number of documents, elements, and update cycle, as shown in Table 4.

In this experiment, three digital English teaching resource datasets including Read Write Think, Many Things, and Reading Rows were selected, which come from different fields of digital English teaching resources and have different characteristics and structures, representing different application fields. Therefore, there are certain differences

Table 4. Characteristics of the Instructional Resource Dataset

Data set	GNS	GeoNames	OSM
File format	Txt	Txt	XML
Number of elements before processing	975	1365	2650
Average depth	4.44	4.76	7.14
Fuzzy DTD number	50	50	50
Number of documents	120	140	160
Number of processed elements	810	1246	2416
Renewal cycle	Minimum week	everyday	everyday

in file format, number of elements, average depth, number of fuzzy DTDs, number of documents, and element update cycle.

3.2 Analysis of Clustering Effects

According to the experimental environment set up in this experiment, three digital English teaching resource clustering methods were used for comparative testing, with F1 score as the indicator. The F1 score combines two indicators, Precision and Recall, to measure the performance of clustering methods. The F1 score is a numerical value between 0 and 1, where 1 indicates a perfect model and 0 indicates a complete failure of the model.

The formula for calculating F1 score is:

$$F1\ Score = 2 \times \frac{Precision \times Recall}{Precision + Recall} \tag{5}$$

In the formula, *Precision* represents accuracy, and *Recall* represents recall.

The F1 score test results of the three methods are shown in Fig. 5.

From Fig. 5, it can be seen that the three experimental datasets selected this time, compared to Group B in clustering digital English teaching resources, obtained the lowest F1 score among the three clustering methods; Although the control group A is higher than the control group B, it is significantly lower than the experimental group; The experimental group clustered the digital English teaching resource data from three datasets. Although the F1 scores obtained after clustering fluctuated irregularly due to the differences in the three datasets, the F1 scores obtained were significantly higher than those of control group B and control group A. It can be seen that the clustering method for digital English teaching resources designed this time can obtain higher F1 scores, and the clustering quality is higher for digital English teaching resources.

On the basis of the first group of experiments, the second group of experiments was conducted. In the first set of experiments, the time required to cluster the three data sets of digitized English teaching resources in the three clustering methods was counted, and the length of clustering time in the three clustering methods was compared. The experimental results are shown in Fig. 6.

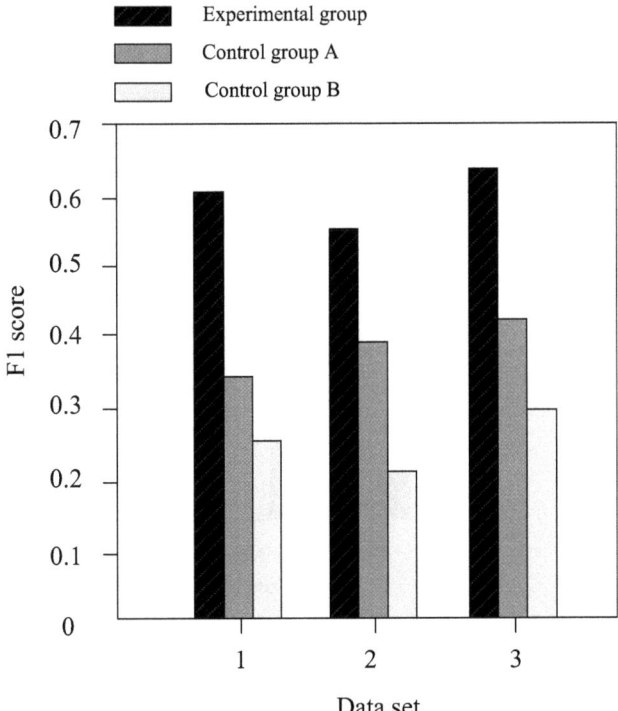

Fig. 5. Results of the quality comparison of the clustering of teaching resources

The data presented in Fig. 6 illustrates distinctive trends among the control and experimental groups regarding clustering quality and time. Notably, while control group B exhibits the poorest clustering quality, its clustering time significantly exceeds that of control group A. Control group A, conversely, demonstrates the lengthiest clustering time compared to the three methods analyzed. In contrast, the experimental group emerges as the most efficient, completing the clustering process in the shortest time frame. These findings, derived from an evaluation of document elements and data set sizes, emphasize the efficiency of the digital English teaching resource clustering approach explored in this research.

Based on the above two groups of experiments, it can be seen that the clustering method of digital English teaching resources in this study can use the fastest speed to get the highest F-value measurement, that is, the higher clustering quality of digital English teaching resources.

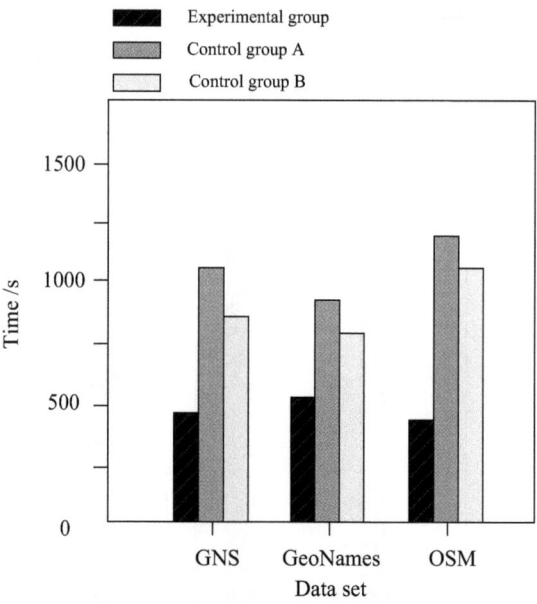

Fig. 6. Results of the time comparison of the clustering of teaching resources

4 Conclusion

The extensive development of digital English teaching resources in the era of rapid technological advancement has raised concerns about the effective management and utilization of these resources to enhance English teaching practices. Consequently, this research focuses on a deep learning-based clustering approach for digital English teaching resources, aiming to organize and provide precise recommendations for these resources through scientific clustering methods. The study delves into the application of deep learning technology in clustering and devises a practical clustering scheme by incorporating the unique attributes of digital English teaching resources. By developing a deep learning model, the automatic classification and clustering of teaching resources are successfully achieved, allowing for the efficient grouping of similar resources and enhancing user accessibility and utilization. The innovation lies in addressing the limitations of traditional clustering methods when dealing with complex and high-dimensional data. Additionally, the model is optimized and enhanced to align more closely with the practical application scenarios of English teaching resources. However, this study still needs further optimization of the model's parameters and structure to improve clustering accuracy and efficiency. Furthermore, exploring the integration of clustering results with actual teaching requirements is essential for delivering personalized and precise resource recommendation services to users.

Looking ahead, the deep learning-based clustering method for digital English teaching resources has a broad application prospect. With the continuous progress of technology and the accumulation of data, the clustering model can be further improved and optimized to achieve a higher level of resource management and utilization. At the same time, the clustering method can be combined with other related technologies, such as natural language processing and recommendation algorithms, to build a more intelligent and efficient digital English teaching platform.

References

1. Peng, C., Zhou, X., Liu, S.: An introduction to artificial intelligence and machine learning for online education. Mob. Networks Appl. **27**(3), 1147–1150 (2022)
2. Yang, Y.: Integration and utilization strategy of university English teaching information resources based on fuzzy clustering algorithm. Mob. Inf. Syst. **22**(5), 1321–1327 (2022)
3. Pérez-Paredes, P.: Review of Viana (2022): teaching English with corpora: a resource book. Int. J. Corpus Linguistics **29**(1), 116–122 (2024)
4. Liu, C., Xia, J., Leung, M.F.: Research on optimization and allocation of English teaching resources. Math. Probl. Eng. Theory Meth. Appl. **45**(24), 1–8 (2022)
5. Liu, M, Li, Q.: Intelligent integration method of AI English teaching resource information under multi-agent collaboration. Adv. Multimedia **12**(3), 1–11 (2022)
6. Li, S., Yuan, D.K., Zhang, L.P., et al.: Teaching resource recommendation method based on multidimensional incomplete classification tree. J. Langfang Normal Univ. (Nat. Sci. Ed.) **23**(04), 39–42+65 (2023)
7. Wu, C.Q., Liu, M.: Teaching resource recommendation algorithm based on kernel canonical correlation analysis. J. Univ. Sci. Technol. Liaon. **44**(01), 62–66 (2021)
8. Yue, P., Zhang, H.: Personalized recommendation system of English teaching resources based on deep learning. Inf. Technol. (06), 149–153+160
9. Zhao, F.: Research on information fuzzy retrieval of network multimedia physical education teaching resource database. Inf. Technol. (07), 29–33 (2023)
10. Ma, M.C.: Intelligent retrieval method of computer professional teaching resources based on knowledge graph. Inf. Comput. **35**(19), 22–24 (2023)
11. Li, X.H.: Real-time sharing system of English micro-course teaching resources based on digital platform. Microcomput. Appl. **39**(04), 153–155 (2023)
12. Hou, Y.C.: English learning resource sharing system based on Internet data. Inf. Technol. **47**(02), 129–133+139 (2023)
13. Zhong, Y., Yuan, W.W., Guan, D.H.: Weighted double q-learning algorithm based on softmax. Comput. Sci. **51**(S1), 58–62 (2024)
14. Wang, Y., He, Y.M., Chen, H.X., et al.: RHS-CNN: a CNN text classification model based on regularized hierarchical softmax. J. Chongqing Univ. Technol. Nat. Sci. **34**(05), 187–195 (2020)
15. Zhang, X.B., Zhu, Z.N., Zhang, W.W.: Combining deep learning and softmax's for part defect identification. Comput. Dig. Eng. **50**(05), 1142–1146 (2022)

An Approach to Integrating Micro-video English Teaching Resources Based on Improved Deep Learning

Ping Huang[✉] and Yalin Sun

School of Economics and Business Foreign Languages, Wuhan Technology and Business University, Wuhan 430065, China
huangping20040701@163.com

Abstract. Some videos may have low resolution, high noise, or blurry images due to technical limitations or poor equipment conditions at the time of initial recording, resulting in poor quality after integrating micro video English teaching resources. However, improving deep learning can improve the quality of videos by training neural networks, converting low resolution videos into high-resolution videos, thereby improving image clarity and details. Therefore, a method for integrating micro video English teaching resources based on improved deep learning is proposed. Extract keyframe images of micro video English teaching resources using the K-means algorithm, enhance the original extracted images, introduce generative adversarial networks in deep learning technology, use dense convolutional neural networks and mixed attention mechanisms to improve the generator in the original generative adversarial network, construct a micro video English teaching resource integration model, input the enhanced keyframe images for classification fusion, obtain the micro video integration result and output, and achieve the integration of micro video English teaching resources. The experimental results show that compared to existing resource integration methods, our method can enhance the quality of English teaching resource images and has significant advantages in objective indicators.

Keywords: Improving Deep Learning · Micro-video · English Language Teaching · Teaching Resources · Resource Integration · Integration Methods

1 Introduction

In today's digital age, online education [1] has become an important branch of the field of education, which breaks the time and space restrictions of traditional education through Internet technology, making learning resources more easily accessible to learners around the world. Micro-video [2], as a new teaching medium, has shown great potential in English teaching with its characteristics of short, concise, content focused, easy to digest and spread. With the deepening of globalization, the importance of English as an international language has become increasingly prominent. The traditional English teaching mode is often limited by teaching materials, teachers and teaching environment, which is difficult to meet the diversified learning needs of modern learners. The rise of online education provides new possibilities for English teaching, especially the integration of micro-video teaching resources, which provides learners with more flexible and personalized learning pathways [3, 4]. Compared with traditional long-form videos or text textbooks, micro-videos can attract students' attention more and stimulate their interest in learning. At the same time, micro-video is extremely portable, and students can learn anytime and anywhere through mobile phones, tablets and other devices. This fragmented learning method greatly improves the flexibility and efficiency of learning. In addition, micro-video can also help students better understand language knowledge and cultural background through intuitive pictures and vivid scenes, so as to improve their language application ability and cross-cultural communication ability.

Reference [5] proposes a teaching resource integration method based on digital platforms, which utilizes the convenience of collecting, processing, and storing teaching resources on digital platforms, collects and processes teaching resource data, and stores it on digital platforms. Calculate the resolution and clarity of images processed by the digital platform, and determine the effectiveness of teaching resource image processing. Classify teaching resource data, establish a dataset buffer based on the category of teaching resource data, match the similarity of data one by one, and complete the integration of teaching resources. This method highly relies on the technical performance of digital platforms, including the ability to collect, process, and store data. If the platform technology is not mature enough or malfunctions, it may affect the integration effect of teaching resources. Reference [6] proposes a method for integrating English teaching resources based on XML technology. By analyzing the data characteristics and organizational structure of resources, they are divided into two types: e-book resources and digital video resources. Using XML technology to construct an English education resource integration model, relying on FC fiber optic switches to connect two types of teaching resources with a disk array. By using an XML resource integration model, data can be transmitted at high speed between relevant servers and backend disk arrays, and actively added to model nodes with lower loads, thereby achieving the integration of English education resources. The method relies on hardware devices such as FC fiber optic switches and disk arrays, which have high costs and may have compatibility issues. In addition, hardware maintenance and upgrades may also bring additional costs and troubles.

This article proposes a micro video English teaching resource integration method based on improved deep learning to address the existing research findings. Improving the application of deep learning in the integration of micro video English teaching resources not only enables intelligent understanding and classification of speech, image, and text information in micro videos through automated content analysis technology, simplifying the complexity of resource integration, but also enhances the accuracy of personalized recommendation systems. Based on students' learning habits and preferences, personalized learning content is tailored for them, thereby improving learning efficiency and interest.

2 Extracting Key Frame Images of Micro-Video English Teaching Resources

Micro video is composed of a series of image frames, and the duration is usually ten or ten seconds. Therefore, the integration of micro video English education resources is the classification and fusion of micro video image frames. The first step of this paper is to extract the key frame images of the micro video English teaching resources to be fused. Although English teaching resources are in the form of micro videos, the number of video frames is still considerable, and the content of adjacent video frames is often very similar, with a lot of redundancy. The efficiency and simplicity of the K-means clustering algorithm enable it to quickly and effectively filter out key frames from a large number of micro video frames, while its automation and scalability ensure the speed and consistency of the processing process, while reducing the need for manual intervention. In addition, the K-means algorithm can identify representative keyframes, which usually contain important information about video content, helping users quickly browse and understand video themes. Due to its strong adaptability, the K-means algorithm can flexibly adjust according to different teaching needs and video content, thereby extracting diverse keyframes, improving the efficiency of teaching resource utilization and learning experience.

K-means clustering [7, 8], an unsupervised technique, seeks to maximize intra-class similarity and minimize inter-class similarity. This results in samples within the same cluster being similar, while samples in different clusters are dissimilar. Typically, the algorithm uses Euclidean distance to measure sample similarity. However, data that does not conform to Euclidean distance requirements must undergo suitable conversion processes. Assume that the original micro video English teaching resource video frame data is $X = x_1, x_2, \cdots, x_n$, of which n is the number of video frame data samples, and the number of K-means clustering is k, of which x_i indicates that the first i. If there are video frame data samples, the specific process of K-means clustering algorithm to extract key frames of micro video English teaching resources is as follows:

(1) Initialize the clustering center of mass. Using random initialization and other methods to obtain that k initial center of mass, denoted as:

$$C = \{c_1, c_2, \cdots, c_k\} \tag{1}$$

In the formula, the C is the initial set of centers of mass of the initial micro-video English teaching resource video frame data samples; the $c_1, c_2, \cdots, c_k$ are the initial center of mass.

(2) Calculate the distance between the sample and each center of mass and divide it into the cluster with the smallest distance until all the samples have been divided, and the formula for calculating the distance between the sample and the center of mass is shown in the following equation:

$$D(x_i, c_i) = \sqrt{\sum_{i=1}^{n} (x_i - c_i)^2} \qquad (2)$$

The formula represents $D(x_i, c_i)$ as the Euclidean distance between a sample image of a video frame from micro-video English teaching resources x_i and the initial center of mass c_i. Within the K-means algorithm, Eq. (2) utilizes Euclidean distance to compute the proximity of each data point to all cluster centers. Consequently, this facilitates the allocation of each data point to the cluster containing the nearest cluster center [9, 10].

(3) Enhance the centroid positioning of each cluster by determining the arithmetic mean of all constituent samples. Designate this calculated mean as the revised centroid for the respective cluster, as stipulated in the subsequent mathematical expression:

$$c_{j'} = \frac{1}{N_j} \sum_{i=1}^{N} x_i \cdot \gamma(x_i, j) \qquad (3)$$

In the formula, the $c_{j'}$ indicates the j new clustering centers for clusters; the N_j denotes the cluster j number of in-sample points. $\gamma(x_i, j)$ is an indicator function, when the sample point, the x_i is assigned to clusters j, the value is 1, otherwise the value is 0.

(4) Repeat steps (2) and (3) until the center of mass is essentially unchanged.

In summary, this paper utilizes the video frame data from micro-video English teaching resources as the initial samples. These initial samples are first obtained by partitioning into k clusters, and the centroids of these k clusters are then utilized as the initial centroids for K-means clustering to re-cluster the sample data. Subsequently, the video frames that closely resemble the centroids in each cluster are identified as key frames, and their corresponding images are extracted. This process offers valuable data support for the subsequent integration of teaching resources.

The K-means algorithm table is shown in Table 1.

3 Enhanced Processing of Micro-video English Teaching Resource Keyframe Images

In order to improve the quality and clarity of the keyframe images, this paper adopts the image enhancement technology [11, 12] to process the keyframe images of micro-video English teaching resources extracted from the above content. First of all, the original key frame image needs to be gray scale transformed, in the image enhancement processing technology, gray scale transformation is a very basic spatial domain image processing method. In image enhancement processing techniques, gray scale transformation is a very basic spatial domain image processing method. Gray scale transformation involves adjusting the gray scale values of individual pixels in an image using a specified transformation function. This manipulation alters the image's dynamic range, enhancing contrast

Table 1. K-means Algorithm Table

Step	Describe
1	Function getmaxdistance()
2	D_1, D_2 ;//Used to store the longest distance
3	C_1, C_2 ;//Storage initial center
4	index 1, index 2;//Store index

5.	for(i=0;i<n;i++)
	 {
	     if $d(t,i) > D_1$
	     $D_1 = d(t,i), index1 = i$
	 }
	 for(i=0;i<n;i++)
	 {
	     if $d(index1,i) > D_2$
	     $D_2 = d(index1,i), index2 = i$
	 }
	 if $t = index2$
	     $C_1 = x_t$, $C_2 = x_{index2}$
	 else if $D_1 > D_2$
	     $C_1 = x_{index1}$, $C_2 = x_{index2}$

and clarity to achieve a more distinct visual representation. By expanding the dynamic range of the image, the transformation method significantly improves visual quality. The transformation process can be succinctly depicted using the following formula:

$$T_1(x, y) = F[T_0(x, y)] \tag{4}$$

The formula includes $T_0(x, y)$ and $T_1(x, y)$, representing the keyframe images of micro-video English teaching resources before and after gray-scale transformation. (x, y) represents the pixel point coordinates, while F is the function for gray-scale transformation, indicating the relationship between input and output gray-scale values.

The histogram enhancement method is one of the practical and effective processing methods in image enhancement. For a grayscale histogram, it is a graph that reflects the relationship between the grayscale levels in an image and the probability of such grayscale levels appearing. Histogram enhancement includes two types: histogram equalization and histogram normalization. This article combines the characteristics of key frame images in micro video English teaching resources and selects histogram equalization as the enhancement method. The specific steps of this method are as follows:

(1) Calculate the probability density function $p(h_i)$ of the original keyframe image, according to Eq. (5):

$$p(h_i) = \frac{n_i}{S} (i = 0, 1, 2, \cdots, L-1) \tag{5}$$

In the formula, h_i represents the gray scale of i; n_i stands for the number of pixels in the gray level i; S is the total number of pixels in the keyframe image; and L indicates the number of gray levels in the keyframe image.

(2) Calculate the cumulative distribution function q_i of the original keyframe image according to Eq. (6):

$$q_i = \sum_{j=0}^{i} p(h_j) \tag{6}$$

(3) Find the corresponding gray level after grayscale transformation according to Eq. (7):

$$h_{i'} = \text{int}\left[(L-1) \cdot q_i + 0.5\right] \tag{7}$$

In the formula, the $h_{i'}$ is the transformed the i gray scale.

(4) According to the gray scale transformation table, the original key frame image gray level mapping to the new gray level, complete the histogram equalization. In practice, some micro-video English teaching resources key frame image due to a variety of reasons makes the gray level distribution is more concentrated, the use of histogram equalization technology to change the distribution of gray levels of the image, so that the spacing of the various gray levels to expand the contrast of the image to expand and promote the distribution of gray levels tends to be uniform, the distribution of the area increases, so that the details of the image becomes clear, to achieve the effect of image enhancement.

(5) It is necessary to smooth the keyframe image after equalization. During the process of image acquisition and transmission, the image is often affected by various noises, which can reduce the quality of the image to a certain extent. Image smoothing is an image processing technique that weakens various interference noises. The basic idea of linear smoothing is to use the weighted average of the current observation value

and the previous smoothing value to calculate the new smoothing value. The calculation formula is as follows:

$$R_t = \lambda \cdot U_t + (1 - \lambda) \cdot R_{t-1} \tag{8}$$

In the formula, the R_t, R_{t-1} are respectively at the time of t and $t-1$ the keyframe image smoothing value; the λ is the smoothing coefficient, usually a value between 0 and 1. U_t is the keyframe image observations of time t.

In the integration of micro-video English teaching resources, high-quality keyframe images need to be integrated to ensure that the subsequent integration of good results.

4 Constructing a Model for Integrating Micro-video English Teaching Resources Based on Improving Deep Learning

Deep Learning is a new research direction in the field of machine learning, mainly by learning the inherent rules and representation levels of sample data, enabling machines to have analytical learning abilities similar to those of humans. Among them, Generative Adversarial Networks (GANS) is one of the representative algorithms of deep learning, which can automatically extract features and classify images. Therefore, this paper introduces Generative Adversarial Networks to classify and fuse extracted micro video keyframe images. However, the original GANs have problems such as loss of high-frequency texture information in the image, weak ability of the model to segment defocused areas in the image, vanishing gradients, and unsatisfactory fusion image quality. Therefore, this paper improves the generator network within the original GANs structure, and designs an improved Generative Adversarial Network model for classifying and fusing English teaching resource keyframe images. The specific structure is shown in Fig. 1:

As shown in the figure above, the overall structure of the improved GANs model is simple, and the core module is the improved generator network and discriminator module.

(1) To resolve the challenges of limited segmentation capabilities for the focused and defocused regions in multifocus images, as well as the issue of vanishing gradients in the network, we introduce enhancements to the original generator by leveraging a dense convolutional neural network and a hybrid attention mechanism. Firstly, we integrate a dense convolutional neural network across each path of the generator. This dense connectivity structure connects each layer of the generator network to its preceding layers, enabling direct access to the feature maps of all prior layers. This approach not only enhances feature reuse and propagation but also helps to mitigate the problem of vanishing gradients. Secondly, we introduce a hybrid attention module following the convolution operation of each layer in the generator network. This module ensures that the generator network pays increased attention to the focal regions of the image during feature extraction. This improvement not only boosts the model's spatial accuracy but also significantly enhances its segmentation capabilities for both the focused and defocused areas of multifocus images. After improving the original generator network with dense convolutional neural network and mixed attention module, this chapter uses two

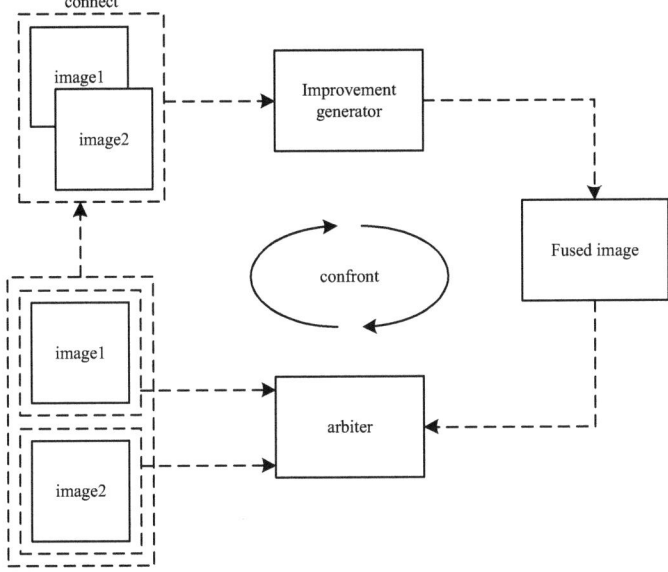

Fig. 1. Improved generative adversarial network model

channels to extract image features. Each path of the generator contains four convolution layers. The first convolution layer uses 5 × 5 convolution cores, and the other three use 3 × 3 convolution cores. In addition, use Concurrent Spatial and Channel Squeeze and Channel Exception (SCSE) to enhance the feature map after each convolution operation. The detailed network parameters of the generator are shown in Table 2:

Table 2. Improved generator model network parameters

Layer_Name	Size	Stride	Channel (Input)	Channel (Output)
Conv (Layer1)	5 × 6	1	1	16
Conv (Layer2)	3 × 3	1	16	16
Conv (Layer2_3)	3 × 3	1	32	16
Conv (Layer3)	3 × 3	1	48	16
Conv (Layer3_4)	1 × 1	1	32	16
Conv (Layer4)	3 × 3	1	64	16
Conv (Layer5)	1 × 1	1	128	1

Finally, use the Relu function to activate the characteristic matrix to prevent over fitting, and its expression is as follows:

$$f_0(x) = \max(0, x) \qquad (9)$$

In the formula, the $f_0(x)$ represents the ReLU function, where x is the model input sample. As shown in Eq. (9), the image of this function is a segmented linear function, when the input values x is less than 0, the output is 0; When the input value x is greater than or equal to 0, the output is equal to the input value x.

(2) Discriminator module. In the generation countermeasure network architecture, the role of discriminator is to maximize the difference between the real image and the generated image, and to distinguish the real image and the generated image as accurately as possible. In essence, it is a neural network model with powerful classification function. The discriminator consists of four convolution layers and a linear layer. In the designed model, the gradient map Real Map generated based on the original image and the gradient map FakeMap output by the GAN generator are sent to the discriminator for identification. The convolution kernel size of the discriminator's convolution layer is 3×3, and there is a normalization layer after each convolution operation. Finally, use the Relu function shown in Eq. (9) to activate. The step of the convolution layer is set to 2. The last layer is the linear layer used to find the classification probability. At the same time, the discriminator loss function enables it to accurately distinguish between true and false samples, because the discriminator input is the fusion image generated by the generator Z_1 and the fused image obtained according to the principle of maximum pixel selection Z_2, then the loss function of the discriminator, the *Loss* The expression is shown in the following equation:

$$Loss = \frac{1}{M} \sum_{m=1}^{M} \left(\left(A(Z_1^n) - B_1 \right)^2 + \left(A(Z_2^n) - B_2 \right)^2 \right) \tag{10}$$

Among them M is the number of fused images during training; the A is a function to compute the pixel average; the B_1 indicates the probability that the discriminator is expected to recognize the real data, setting $B_1 = 1$; B_2 denotes the probability that the discriminator is expected to be able to recognize the false data, setting $B_2 = 0$. With this constraint, the discriminator can continuously improve its ability to identify true and false data, guiding the generator to produce strongly textured fused images.

Therefore, in the above improved generative adversarial network model, the enhanced keyframe images are inputted for classification and fusion, the classification and fusion results are outputted, and then the keyframe images of the same type are combined together to form multiple new microvideos, thus completing the integration of microvideo English teaching resources.

5 Experimental Analysis

5.1 Experimental Setup

To assess the feasibility and effectiveness of the proposed method for integrating micro video English teaching resources through enhanced deep learning, this section conducts a comparative simulation experiment using the UCF101 dataset, a popular source for behavior recognition collected from YouTube. The UCF101 dataset consists of 101 behavior types, totaling 13,320 short videos, ensuring data diversity by capturing various aspects such as camera movement, subject appearance changes, target scale alterations, background lighting fluctuations, and shifts in perspective. Each video type is segmented into 25 groups, each containing 4 to 7 short videos that share common attributes like backgrounds and perspectives. Videos are broadly categorized into human-object interaction, human-human interaction, bodily movements, musical instrument performance, and sports, with specific subcategories including children's crawling, archery, basketball, tooth brushing, high jump, cycling, and dog walking. Select categories are illustrated in Fig. 2:

Fig. 2. Representative Categories of UCF101 Dataset

Randomly select a poor quality English teaching resource image from the dataset and use the method proposed in this article to enhance it. The original image of English teaching resources is shown in Fig. 3.

The image of English teaching resources enhanced by the method in this article is shown in Fig. 4.

By comparing the image enhancement effects of English teaching resources shown in Figs. 3 and 4, it can be seen that the clarity of the English teaching resource images is significantly improved after the method proposed in this paper. The resources in the images can be clearly observed, which helps to improve the learning quality of students. Therefore, it indicates that the English teaching resource image enhancement effect of the method proposed in this article is good.

5.2 Evaluation Indicators

In order to objectively verify the effectiveness of the proposed micro video English teaching resource integration method, 10 groups of short videos of a certain category

Fig. 3. Image of Original English Teaching Resources

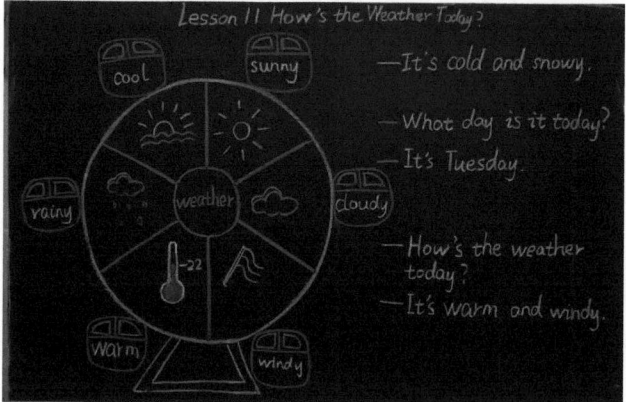

Fig. 4. Enhanced image of English teaching resources

were randomly selected from the UCF101 dataset for resource integration. The average gradient, information entropy, spatial frequency, visual fidelity, and differential correlation were used as indicators to compare and test the method proposed in this paper with the methods in reference [5] and reference [6].

(1) Average gradient η_1.

η_1, being the average gradient of the microvideo frame map, assesses the sharpness of the integrated video. Its greater value indicates enhanced clarity, and the exact computation is outlined in the equation below.

$$\eta_1 = \frac{1}{(H-1)(W-1)} \sum_{i=0}^{H-1} \sum_{j=0}^{W-1} \sqrt{\frac{(I_{i,j+1} - I_{i,j})^2 + (I_{i+1,j} - I_{i,j})^2}{2}} \quad (11)$$

In the formula, the H, W denote the height and width of the frame map in the microvideo integration result; The I denotes the frame map in the microvideo integration result; The i, j represent the rows and columns of frame pixels in the microvideo integration result.

(2) Information entropy η_2.

η_2 is used to measure the information content of micro-video frame map, the larger its value is, the more information is contained in the integration result, the specific calculation formula is shown in the following equation.

$$\eta_2 = -\sum_{i=0}^{255} P_i \log P_i \tag{12}$$

In the formula, the P_i denotes the gray value of the pixel point on the frame map in the microvideo integration result i corresponding normalized probabilities.

(3) Spatial frequency η_3.

η_3 is used to measure the structural texture of the microvideo frame map, the larger its value is, the richer the edges and texture of the integrated result frame map, the specific calculation formula is shown in the following equation.

$$\eta_3 = \sqrt{\alpha^2 + \beta^2} \tag{13}$$

In the formula, the α, β denote the row frequency and column frequency of the pixels on the frame map in the microvideo integration result, respectively.

(4) Visual fidelity η_4.

η_4 is used to measure the information fidelity of the microvideo frame map, similar to the human visual system, the larger its value, the better the visual effect of the integration result, the specific formula is shown in the following equation.

$$\eta_4 = \omega_1 I_1 + \omega_2 I_2 + \omega_3 I_3 \tag{14}$$

In the formula, the I_1, I_2, I_3 respectively represent the VIF values of the micro video integration result frame diagram, the original micro video frame to be integrated Fig. 1, and the original micro video frame to be integrated Fig. 2; $\omega_1, \omega_2, \omega_3$ are weighting factors, respectively.

(5) Differential relevance η_5.

η_5 quantifies the correspondence between the original microvideo frame map information and the integrated frame map data, effectively gauging the degree of spurious information within the integrated outcome. A higher Q value signifies superior integration quality and reduced spurious information. The detailed computation is provided in the following equation.

$$\eta_5 = f(I_2, I_1) + f(I_3, I_1) \tag{15}$$

In the formula, the f denotes the correlation coefficient function.

As shown in the above formula, this experiment adopts five objective and intuitive indicators to assess the overall quality of microvideo integration results, and the evaluation results are more comprehensive.

5.3 Results and Conclusions

Following the completion of microvideo integration using both the experimental and control group methods, the evaluation indices are separately calculated for each method. The specific comparison results can be found in the table below (Table 3).

Table 3. Experimental results of five indicators of different methods on UCF101 dataset

Methods	η_1	η_2	η_3	η_4	η_5
Reference [5] method	5.4379	5.9876	12.6965	0.1057	0.2718
Reference [6] method	5.5594	6.2441	13.9536	0.4127	0.2859
Proposed method	8.2129	8.3467	20.1745	0.6159	0.6694

The experimental results of five key indicators using this method on the UCF101 dataset showed significant advantages. In terms of average gradient, information entropy, spatial frequency, visual fidelity, and differential correlation, the performance of our method is superior to the methods in references [5] and [6]. Specifically, the average gradient value of the method in this article is 8.2129, the information entropy value is 8.3467, the spatial frequency value is 20.1745, the visual fidelity value is 0.6159, and the difference correlation value is 0.6694, all of which are higher than the other two methods. This indicates that the method proposed in this article can more effectively preserve the clarity, information content, texture details, visual quality, and consistency before and after image processing, thereby demonstrating higher practical value and application potential in the field of image processing and analysis.

6 Conclusion

This paper delves into the integration methodology of micro-video English teaching resources, presenting an enhanced deep learning model applied effectively in key processes like key frame extraction, image enhancement, and resource fusion. However, some limitations remain, including the need to bolster the model's generalization capacity and address the efficiency of handling large datasets. Future efforts will concentrate on refining the model structure for improved performance in complex settings. Moreover, exploration will extend to utilizing this approach across various languages and educational domains. As technology advances and research progresses, the enhanced deep learning-based integration method for micro-video English teaching resources is poised to substantially uplift teaching standards and enhance learning efficiency.

Acknowledgment. 1. The Scientific Research Team Plan of Wuhan Technology and Business University (WPT2023041);

2. Research on the Teaching Model Construction and Practice of College English "Golden Course" based on OBE (2019ZA10), one of the key projects supported by the Educational Science Planning Program of Hubei Province with Special Funding, 2019;

3. Teaching Reform Research Project of Wuhan Technology and Business University- "Design and Application of Blended Teaching for College English from the Perspective of Learning Experience" (2022Y01);

4. One of the First Class Undergraduate Courses of Hubei Province-- College English II of Wuhan Technology and Business University.

References

1. Huang, Y., Peng, C., Liu, S.: Empirical research of classroom behavior based on online education: a systematic review. Mobile Networks Appl. online first, https://doi.org/10.1007/s11036-023-02251-2
2. Chen, C.: Research on micro-video function design of the practical teaching system of ideological and political course in colleges and universities. Techniques Autom. Appl. **42**(06), 139–142+159 (2023)
3. Shi, N.J.: Research on personalized English learning system based on MOOC. Inf. Technol. (07), 34–37+42 (2023)
4. Ye, L., Chen, J., Le, H.P.: Design of personalized English teaching system based on big data. Microcomput. Appl. **38**(01), 41–44 (2022)
5. Zhang, J.: Research on integration method of history teaching resources based on digital platform. Techniques Autom. Appl. **42**(07), 78–81+94 (2023)
6. Zhang, J.: A model of distributed English educational resources integration based on XML technology. Inf. Technol. **47**(04), 112–116 (2023)
7. Cao, J., Zhou, H., Dai, Z.B., et al.: A self-verification and self-classification method for FITS images based on K-means. Electron. Des. Eng. **31**(23), 180–183+195 (2023)
8. Xu, W.J., Guan, K.H., Ma, Y., et al.: Track data hot spot mining algorithm based on k-means. Comput. Modernization (10), 23–28+34 (2021)
9. Ning, L.L.: Extraction of Web user interest feature factor based on K-means algorithm in the context of big data. Autom. Instrument. (09), 41–45 (2023)
10. Liu, W.M., Cui, Y., Mao, Y.M., et al.: Parallel K-means algorithm based on MapReduce and MSSA. Appl. Res. Comput. **39**(11), 3244–3251+3257 (2022)
11. Zhu, X.K., Wang, K., Yun, L.J., et al.: Research on image enhancement technology under the low-light condition based on multi-scale fusion. J. Yunnan Normal University: Nat. Sci. Ed. **43**(03), 30–35 (2023)
12. Kang, L.J., Chen, X.Q.: Image contrast enhancement technology based on multi-level histogram shape segmentation. Comput. Appl. Software **39**(03), 207–212+321 (2022)

Intelligent Optimization Method for Charging Power of Electric Vehicle Charging Station Based on VSM

Jiatong Wei[1], Chunhua Kong[1(✉)], Shujiang Song[2], and Lei Ma[3]

[1] College of Automotive Engineering, Jilin Communications Polytechnic, Changchun 130012, China
`13596157766@163.com`
[2] Network Management Department, FAW-Volkswagen Sales Co., Ltd., Changchun 130031, China
[3] Beijing Polytechnic, Beijing 100176, China

Abstract. To ensure stable output of charging power at the charging station, this study designed an intelligent optimization method for charging power of electric vehicle charging stations based on VSM. Firstly, the circuit topology of the electric vehicle is analyzed in depth, and the power demand and variation characteristics during the charging process are clarified. Then, based on VSM theory, an optimization model of electric vehicle charging power is established. This model can comprehensively consider the operating parameters of the charging station and the charging demand of the electric vehicle, and realize the intelligent distribution of charging power. Finally, a three-phase balance optimization charging power regulation strategy is proposed to ensure the balance of three-phase current during the charging process, and to minimize resource waste by improving charging efficiency. Through testing and verification, it was found that the deviation of grid frequency was significantly reduced after applying this method. When the frequency deviation of the power grid increases from 0 to -1.5 Hz, the active power absorbed by the power grid decreases from 15.0 kW to 13.5 kW, a decrease of 10%. However, during this period, the output power of the load remained at 15.0 kW, indicating that this method can maintain stable power output under different input voltages.

Keywords: VSM · Electric vehicle charging station · Charging power · Power optimization

1 Introduction

Electric vehicles, with their clean, efficient and low-carbon characteristics, are gradually becoming an important part of future urban transportation, replacing traditional fuel vehicles. This shift not only represents technological progress, but also reflects the human society's pursuit of environmental protection, energy saving and sustainable development. However, with the increasing position of the electric vehicle market, the construction and operation of charging stations are gradually emerging, becoming a key factor restricting the further development of electric vehicles.

As the energy supply station for electric vehicles, the performance of charging station is directly related to the charging efficiency, cost and user's charging experience [1]. The distribution and scheduling of charging power is the core of charging station operation and management. How to intelligently adjust and optimize the charging power according to the actual condition of the charging equipment, the user's demand and the load condition of the power grid has become a key issue to be solved in the field of electric vehicles.

Traditional charging power optimization methods are often based on fixed algorithms or rules, which lack flexibility and adaptability and are difficult to cope with the complexity and uncertainty in the charging process of electric vehicles. Therefore, we need a more intelligent and efficient method to solve this problem.

Vector Space Model (VSM) is a typical text representation method that can be applied to tasks such as data retrieval and processing. VSM transforms text into points in vector space, reveals the internal relationship between texts by using similarity and distance measures between vectors, and realizes accurate text matching and efficient retrieval. The idea of simplifying complex problems into vector operations provides us with a new optimization design approach. The intelligent optimization method of charging power of electric vehicle charging station based on VSM aims to realize intelligent allocation and scheduling of charging power by constructing a vector space model of charging power of charging station. Compared with the traditional optimization methods, the intelligent optimization method of charging power based on VSM has higher flexibility and adaptability. It can dynamically adjust the charging power distribution scheme according to real-time data, and effectively deal with the uncertainty and dynamic changes in the charging process of electric vehicles. At the same time, this method can also make full use of the remaining capacity of the power grid, realize the collaborative optimization between the power grid and the charging station, and further improve the energy utilization efficiency.

Therefore, this paper deeply studies the intelligent optimization method of charging power of electric vehicle charging stations based on VSM.

2 Analysis of Electric Vehicle Circuit Topologies

Analysis of electric vehicle circuit topology is a key step in optimizing the charging power of electric vehicle charging stations. The circuit topology determines the current flow path and the connection between components, which has an important impact on the performance and charging efficiency of electric vehicles [2].

In electric vehicles, the circuit topology typically includes the interrelationships between components such as battery management systems (BMS), motor controllers, on-board chargers (OBC), DC-DC converters, etc. [3]. These parts are connected together through specific circuits to form a complete circuit system. Among them:

① OBC is a device that converts alternating current (AC) to direct current (DC) for battery charging. The efficiency of OBC directly affects the charging efficiency. Efficient OBC can reduce losses during energy conversion and improve charging efficiency.

② BMS is responsible for monitoring and managing the status of batteries, including State of Charge (SOC), State of Health (SOH), and temperature. An efficient BMS can ensure that the battery is charged at its optimal state, avoiding overcharging or undercharging, thereby improving charging efficiency and battery life.
③ In some electrical topologies, DC-DC converters are used to convert the high voltage of the battery into the low voltage required by other systems of the vehicle. This conversion process will also affect the overall energy efficiency.

In this paper, we first analyze the topology of the main electric vehicle charging/discharging circuit in detail, and Fig. 1 shows the topology of the main electric vehicle charging/discharging circuit.

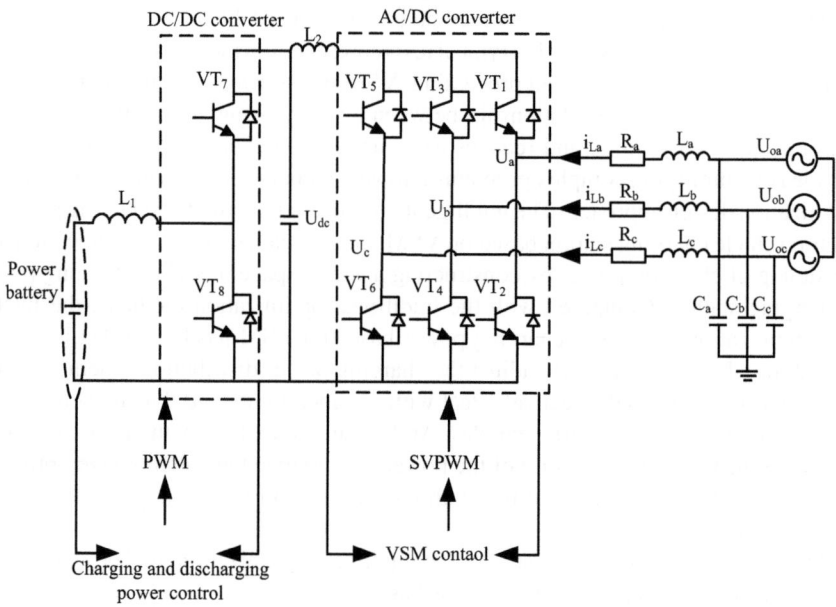

Fig. 1. Charging and discharging circuit structure

In Fig. 1: U_{oa}, U_{ob} and U_{oc} are the three-phase voltage of AC side bus; L and C form LC filter; R is the equivalent filter resistance; U_a, U_b and U_c correspond to the three-phase voltage on the AC side of the converter, respectively; i_{La}, i_{Lb} and i_{Lc} correspond to the current on the input AC side respectively; U_{dc} is the DC bus voltage.

In Fig. 1, the main circuit topology is divided into two parts: DC/DC converter topology on the left and bidirectional AC/DC converter topology on the right. Among them, the electric vehicle DC converter is composed of DC/DC non isolated bidirectional half bridge circuit, also known as bidirectional Boost/Buck circuit. The bidirectional DC/DC converter is shown in Fig. 2.

In Fig. 2: Switching tubes, S_1 and S_2 simultaneou action and opposite direction of action; series internal resistance: represents the energy storage element. D_1 and D_2 are freewheeling diode; R is the equivalent load on the busbar side. By adjusting the status

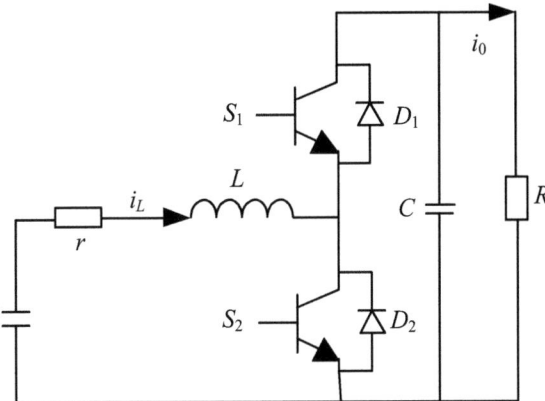

Fig. 2. Bidirectional DC/DC Converter

of the switch, power can flow between high and low voltage sides. The circuit structure is simple, the control is convenient, and the reliability is high.

In the process of interconnection between the charging station and the power grid, the bidirectional AC/DC three-phase converter is used as the main converter, and Fig. 3 is its structural diagram. The DC side is the capacitor at the high-voltage side of the electric vehicle, and the main circuit topology uses a two-level inverter with LC filtering.

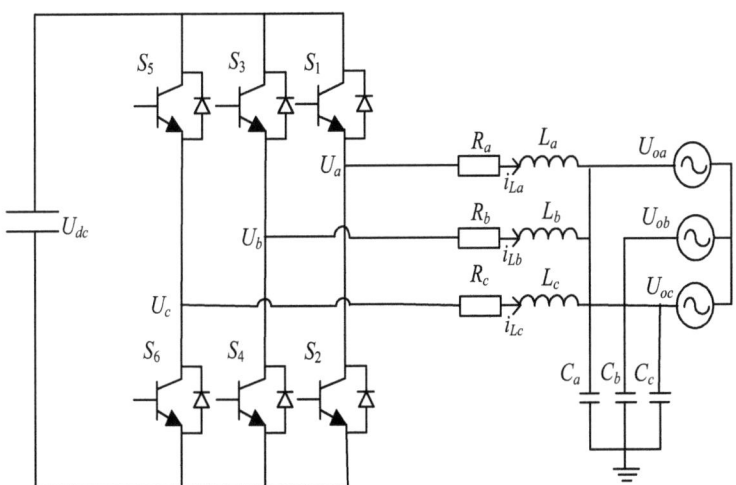

Fig. 3. Bidirectional AC/DC converter

In Fig. 3, R, L and C are the equivalent filter resistance, inductance and capacitance on the AC side of the converter respectively. The switching function defining the converter is shown in Formula (1):

$$S_k = \begin{cases} 1, \text{ Upper end connectivity, lower end disconnection} \\ 0, \text{ Lower end connectivity, upper end disconnection} \end{cases} \quad (1)$$

where: S_k is the converter switching function.

According to Kirchhoff's law, the dynamic equation of AC side voltage of bidirectional AC/DC converter is as shown in Eq. (2):

$$L = \frac{di_{Labc}}{dt} = u_{abc} - u_{oabc} - Ri_{Labc} \quad (2)$$

where: u_{abc} is the three-phase voltage on the AC side of the converter; i_{Labc} is the three-phase current of the AC side of the converter; u_{oabc} is the AC bus three-phase voltage. Where, $u_{abc} = S_k u_{dc}$. Through Clark transformation processing, a two-phase stationary model can be established [4], and the transformation process is as follows:

$$T = \frac{2}{3} \begin{bmatrix} 1 & -\frac{1}{2} & -\frac{1}{2} \\ 0 & \frac{\sqrt{3}}{2} & -\frac{\sqrt{3}}{2} \end{bmatrix} \quad (3)$$

On this basis, T is further transformed into the following form:

$$T\prime = \begin{bmatrix} \cos\theta & \sin\theta \\ -\sin\theta & \cos\theta \end{bmatrix} \quad (4)$$

By combining the above two formulas, the transformation matrix can be obtained as follows:

$$T'' = \frac{2}{3} \begin{bmatrix} \cos\theta & \cos(\theta - \frac{2\pi}{3}) & \cos(\theta + \frac{2\pi}{3}) \\ -\sin\theta & -\sin(\theta - \frac{2\pi}{3}) & -\sin(\theta + \frac{2\pi}{3}) \\ \frac{1}{2} & \frac{1}{2} & \frac{1}{2} \end{bmatrix} \quad (5)$$

A transformation of Eq. (5) yields.

$$\frac{d}{dt}\begin{bmatrix} i_{Ld} \\ i_{Lq} \end{bmatrix} = A\begin{bmatrix} i_{Ld} \\ i_{Lq} \end{bmatrix} + B\begin{bmatrix} u_d \\ u_q \end{bmatrix} + C\begin{bmatrix} u_{od} \\ u_{oq} \end{bmatrix} = A\begin{bmatrix} i_{Ld} \\ i_{Lq} \end{bmatrix} + B\begin{bmatrix} S_d u_{dc} \\ S_q u_{dc} \end{bmatrix} + C\begin{bmatrix} u_{od} \\ u_{oq} \end{bmatrix}$$

$$A = \begin{bmatrix} -\frac{R}{L} & \omega \\ -\omega & -\frac{R}{L} \end{bmatrix}, B = \begin{bmatrix} \frac{1}{L} & 0 \\ 0 & \frac{1}{L} \end{bmatrix}, C = \begin{bmatrix} -\frac{1}{L} & 0 \\ 0 & -\frac{1}{L} \end{bmatrix} \quad (6)$$

Through the above analysis, the dynamic equation of AC side voltage of bidirectional AC/DC converter is obtained, which is convenient for subsequent control and optimization.

3 Establish VSM Electric Vehicle Charging Power Optimization Model

After defining the circuit topology, a mathematical model needs to be established to optimize the charging power. This model will consider many factors to find an optimal charging power distribution scheme, so as to maximize the charging efficiency and reduce the impact on the power grid. VSM is a power system control strategy, which is mainly used in the control of power converters and grid connected converters. VSM simulates the dynamic behavior of traditional synchronous generators, making charging stations behave more similar to traditional power plants in the power grid. This simulation includes the inertia response of synchronous generators and the behavior of speed controllers, enabling charging stations to provide better frequency support when the grid frequency changes, thereby stabilizing the grid frequency. In addition, VSM introduces inertia and damping parameters into the control strategy of charging stations, enabling them to provide necessary inertia support and damping when the grid frequency changes, reducing frequency deviation.

The core idea of VSM is to control the output of the converter by simulating the dynamic response voltage of the power system, so as to make it behave like a power system with specific inertia and damping. The mathematical model is as follows:

$$\begin{cases} \delta\theta = \alpha + K(P_r - P_m) \\ \alpha = \frac{1}{M}(P_{in} - P_{out} - D\delta\theta) \end{cases} \quad (7)$$

where, $\delta\theta$ is the phase angle deviation; the α is the frequency deviation of the system; the P_r and P_m reference power and measured power, respectively; the P_{in} and P_{out} are respectively the input and output power of the converter; M is the virtual inertia constant; D is the virtual damping coefficient [5]. It can be said that VSM has the technical characteristics of simulating traditional power system, high adaptability, high precision control, and configurability. It has strong application advantages in electric vehicle charging pile charging power optimization control.

Based on the above content, before establishing the VSM electric vehicle charging power optimization model, the operating mode of the electric vehicle charging pile converter is first analyzed. The process is as follows:

Step 1: Modal analysis of electric vehicle charging pile converter operation.

Considering the voltage stability and control accuracy of the electric vehicle charging pile converter, the following is a comprehensive analysis of the operating modes of the converter, in order to provide detailed theoretical support for the optimal control of the charging power of the converter.

The particular DC voltage state will affect the mode of operation of the converter [6]. The main operating modes of the electric pile converter at different voltage conditions are as follows.

① The DC voltage state of the front stage is 0–320 V, and the converter is in the current limiting mode: within this voltage range, the converter will limit its current output to ensure that it will not be overloaded, which is usually limited to below 20 A.
② The DC voltage state of the front stage is 320–380 V, and the converter is in the standard operation mode: within this voltage range, the converter will work normally, providing a maximum current output of 50 A.
③ The DC voltage state of the front stage is 380–400 V, and the converter is in the efficient mode: when the voltage is close to the maximum value, the converter will enter the efficient mode, optimize the power conversion efficiency, and the output current can reach 55 A.

④ The DC voltage status of the front stage is >400 V, and the converter is in the protection mode: if the voltage exceeds the safety range, the converter will automatically enter the protection mode, limiting or cutting off the output to protect the internal circuit.

Step 2: Electric vehicle charging power VSM model.

The VSM allows the EV Charge Point Converter to be simulated as a synchronous generator. To achieve this simulation, the following relationships can be used:

$$P_d = V_d \times I_d \tag{8}$$

In the formula, the P_d is the DC power, the V_d is the DC voltage, the I_d is DC current. According to the VSM principle, the relationship between the voltage and the current can be described as:

$$T_m - T_e = J\frac{d\alpha}{dt} \tag{9}$$

In the formula, the T_m and T_e are mechanical torque and electrical torque respectively, and J is inertia constant. Then, in order to maintain the stability of the DC side voltage, the following front-end DC port voltage controller is designed.

① Calculate the difference between the set voltage and the actual voltage:

$$e(t) = V_{dc1} - V_{dc2} \tag{10}$$

where: V_{dc1} is the set voltage. V_{dc2} is the actual voltage.

② PID control algorithm is introduced. Considering that the controller is real-time, easy to implement and adjust, PID algorithm is selected for voltage control, and the formula is as follows:

$$u(t) = K_p e(t) + K_i \int e(t)dt + K_d \frac{de(t)}{dt} \tag{11}$$

In the formula, the K_p is the proportional gain, the K_i is the integration gain, the K_d is the differential gain, the t is time. For the error $e(t)$, the algorithm can provide proper control inputs to the controller [7].

③ Controller programming. In order to realize the design of the formula described above, the internal logic of the controller is accurately programmed, and the following pseudo-code is simplified.

Initialize:K_p, K_i, K_d, V_dc_ setpoint
Pr electric vehicle ious_error=0
Integral=0
Loop:
V_dc_actual=read_dc_voltage()
Error=V_dc_setpoint—V_dc_actual
Integral=Error*dt
Derivative=(Error-Pr electric vehicle ious_error) /dt
Control_input=K_p*Error+K_i*Integral+K_d*Derivative
adjust current (Control_ input)
Pr electric vehicle ions_error=Error

wait(dt)
End Loop

In conclusion, this design logic can ensure the stability of DC voltage of electric vehicle charging pile converter in the charging process, so as to improve the charging efficiency of the equipment and maintain the life and safety of electric vehicles and charging piles.

4 Three-Phase Balanced Optimized Charging Power Regulation

After establishing the charging power optimization model, it is necessary to further realize the optimal regulation of charging power. Three-phase balance is an important consideration, which involves the stability and safety of the power grid [8]. During the charging process, it is necessary to ensure the balance of the three-phase currents to avoid excessive burden on the grid or safety problems.

This paper establishes a two-phase optimization strategy to optimize EV charging loads based on real-time three-phase currents, and divides the optimization day into 96 time segments. When the i th electric cars are in time of t entering the cluster charging area at any time, the control center collects the battery status and expected charging duration of the vehicle. For each electric vehicle that enters the charging area for charging, charging position optimization is carried out; at the end of each time segment, charging power control optimization is carried out for all vehicles. The execution sequence of the two-stage optimization is shown in Fig. 4.

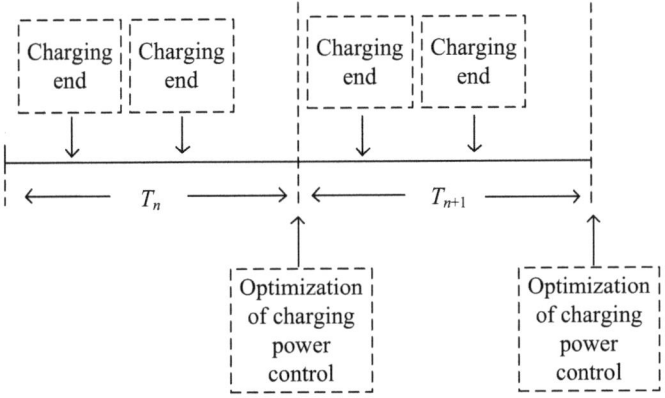

Fig. 4. Two-stage optimization of execution order

(A) Phase I Electric Vehicle Access Optimization.

Electric vehicle charging load has the flexible load characteristics of optional access point, when the electric vehicle enters the charging area, it should be optimized according to the operation of charging area power supply line trend to optimize the selection of its access location, to achieve the optimization of electric vehicle load access. For the i electric vehicle the access optimization process is shown in Fig. 5.

The idea of optimizing access is: in the presence of the N transformer, it is calculated that after connecting the electric vehicle to each phase of each transformer, the corresponding $3N$ neutral current values for each case. Starting from the smallest neutral current case, if the charging facility corresponding to this case has free parking spaces and the connection of EVs will not cause overloading of the transformer, then the optimized connection is completed. Otherwise, the next smallest neutral current case is examined until the judgment is completed for all cases.

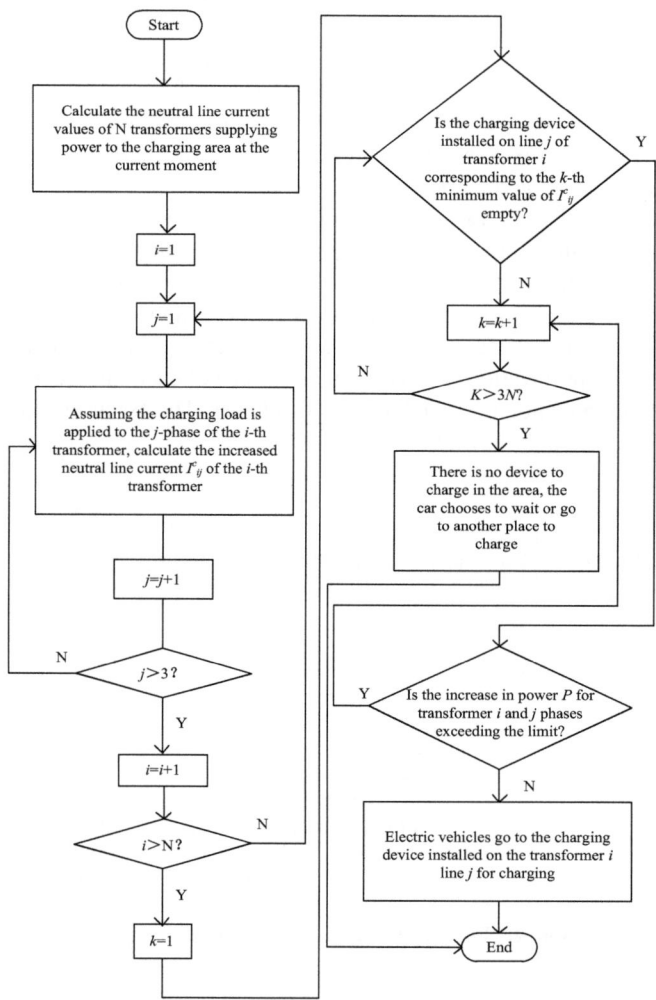

Fig. 5. Optimized access process for electric vehicles

(B) Stage II Electric vehicle charging power optimization.

In the second stage, i.e. EV charging power optimization stage, power control decisions are made for EVs still charging in the charging area according to the system three-phase operation status and conventional load change trend. This process involves the analysis of neutral current confidence coefficients and historical data to ensure that the decision is scientific and reasonable. Once a decision is made to power control an EV, its charging power is adjusted according to predetermined rules and steps to achieve three-phase balance and maximize charging efficiency [9].

After the end of the chronological segment t, the optimization center, according to the system three-phase operation status, conventional load change trend, the next time segment is still in the charging area for charging electric vehicles to decide whether to take power control, power control flow chart shown in Fig. 6.

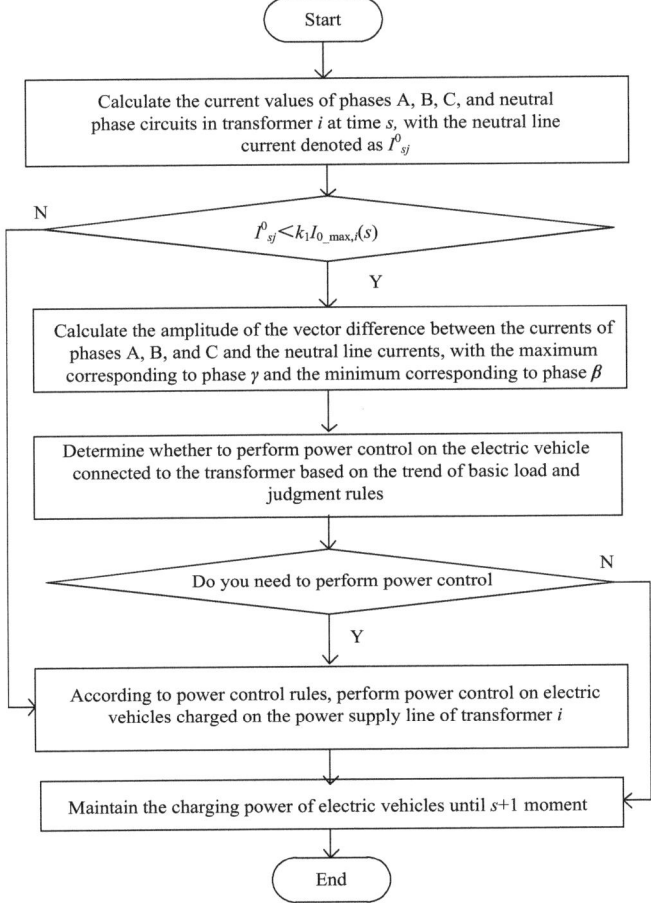

Fig. 6. Charging power optimization process for electric vehicles

In Fig. 6 k_1 is the neutral current confidence factor; and $I_{0-\max,i}(s)$ is the historical data, the maximum neutral current of transformer i at s moment.

The decision rules for deciding whether or not to power control the EV based on the trend eigenvalues and the tidal operating state are shown in Table 1.

Table 1. Power control execution criteria

$\delta_\gamma(t)$	$\delta_\beta(t)$		
	−1	0	1
−1	No Execute	No Execute	Execute power control
0	No Execute	No Execute	No Execute
1	No Execute	No Execute	No Execute

In Table 1, $\delta_\gamma(t)$, $\delta_\beta(t)$ are the trend eigenvalue of the phase γ and β at time t. The power optimization control steps are as follows.

① According to power variation constraints, all electric vehicles available for power control on A, B and C phases are selected. Determine which electric vehicle charging power can be adjusted to optimize the power distribution in the entire charging area. In the screening process, various factors are considered, such as the charging state of electric vehicles, the maximum and minimum power limits of charging equipment, and the power supply capacity of the grid [10].

② Subject to the fulfillment of step ①, γ is for electric vehicles powered by phase transformer, select the vehicles whose transformer power does not exceed the limit after increasing the power by one gear. All increased by 4P. (If the charging power of the vehicle has reached the maximum power value of the charging equipment, the power will remain unchanged).

③ Subject to the fulfillment of step ①, β is for electric vehicles supplied by the phase house are selected to meet the charging completion guarantee constraints, and all vehicles are reduced the charging power ΔP_c of the vehicle (if the vehicle's charging power has reached the minimum power value of the charging equipment, the power will remain unchanged). These vehicles are usually those electric vehicles that are about to complete their charging tasks, and their charging power can be appropriately reduced to release more power resources for other vehicles that need to be charged. For these vehicles, their charging power is uniformly reduced, and similarly, if the charging power of a vehicle has reached the minimum power value of the charging equipment, its power will be kept unchanged to ensure that the charging process can be carried out properly.

Through the fine control of the above three steps, the charging power of electric vehicles can be optimized. It should be noted that, in the actual application, it is necessary to flexibly adjust and optimize according to the actual operation of the power grid, the distribution of electric vehicles and the technical parameters of the charging equipment, in order to achieve the best charging power control effect.

5 Experiments and Analysis of Results

This simulation experiment aims to verify the effectiveness and feasibility of the charging power optimization control strategy of the VSM based electric vehicle charging pile converter in practical application scenarios. Through the 15k (15 kW power scale) simulation model in MATLAB/SIMULINK environment, it is expected to clarify the impact of this strategy on charging power regulation, and provide a reference for subsequent large-scale applications.

5.1 Experimental Preparation

In this experiment, a 15 kW electric vehicle charging pile converter model is built on the MATLAB/SIMULINK platform, and the charging process of different charging stations and electric vehicles is simulated.

Create the charging pile converter model in MATLAB/SIMULINK. The main parameters of the model have been given, including the input voltage of 400 V, DC side capacitance 500 μF, inductance is 2 mH, and the switching frequency of the converter is 10 kHz. These parameters will serve as the basis for model construction to ensure the accuracy and reliability of the model.

In order to improve the richness of data in the experiment, different models and specifications of charging stations and electric vehicles were applied: small charging stations can accommodate 12 electric vehicles, medium charging stations can accommodate 120 electric vehicles, and large charging stations can accommodate 1200 electric vehicles. This setting can simulate charging stations of different scales to test the adaptability and scalability of the model.

At the same time, the models of electric vehicles will be divided according to the upper limit of charging power. The upper limit of charging power of each small electric vehicle will be randomly generated from the uniform distribution on the [5, 10] kW interval, the upper limit of charging power of each medium-sized electric vehicle will be randomly generated from the uniform distribution on the [30, 40] kW interval, and the upper limit of charging power of each large electric vehicle will be randomly generated from the uniform distribution on the [80, 100] kW interval. This setting can simulate electric vehicles with different charging requirements to test the performance of the model under various charging scenarios.

5.2 Experimental Results and Analysis

The experimental results are shown in Table 2.

According to Table 2, under different input voltage conditions, when there is no deviation in the grid frequency, the active power absorbed by the grid is the same as the load output power. At 400 V and 420 V, the active power absorbed by the power grid is 15.0 kW and 15.5 kW respectively, which is consistent with the load output power, proving the effect of the optimization strategy under the standard grid frequency.

When there is power grid frequency deviation, the load output power is always stable even if the input voltage changes. Taking 400 V input voltage as an example, when the frequency deviation of the power grid increases from 0 to −1.5Hz, the active power absorbed by the power grid decreases from 15.0 kW to 13.5 kW, reducing by 10%. However, during this period, the load output power is always maintained at 15.0 kW, indicating that the method proposed in this paper has good stability. When the frequency deviation continues to increase, the active power decreases, but the load output power remains basically constant. Under the condition that the grid frequency deviation is −

Table 2. Simulation experimental results of electric vehicle charging pile

INPUT VOLTAGE (V)	INPUT VOLTAGE (V)	Active power absorbed by the power grid (kW)	Load output power (kW)
400	0	15.0	15.0
	−0.5	14.5	15.0
	−1.0	14.0	15.0
	−1.5	13.5	15.0
420	0	15.5	15.5
	−0.5	15.0	15.5
	−1.0	14.5	15.5
	−1.5	14.0	15.5
380	0	14.5	14.5
	−0.5	14.0	14.5
	−1.0	13.5	14.5
	−1.5	13.0	14.5

1Hz, the input voltage increases from 380 V to 420 V, and the active power absorbed by the grid increases from 13.5 kW to 14.5 kW. However, the load output power always remains at its corresponding standard value, 14.5 kW at 380 V and 15.5 kW at 420 V.

It can be seen that the optimization method based on VSM can not only effectively combat the power grid frequency deviation, but also maintain stable output under different input voltages. This provides high stability and reliability for reasonable control of charging power of electric vehicle charging pile in practical application. This is because the method proposed in this article uses VSM theory to simulate the dynamic behavior of traditional synchronous generators, making electric vehicle charging stations behave more similar to traditional power plants in the power grid. This simulation helps to improve the inertia and damping of the system, thereby providing better frequency support and reducing frequency deviation when the grid frequency changes. Meanwhile, the three-phase balance optimization charging power regulation strategy proposed in this study ensures the three-phase current balance during the charging process. Three phase balance is the key to stable operation of power systems. By optimizing the distribution of three-phase power, additional losses and voltage fluctuations caused by unbalanced currents can be reduced.

6 Conclusion

The VSM based intelligent optimization method for charging power of electric vehicle charging stations not only significantly improves the operating efficiency of charging stations, reduces the waste of resources caused by improper distribution of charging power, but also provides more convenient and efficient charging services for electric vehicle users, and improves the user experience. However, although the VSM based charging power optimization method has achieved remarkable results, there are still some shortcomings in practical applications. First, when dealing with large-scale charging station networks, the computational complexity of this method is high, which may lead to a decline in real-time response speed and affect the overall performance of the system. Secondly, there are significant differences in the charging demand in different regions and different periods of time, and the adaptability of the current method to these changes still needs to be further strengthened to better meet the actual charging demand.

In a word, the intelligent optimization method of charging power of electric vehicle charging station based on VSM has broad application prospects and huge development potential. We will continue to work hard to improve this method and provide more efficient and intelligent solutions for the operation management and user service of electric vehicle charging stations.

Acknowledgment. 1. Scientific Research Project of Provincial Department of Education in 2024, Application Research of Green Energy Saving and Emission Reduction Engine Oil Additives in Cold Environment, No. JJKH20241152KJ.

2. 2022Jilin Communications Vocational and Technical College, New Energy Vehicle Charging Technology Innovation Team in High Latitude Cold and Humid Environment.

References

1. Chowdhury, R., Mukherjee, B.K., Mishra, P., et al.: Optimum positioning of electric vehicle charging station in a distribution system considering dependent loads. In: Bansal, H.O., Ajmera, P.K., Joshi, S., Bansal, R.C., Shekhar, C. (eds.) BITS-EEE-CON 2022. LNNS, vol. 641, 469–480. Springer, Singapore (2022). https://doi.org/10.1007/978-981-99-0483-9_38
2. Tian, X., Zha, H., Tian, Z., et al.: Carbon emission reduction capability assessment based on synergistic optimization control of electric vehicle V2G and multiple types power supply. Energy Rep. **11**, 1191–1198 (2024)
3. Srilakshmi, K., Teja Santosh, D., Ramadevi, A., et al.: Development of renewable energy fed three-level hybrid active filter for EV charging station load using Jaya grey wolf optimization. Sci. Rep. **14**(1), 4429–4429 (2024)
4. Li, K., Gong, A., Liu, Q., et al.: A MCR-WPT system based on LCL-S/P hybrid compensation network with CC/CV and maximum power optimization suiting for battery charging. IET Power Electron. **17**(3), 351–363 (2024)
5. Chen, J.B., Bei, G.Y., Zhang, Q.Z., et al.: Integrated reactive power optimization for distribution systems considering electric vehicle dis-/charging support. Energy Rep. **9**(S7), 1888–1896 (2023)
6. Tian, R., Ling, Y.S., You, J.W., et al.: Soft start of full bridge LLC resonant converter of charging module control strategy research. Chin. J. Power Sources **45**(06), 809–813 (2021)
7. Guo, J.J., Ding, H., Chang, C.Y., et al.: A wireless charging system control strategy based on adaptive genetic tuning PID control. Chin. J. Electron Dev. **46**(01), 157–162 (2023)
8. Abid, M.S., Apon, H.J., Alavi, A., et al.: Mitigating the effect of electric vehicle integration in distribution grid using slime mould algorithm. Alex. Eng. J. **64**, 785–800 (2023)
9. Liu, Y., Wang, J.M.: Simulation of orderly charging control method for guided electric vehicle in charging station. Comput. Simul. **37**(02), 149–153 (2020)
10. Swamy, P.S.P.R., Jayanthi, R., Sai Veerraju, M.: An intelligent optimal charging stations placement on the grid system for the electric vehicle application. Energy **16**(5), 285–296 (2023)

Author Index

A
Ai, Ziyu 143

C
Cao, Shaoyong 51
Che, Xue 186

D
Dai, Xuedong 81

G
Gao, Shan 214
Guo, Jiangtao 3

H
Han, Huanqing 51
Hong, Xiuyan 229
Huang, Ping 157, 344

J
Jia, Junqiang 3
Jiang, Jianhua 143

K
Kong, Chunhua 358
Kong, He 285, 300

L
Li, Degao 3
Li, Hui 95
Li, Jingyu 201
Li, Xiaodi 285, 300
Li, Xintao 3
Liang, Zheheng 174
Liu, Ce 243
Liu, Lulu 3
Lu, Zhengfang 285

M
Ma, Lei 201, 358

S
Shen, Wuqiang 174
Shi, Xiangjun 214
Song, Cao 51
Song, Chao 65
Song, Shaoyong 65
Song, Shujiang 358
Sun, Yalin 157, 344

W
Wang, Fang 243
Wang, Hongwei 81
Wang, Tao 3
Wang, Tingwen 272
Wang, Yuefei 186
Wei, Jiatong 201, 358
Wu, Niyan 18

X
Xu, Hui 330

Y
Yang, Hongxue 201
Yao, Chaosheng 174

Z
Zhan, Man 112, 127
Zhang, Weiwei 95
Zhang, Yuanyuan 229, 317
Zhao, Jingtao 65
Zhu, Xiaorong 330
Zuo, Liwen 32, 258

MIX
Papier aus verantwortungsvollen Quellen
Paper from responsible sources
FSC® C105338

If you have any concerns about our products,
you can contact us on
ProductSafety@springernature.com

In case Publisher is established outside the EU,
the EU authorized representative is:
**Springer Nature Customer Service Center GmbH
Europaplatz 3, 69115 Heidelberg, Germany**

Printed by Libri Plureos GmbH
in Hamburg, Germany